THE MIND *AND* THE BRAIN

ALSO BY

JEFFREY M. SCHWARTZ, M.D.

Brain Lock

(with Beverly Beyette)

A Return to Innocence

(with Annie Gottlieb and Patrick Buckley)

THE MIND *AND* THE BRAIN

NEUROPLASTICITY AND THE POWER OF MENTAL FORCE

JEFFREY M. SCHWARTZ, M.D., AND SHARON BEGLEY

HARPER PERENNIAL

NEW YORK • LONDON • TORONTO • SYDNEY

HARPER PERENNIAL

A hardcover edition of this book was published in 2002 by HarperCollins Publishers.

HarperCollins books may be purchased for educational, business, or sales promotional use. For information please write: Special Markets Department, HarperCollins Publishers, 10 East 53rd Street, New York, NY 10022.

FIRST HARPER PERENNIAL PAPERBACK EDITION PUBLISHED 2003.

The Library of Congress has cataloged the hardcover edition as follows:

Schwartz, Jeffrey, 1951–
The mind and the brain : neuroplasticity and the power of mental force / Jeffrey M. Schwartz and Sharon Begley.
p. cm.
Includes index.
ISBN 978-0-06-039355-7
1. Brain. 2. Neuroplasticity. 3. Neuropsychology. 4. Cognition. I. Begley, Sharon, 1956– II. Title.

QP360 .S377 2002
153—dc21

2002069700

ISBN 978-0-06-098847-0 (pbk.)

08 09 10 11 12 QK/RRD 21 20 19 18 17

To my parents, who never stopped believing in me;
and to Ned, Sarah, and Daniel, for enduring.
—Sharon Begley

To the Venerable U Silananda Sayadaw
on the occasion of his seventy-fifth birthday
—Jeffrey M. Schwartz

May all beings be well, happy, and peaceful

When he speaks of "reality" the layman usually means something obvious and well-known, whereas it seems to me that precisely the most important and extremely difficult task of our time is to work on elaborating a new idea of reality. This is also what I mean when I always emphasize that science and religion *must* be related in some way.

—Wolfgang Pauli, letter to M. Fierz, August 12, 1948

It is interesting from a psychological-epistemological point of view that, although consciousness is the only phenomenon for which we have direct evidence, many people deny its reality. The question: "If all that exists are some complicated chemical processes in your brain, why do you care what those processes are?" is countered with evasion. One is led to believe that . . . the word "reality" does not have the same meaning for all of us.

—Nobel physicist Eugene Wigner, 1967

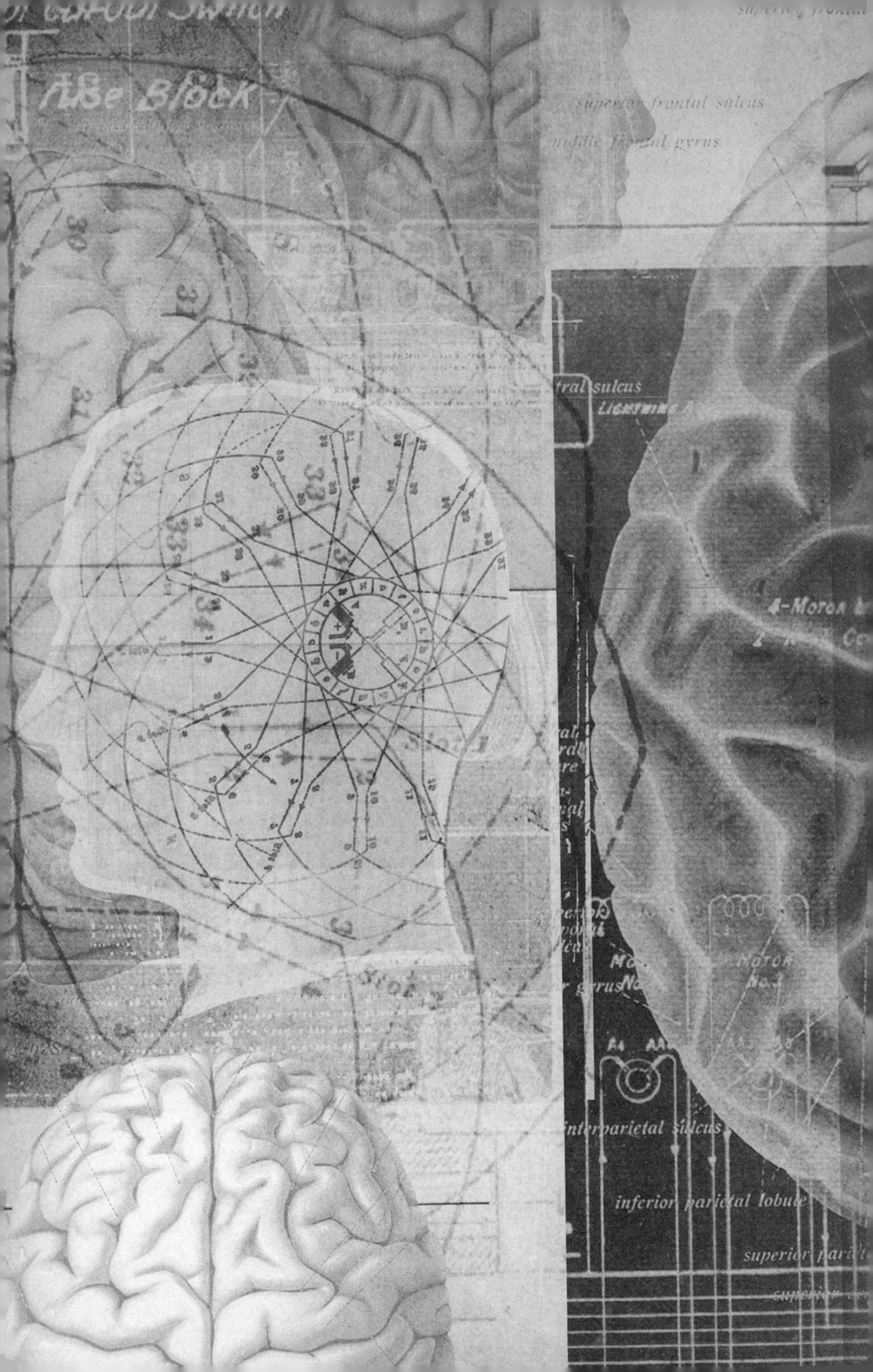

Fuse Block
superior frontal sulcus
middle frontal gyrus
interparietal sulcus
inferior parietal lobule

CONTENTS

{ACKNOWLEDGMENTS}

This book has a virtual, third coauthor: Henry Stapp, whose research into the foundations of quantum mechanics provided the physics underpinning for JMS's theory of directed mental force. For that, and for the countless hours he spent with the authors explaining the basics of quantum theory and reviewing the manuscript, we owe our deepest gratitude.

For more than a decade the Charles and Lela Hilton Family provided donations to support the academic career of JMS at UCLA.

Scores of scientists and philosophers gave tirelessly of their time to discuss their research or review the manuscript, and often both. Our heartfelt thanks to Floyd Bloom, Joseph Bogen, David Burns, Nancy Byl, David Chalmers, Bryan Clark, Almut Engelien, John Gabrieli, Fred Gage, Eda Gorbis, Phillip Goyal, Ann Graybiel, Iver Hand, J. Dee Higley, William Jenkins, Jon Kaas, Nancy Kanwisher, Michael Kozak, Patricia Kuhl, James Leckman, Andrew Leuchter, Benjamin Libet, Michael Merzenich, Steve Miller, Ingrid Newkirk, Randolph Nudo, Kevin Ochsner, Don Price, Alvaro Pascual-Leone, John Piacentini, Greg Recanzone, Ian Robertson, Cary Savage, John Searle, Jonathan Shear, David

Silbersweig, Edward Taub, John Teasdale, Max Tegmark, Elise Temple, Xiaoqin Wang, Martin Wax and Anton Zeilinger. We thank Christophe Blumrich for the care he took in producing the compelling artwork and, most of all, Judith Regan, Susan Rabiner, and Calvert Morgan for their commitment to this project. To those whom we have forgotten to mention (and we know you're out there), our apologies.

{INTRODUCTION}

Hamlet: My father, methinks I see my father.
Horatio: O! where, my lord?
Hamlet: In my mind's eye, Horatio.
—*William Shakespeare*

Every Tuesday, with the regularity of traffic jams on I-405, the UCLA Department of Psychiatry holds grand rounds, at which an invited researcher presents an hour-long seminar on a "topic of clinical relevance." One afternoon in the late 1980s, I saw, posted on a bulletin board at the Neuropsychiatric Institute, an announcement that stopped me cold. One of the nation's leading behavior therapists was scheduled to discuss her high-profile and hugely influential work with obsessive-compulsive disorder (OCD), the subject of my own research as a neuropsychiatrist. OCD is a condition marked by a constant barrage of intrusive thoughts and powerful urges, most typically to wash (because patients are often bombarded with thoughts about being dirty and contaminated with deadly pathogens) and to check (because of irresistible urges to make sure an appliance has not been left on, or a door left unlocked, or to satisfy oneself that something else is not amiss). I had a pretty good idea of what to expect—the speaker was widely known in medical circles for her application of rigorous behaviorist principles to psychological illnesses. "Rigorous," actually, hardly did the behaviorist approach justice. The very first paragraph of the very first paper that formally announced the behaviorist creed—John B. Watson's 1913 classic, "Psychology as the Behaviorist Views It"—managed, in a single throw-down-the-gauntlet statement, to deny man's humanity, to dismiss the significance of a mind capable of reflection, and to deny implicitly the existence of free will: "The behaviorist," declared Watson, "recognizes no dividing line between man and brute."

Rarely in the seventy-five years since Watson has a secular discipline adhered so faithfully to a core principle of its founder. Behaviorists, ignoring the gains of the cognitive revolution that had been building momentum and winning converts throughout the 1980s, continued to believe that there is no need for a therapist to acknowledge a patient's inner experiences while attempting to treat, say, a psychological illness such as a phobia; rather, this school holds that all desired changes in behavior can be accomplished by systematically controlling relevant aspects of a patient's environment, much as one would train a pigeon to peck particular keys on a keyboard by offering it rewards to reinforce correct behavior and punishments to reverse incorrect behavior. The grand rounds speaker, faithfully following the principles of behaviorist theory, had championed a particular method to treat obsessive-compulsive disorder known as "exposure and response prevention."

Exposure and response prevention, or ERP, was a perfect expression of behaviorist tenets. In ERP therapy sessions as routinely practiced, the OCD patient is almost completely passive. The therapist presents the patient with "triggers" of varying intensity. If, for instance, an OCD patient is terrified of bodily secretions and feels so perpetually contaminated by them that he washes himself compulsively, then the therapist exposes him to those very bodily products. The patient first ranks the level of distress various objects cause. Touching a doorknob in the therapist's office (which the patients believes is covered with germs spread by people who haven't washed after using the bathroom) might rate a 50. Touching a paper towel dropped in the sink of a public rest room might rate a 65; a sweaty T-shirt, 75; toilet seats at a gym, 90; a dollop of feces or urine, 100. Presenting one of these triggers constitutes the "exposure," the first half of the process. In the second half, the "response prevention," the therapist keeps the patient from reacting to the trigger with compulsive behaviors—in this example, washing. Instead of allowing him to run to a sink, the therapist waits for the intensity of the patient's distress to return to preexposure levels. During this waiting period, the patient is typically quite passive, but hardly calm or relaxed. Quite the contrary: patients suffer unpleasant, painful, intense anxiety in the face of the triggers—anxiety that can take hours to dissipate.

The theoretical basis of the approach, to the extent that there is one, involves the rather vague notion that the intense discomfort will somehow cause the symptoms to "habituate," much as the intense feeling of cold one feels after jumping into the ocean fades in a few minutes. During these treatment sessions, if a patient asks about the possible risks of exposure and response prevention he is usually rebuffed for "seeking reassurance," which supposedly undermines the efficacy of the treatment. And yet examples abound in which the risks endured by patients were only too real. In the United States, therapists in the forefront of developing these techniques have had patients rub public toilet seats with their hands and then spread—well, then spread whatever they touched all over their hair, face, and clothes. They have had patients rub urine over themselves. They have had patients bring in a piece of toilet paper soiled with a minuscule amount of their fecal material and rub it on their face and through their hair during the therapy session—and then, at home, contaminate objects around the house with it. In other cases, patients are prevented from washing their hands for days at a time, even after using the bathroom.

To me, this all seemed cruel and distasteful in the extreme—but it also seemed unnecessary. At the time, my UCLA colleague Lewis Baxter and I had recently begun recruiting patients into what was probably one of the first organized, ongoing behavior-therapy groups in the United States dedicated solely to the study and treatment of OCD. The study would examine, through the then-revolutionary brain imaging technique of positron emission tomography (PET), the neurological mechanisms underlying the disease. The group therapy sessions held in conjunction with the study would allow us to offer treatment to the study participants, of course. But the therapy sessions also presented what, to me, was an intriguing opportunity: the patients whom Baxter and I would study for clues to the causes of OCD might also tell us something about the relative efficacy of different treatments and treatment combinations. Our UCLA group had decided to study the effects of both drug and behavior therapy. I wasn't interested in doing research on the first of these, but I was extremely curious about the effects of psychologically oriented drug-free treatments on brain function. I didn't have much competition: by the late 1980s drugs were where the glamour was in major academic research centers. My offer to lead the behavior-therapy research group was accepted gladly.

I was becoming increasingly convinced of what was then a heresy in the eyes of mainstream behaviorists: that a patient undergoing behavior therapy need *never* do anything that a normal, healthy person would object to doing. I believed, too, on the basis of preliminary clinical research, that OCD might be better treated by systematically activating healthy brain circuits, rather than merely letting the pathological behaviors and their associated circuits burn themselves out, as it were, while the patient's distress eventually dissipated in a miasma of pain and anxiety.

My quest for an alternative treatment grew in part from my discomfort with exposure and response prevention treatment, which is based on principles gleaned almost solely from research on animal behavior. The difference between the techniques used in animal training and those applied to humans was negligible, and I had come to suspect that, in failing to engage a patient's mental faculties, behavior therapy was missing the boat. Treatments based on the principles of behaviorism denied the need to recognize and exploit the uniquely human qualities that differentiate humans from animals. If anything, such treatments are imbued with an obstinate machismo about not doing so; the behaviorists seemed to take a perverse pride in translating their work directly from animals to humans, allowing their theoretical preconceptions to displace common sense.

But exposure and response prevention, with its visits to public toilets and patients' wiping urine-impregnated paper over themselves, was claiming success rates of 60 to 70 percent. (Only years later would I discover that that percentage excluded the 20 to 30 percent of patients who refused to undergo the procedure once they saw what it entailed, as well as the 20 percent or so who dropped out.) Clearly, any alternative would face an uphill battle.

When I walked alone into the grand rounds auditorium that afternoon, I had a pretty clear idea of the techniques the speaker had applied to her OCD patients. Still, it was a welcome opportunity to hear directly from an established behaviorist about her methods, her theories, and her results. The audience settled down, the lights dimmed, and the speaker began. She had the tone and demeanor of someone on a mission. After explaining her diagnostic techniques—she was well known for a detailed questionnaire she had developed to pinpoint patients' fears, obsessions, and compulsions—she launched into a description of the behavioral treatment she used in the case of one not-atypical OCD sufferer. When this patient hits a bump in the road while driving, she explained, he feels he has run over someone and so looks obsessively in the rearview mirror. He frequently stops the car and gets out or drives around for hours looking desperately for a body he anxiously worries must be lying, bleeding and dying, on the pavement. She reported, with what I would come to recognize as her trademark self-assurance, that the key to her treatment of this case was . . . removing the rearview mirror from the car! Just as she made germ-obsessed patients touch toilet seats until their distress evaporated, she had this hit-and-run-obsessed patient drive without his mirror until his urge to check for bodies in the road behind him "habituated."

I was aghast. The potential danger she put the patient in was astonishing—but this apparently made not a whit of difference. The prevailing view among behaviorists was that normal standards of judgment and taste could be set aside during behavioral interventions. I already had qualms about how mechanistic the treatment based on behaviorist principles was, how in thrall to problematic dogma and, indeed, to the cult of scientism itself, which has been described by Jacques Barzun as "the fallacy of believing that the method of science must be used on all forms of experience and, given time, will settle every issue." Imagining the implications of a mainstream treatment that called for a patient to drive around without a rearview mirror, I found it hard to focus on the rest of the talk.

But what I had heard had triggered an epiphany. From then on, I decided, I would commit myself to finding a way to spare OCD patients (as well as patients with other mental disorders) from unnecessary, irresponsible, even brutal treatment by experts who pride themselves on ignoring what patients are feeling, or indeed whether they are even conscious. Surely there is something deeply wrong, both morally and scientifically, with a school of psychology whose central tenet is that people's conscious life experience (the literal meaning of the word *psyche*) is irrelevant, and that the intrinsic difference between humans and "brutes" (as Watson had candidly put it) could be safely ignored. I became determined to show that OCD can be effectively treated without depriving patients of rearview mirrors, without forcing them to touch filthy toilets, without ordering them to use the bathroom without washing their hands afterward—without, in short, forcing them to do anything dangerous, unsanitary, or just plain ridiculous. There is no need to suspend common sense and simple old-fashioned decency to use behavioral interventions successfully, I reasoned, as I walked back to my office. By applying a new and scientifically testable method that would empower OCD patients actively and willfully to change the focus of their attention, I just might help them learn to overcome their disease. But I had a hunch that I might achieve something else, too: demonstrating, with the new brain imaging technology, that patients could systematically alter their own brain function. The will, I was starting to believe, generates a force. If that force could be harnessed to improve the lives of people with OCD, it might also teach them how to control the very brain chemistry underlying their disease.

What determines the question a scientist pursues? One side in the so-called science wars holds that the investigation of nature is a purely objective pursuit, walled off from the influences of the surrounding society and culture by built-in safeguards, such as the demand that scientific results be replicable and the requirement that scientific theories accord with nature. The gravitational force of a Marxist, in other words, is identical to the gravitational force of a fascist. Or, more starkly, if you're looking for proof that science is not a social construct, as so-called science critics contend, just step out the window and see whether the theory of gravity is a mere figment of a scientist's imagination.

That the findings of science are firmly grounded in empiricism is clear. But the *questions* of science are another matter. For the questions one might ask of nature are, for all intents and purposes, without end. Although the methods of science may be largely objective, the choice of what question to ask is not. This is not a shortcoming, much less a fault, of science. It is, rather, a reflection of the necessary fact that science is, at bottom, a human endeavor. Running through both psychiatry and neuroscience is a theme that seemed deeply disturbing to me almost from the moment I began reading in the field as a fifteen-year-old in Valley Stream, Long Island, when my conviction that the inner working of the mind was the only mystery worth pursuing made me vow to become a psychiatrist. What disturbed me was the idea that free will died with Freud—or even earlier, with the materialism of the triumphant scientific revolution. Freud elevated unconscious processes to the throne of the mind, imbuing them with the power to guide our every thought and deed, and to a significant extent writing free will out of the picture. Decades later, neuroscience has linked genetic mechanisms to neuronal circuits coursing with a multiplicity of neurotransmitters to argue that the brain is a machine whose behavior is predestined, or at least determined, in such a way as seemingly to leave no room for the will. It is not merely that the will is not free, in the modern scientific view; not merely that it is constrained, a captive of material forces. It is, more radically, that the will, a manifestation of mind, does not even exist, because a mind independent of brain does not exist.

My deep doubts that human actions can be explained away through materialist determinism simmered just below the surface throughout my years of medical school. But by the time I completed my psychiatric residency at Cedars-Sinai Medical Center in 1984, my research interests had converged on the question of the role of the brain in mental life. After two years conducting brain research under the mentorship of Floyd Bloom at the Salk Institute in La Jolla from 1980 to 1982—investigating a possible role for the endogenous opiate beta-endorphin in manic depression, as well as doing basic research on the functional neuroanatomy of changes in mood states—I was growing ever more curious about the mysterious connection between mental events and the activity of discrete brain structures. The timing was perfect: even then, that area of neuroscience, broadly known as functional neuroanatomy, was achieving gains few even dreamed of. Brain imaging techniques such as PET (and, later, functional magnetic resonance imaging, or fMRI) were, for the first time, allowing neuroscientists to observe the living, working human brain in action. Ordering a forefinger to lift, reading silently, matching verbs to nouns, cogitating on faces, conjuring up a mental image of a childhood event, mentally manipulating blocks to solve the game Tetris—scans were mapping the parts of the brain responsible for each of these activities, and for many more.

But even as Congress declared the 1990s the Decade of the Brain, a nagging doubt plagued some neuroscientists. Although learning which regions of the brain become metabolically active during various tasks is crucial to any understanding of brain function, this mental cartography seemed ultimately unsatisfying. Being able to trace brain activity on an imaging scan is all well and good. But what does it *mean* to see that the front of the brain is underactive in people with schizophrenia? Or that there is a quieting of the frontal "executive attention network" when experienced practitioners of the ancient technique of *yoga nidra* attain meditative relaxation? Or even that a particular spot in the visual cortex becomes active when we see green? In other words, what kind of internal experience is generated by the neuronal activity captured on a brain scan? Even more important, how can we use scientific discoveries linking inner experience with brain function to effect constructive changes in everyday life? Soon after I joined the UCLA faculty in 1985, I realized that obsessive-compulsive disorder might offer a model for these very questions of mind and brain.

At the same time, I was regaining an interest in Buddhist philosophy that I had developed a decade earlier, when a poet friend (who later perished on that ill-fated KAL flight that ran into the wrong end of the cold war) became deeply involved in Buddhist meditation. As a premed philosophy major I always had a healthy dose of skepticism about what my poet friends were into, but I was nevertheless intrigued. The first Noble Truth, Dukkha—or, as it is generally translated, "Suffering"—had an immense intuitive sensibility to me. Life, I already felt, was not an easy undertaking. In addition, Buddhist philosophy's emphasis on the critical importance of observing the Basic Characteristic of Anicca, or Impermanence, appealed to me. As an aspiring psychiatrist in self-directed training,

I was drawn to the practical aspect of Buddhist philosophy: the systematic development and application of a clear-minded observational power, known in the Buddhist lexicon as Mindfulness.

I had first pursued this new direction in earnest during my first year of medical school. I added two self-taught extracurricular courses to my required studies: introductory training in Yoga as expounded in the classic text *Light on Yoga*, by B. K. S. Iyengar, and regular reading of the *Archives of General Psychiatry*, which, of all the leading journals, seemed most focused on the newly developing field of neuropsychiatry (I had already decided that I would specialize in the brain-related aspects of psychiatry). During that first year I arranged to continue these pursuits by setting up a summer clerkship in neuropsychiatry research and enrolling, at the end of the summer, in an intensive retreat in the practice of Buddhist mindfulness meditation. When the second year of medical school began in September 1975, I knew I was setting off on what would become a lifelong quest, to develop and integrate these two fields.

At the core of Buddhist philosophy lies this concept of mindfulness, or mindful awareness: the capacity to observe one's inner experience in what the ancient texts call a "fully aware and nonclinging" way. Perhaps the most lucid modern description of the process comes from the German monk Nyanaponika Thera (his name means "inclined toward knowledge," and *thera* is a title roughly analogous to "teacher"). A major figure of twentieth-century Buddhist scholarship, he coined the term *Bare Attention* to explain to Westerners the type of mental activity required to attain mindful awareness. In his landmark book *The Heart of Buddhist Meditation*, Nyanaponika wrote, "Bare Attention is the clear and single-minded awareness of what actually happens *to* us and *in* us at the successive moments of perception. It is called 'Bare' because it attends just to the bare facts of a perception as presented either through the five physical senses or through the mind . . . without reacting to them." One Buddhist scholar captured the difference between mindfulness and the usual mode of mind this way: "You're walking in the woods and your attention is drawn to a beautiful tree or a flower. The usual human reaction is to set the mind working, 'What a beautiful tree, I wonder how long it's been here, I wonder how often people notice it, I should really write a poem.' . . . The way of mindfulness would be just to see the tree . . . as you gaze at the tree there is nothing between you and it." There is full awareness without running commentary. You are just watching, observing all facts, both inner and outer, very closely.

The most noteworthy result of mindfulness, which requires directed willful effort, is the ability it affords those practicing it to observe their sensations and thoughts with the calm clarity of an external witness: through mindful awareness, you can stand outside your own mind as if you are watching what is happening to another rather than experiencing it yourself. In Buddhist philosophy, the ability to sustain Bare Attention over time is the heart of meditation. The meditator views his thoughts, feelings, and expectations much as a scientist views experimental data—that is, as natural phenomena to be noted, investigated, reflected on, and learned from. Viewing one's own inner experience as data allows the meditator to become, in essence, his own experimental subject. (This kind of directed mental activity, as it happens, was critical to the psychological and philosophical work of William James, though as far as we know he had no more than a passing acquaintance with Buddhist meditation.)

Through the centuries, the idea of mindfulness has appeared, under various names, in other branches of philosophy. Adam Smith, one of the leading philosophers of the eighteenth-century Scottish Enlightenment, developed the idea of "the impartial and well-informed spectator." This is "the man within," Smith wrote in 1759 in *The Theory of Moral Sentiments*, an observing power we all have access to, which allows us to observe our internal feelings as if from without. This distancing allows us to witness our actions, thoughts, and emotions not as an involved participant but as a disinterested observer. In Smith's words:

> *When I endeavor to examine my own conduct . . . I divide myself as it were into two persons; and that I, the examiner and judge, represent a different character from the other I, the person whose conduct is examined into and judged of. The first is the spectator. . . . The second is the agent, the person whom I properly call myself, and of whose conduct, under the character of a spectator, I was endeavoring to form some opinion.*

It was in this way, Smith concluded, that "we suppose ourselves the spectators of our own behaviour." The change of perspective accomplished by the impartial spectator is far from easy, however: Smith clearly recognized the "fatiguing exertions" it required.

For years I had wondered what psychiatric ailment might best lend itself to a study of the effects of mindfulness on brain function. So within a few days of beginning to study the literature on obsessive-compulsive disorder at UCLA, I suspected that the disease might offer an entrée into some of the most profound questions of mind and brain, and an ideal model in which to examine the interface between the two. And soon after I began working intensively with people who had the condition and looked at the PET data being collected on them, I realized I'd stumbled onto a neuropsychiatrist's gold mine.

The obsessions that besiege the patient seemed quite clearly to be caused by pathological, mechanical brain processes—mechanical in the sense that we can, with reasonable confidence, trace their origins and the brain pathways involved in their transmission. OCD's clear and discrete presentation of symptoms, and reasonably well-understood pathophysiology, suggested that the brain side of the equation could, with enough effort, be nailed down.

As for the mind side, although the cardinal symptom of obsessive-compulsive disorder is the persistent, exhausting intrusion of an unwanted thought and an unwanted urge to act on that thought, the disease is also marked by something else: what is known as an ego-dystonic character. When someone with the disease experiences a typical OCD thought, some part of his mind knows quite clearly that his hands are not really dirty, for instance, or that the door is not really unlocked (especially since he has gone back and checked it four times already). Some part of his mind (even if, in serious cases, it is only a small part) is standing outside and apart from the OCD symptoms, observing and reflecting insightfully on their sheer bizarreness. The disease's intrinsic pathology is, in effect, replicating an aspect of meditation, affording the patient an impartial, detached perspective on his own thoughts. As far as I knew, the impartial spectator in the mind of an OCD patient—overwhelmed by the biochemical imbalances in the brain that the disease causes—remained only that, a mere spectator and not an actor, noting the symptoms that were laying siege to the patient's mind but powerless to intercede. The insistent thoughts and images of OCD, after all, are experienced passively: the patient's volition plays no role in their appearance.

But perhaps, I thought, the impartial spectator needn't remain a bystander. Perhaps it would be possible to use mindfulness training to empower the impartial spectator to become more than merely an effete observer. Maybe, just maybe, patients could learn a practical, self-directed approach to treatment that would give them the power to strengthen and utilize the healthy parts of their brain in order to resist their compulsions and quiet the anxieties and fears caused by their obsessions. And then, despite the painful intrusions into consciousness caused by the faulty brain mechanisms, the patient could exercise the power to make a choice about whether the next idea the brain attends to will be "I am going to work in the garden now," rather than "I am going to wash my hands again." Although the passive stream of the contents of consciousness may well be determined by brain mechanism, the mental and emotional response to that stream may not be. The OCD patient, in other words, may have the capacity to focus attention in a way that is not fixed or predestined by the (pathological) brain state.

To my way of thinking, the Buddhist concept of mindfulness offered a guide to what would be a radically new approach to OCD treatment. In what came to be called the Four Steps regimen of cognitive-behavioral therapy for OCD, patients gain insight into the true nature and origin of the bothersome OCD thoughts and urges. They *Relabel* their obsessions and compulsions as false signals, symptoms of a disease. They *Reattribute* those thoughts and urges to pathological brain circuitry ("This thought reflects a malfunction of my brain, not a real need to wash my hands yet again"). They *Refocus*, turning their attention away from the pathological thoughts and urges onto a constructive behavior. And, finally, they *Revalue* the OCD obsessions and compulsions, realizing that they have no intrinsic value, and no inherent power. If patients could systematically learn to reassess the significance of their OCD feelings and respond differently to them through sustained mindful awareness, I reasoned, they might, over time, substantially change the activity of the brain regions that underlie OCD. Their mind, that is, might change their brain.

At first, whenever I tried to discuss these ideas with colleagues, the reaction ranged from mere amusement to frank annoyance. Like all of modern science, the field of psychiatry, especially in its current biological incarnation, has become smitten with *materialist reductionism,* the idea that all phenomena can be explained by the interaction and movements of material particles. As a result, to suggest that anything other than brain mechanisms in and of themselves constitute the causal dynamics of a mental phenomenon is to risk being dismissed out of hand. But there was another problem. For decades, a key tenet of neuroscience held that although the organization and wiring of the infant brain are molded by its environment, the functional organization and structure of the adult brain are immutable. Experiments in rats, monkeys, ferrets, and people showing that the adult brain can indeed change, and change in profound ways, still lay in the future. Since I was arguing that the mind can change the brain, persuading the scientific community that I was right required that scientists accept an even more basic fact: that the adult brain can change at all.

The chapters that follow explore the new vistas in neuroscience opened by the original UCLA work on obsessive-compulsive disorder. We'll survey both historical and current approaches to the mind-brain enigma surrounding how mental phenomena emerge from three pounds of grayish, gelatinous tissue encased in the human skull. We'll also explore the OCD research in further detail. My discovery that mental action can alter the brain chemistry of an OCD patient occurred when neuroscientists were reopening a question that most had thought long settled: can the adult brain change in ways that are significant for its function? Does it, in other words, display an attribute that researchers had thought lost with the final years of childhood—neuroplasticity? *Neuroplasticity* refers to the ability of neurons to forge new connections, to blaze new paths through the cortex, even to assume new roles. In shorthand, neuroplasticity means rewiring of the brain. After chronicling the ongoing discoveries of neuroplasticity in the brain of the developing child—from the first tentative neuronal synapses as they form in fetal life to the wiring of the visual, auditory, and somatosensory systems and higher cortical functions such as cognition and emotions—we will review the notorious tale of the Silver Spring monkeys. The mistreatment of these seventeen macaques at a behavioral psychology institute in the 1970s led to their seizure by federal agents, conviction of the lead researcher on six counts of animal cruelty, and, more than any other single event, the rise of the animal rights movement in the United States. But experiments on the Silver Spring monkeys also demonstrated, for the first time, the massive plasticity of the adult primate brain.

The heart of the book describes the burgeoning field of neuroplasticity—how plasticity is induced by changes in the amount of sensory stimulation reaching the brain. Neuroplasticity can result not only in one region of the brain colonizing another—with remarkable effects on mental and physical function—but also in the wholesale remodeling of neural networks, of which the changes in the brains of OCD patients are only one example. The discovery that neuroplasticity can be induced in people who have suffered a stroke demonstrated, more than any other finding, the clinical power of a brain that can rewire itself.

It was through my informal collaboration with the physicist Henry Stapp, a preeminent scholar of the interpretation of quantum mechanics, that my tentative musings on the causal efficacy of attention and will found a firm basis in physics. Stapp's youthful pursuit of the foundations of quantum physics evolved, in his later years, into an exploration of the mind's physical power to shape the brain. It has long been Stapp's contention that the connection between consciousness and the brain is (*pace,* philosophers) primarily a problem in physics and addressable by physics—but only the correct physics. Though you would hardly know it from the arguments of those who appeal to physics to assert that all mental phenomena can be reduced to the electrochemical activity of neurons, physics has progressed from its classical Newtonian form and found itself in the strange land of the quantum. Once, physics dealt only with tangible objects: planets, balls, molecules, and atoms. Today, in the form of quantum mechanics, it describes a very different world, one built out of what Stapp calls "a new kind of stuff," with properties of both the physical and the mental. The physics that informs the neuroscience I describe is, I think, one of the features that make this book different from those that have come before, for it is in quantum mechanics that the hypotheses born of studies of OCD and the mind find their harmonizing voice. What we now know about quantum physics gives us reason to believe that conscious thoughts and volitions can, and do, play a powerful causal role in the world, including influencing the activity of the brain. Mind and matter, in other words, *can* interact.

One result of my collaboration with Stapp was a short paper we wrote for the *Journal of Consciousness Studies*. In it, we marshaled evidence from neuroscience and quantum physics to argue for the existence and causal efficacy of volition. In his 1890 masterpiece *The Principles of Psychology*, William James argued that the ability to fix one's attention on a stimulus or a thought and "hold it fast before the mind" was the act that constituted "the essential achievement of the will." If the effort to hold an object in one's attention is determined wholly by the properties of that object and its effects on the nervous system, then the case for free will is weak. If, however, one can make more or less effort, as one chooses, and so willfully "prolong the stay in consciousness of innumerable ideas which else would fade away more quickly," then free will remains a real scientific possibility.

James struggled throughout his life to find a rigorous alternative to the reductive determinism that ruled science in his day and has persisted down to ours. Although he rejected determinism on ethical grounds, as a working scientist he had to admit that "the utmost a believer in free-will can *ever* do will be to show that the deterministic arguments are not coercive. That they are seductive, I am the last to deny." But although determinism has indeed seduced both the scientifically sophisticated and the scientifically innocent for well over a century, its arguments are not "coercive." During the reign of classical physics one could be excused for thinking otherwise. But not anymore. Individuals choose what they will attend to, ignoring all other stimuli in order to focus on one conversation, one string of printed characters, or, in Buddhist mindfulness meditation, one breath in and one breath out.

In the last section of the book, we explore this third rail of neuroscience: the existence, character, and causal efficacy of will. There, I propose that the time has come for science to confront the serious implications of the fact that directed, willed mental activity can clearly and systematically alter brain function; that the exertion of willful effort generates a *physical force* that has the power to change how the brain works and even its physical structure. The result is directed neuroplasticity. The cause is what I call directed mental force.

Mainstream philosophical and scientific discourse may remain strongly biased toward a materialist perspective. Yet the simple fact is that the materialism of classical physics offers no intuitively meaningful way of explaining the critical role played by the will in the brain changes seen in OCD patients. The striving of the mind to be free of its inner compulsions—what Buddhists call Right Effort—is much more than just the play of electrochemical impulses within a material construct. In this book, I describe experimental data that support an alternative, offering evidence that the brain truly is the child of the mind.

How? Through the mental act of focusing attention, mental effort becomes directed mental force. "[T]he effort to attend," James believed, may well be a true and genuine "original force." Modern neuroscience is now demonstrating what James suspected more than a century ago: that attention is a mental state (with physically describable brain state correlates) that allows us, moment by moment, to "choose and sculpt how our ever-changing minds will work, [to] choose who we will be the next moment in a very real sense. . . . Those choices are left embossed in physical form on our material selves." If James was speaking metaphorically, he was also speaking with almost eerie prescience. For it is now clear that the attentional state of the brain produces physical change in its structure and future functioning. The seemingly simple act of "paying attention" produces real and powerful physical changes in the brain. In fact, Stapp's work suggests that there *is* no fully defined brain state until attention is focused. That physical activity within the brain follows the focus of attention offers the clearest explanation to date of how my hypothesized mental force can alter brain activity. The choice made by a patient—or, indeed, anyone—causes one physical brain state to be activated rather than another. A century after the birth of quantum mechanics, it may at last be time to take seriously its most unsettling idea: that the observer and the way he directs his attention are intrinsic and unavoidable parts of reality.

Finally, in the Epilogue, we attempt to come to terms with why any of this matters. One important answer is that the materialist-determinist model of the brain has profound implications for notions like moral responsibility and personal freedom. The interpretation of mind that dominates neuroscience is inimical to both. For if we truly believe, when the day is done, that our mind and all that term entails—the choices we make, the reactions we have, the emotions we feel—are nothing but the expression of a machine governed by the rules of classical physics and chemistry, and that our behavior follows ineluctably from the workings of our neurons, then we're forced to conclude that the subjective sense of freedom is a "user illusion." Our sense that we are free to make moral decisions is a cruel joke, and society's insistence that individuals (with exceptions for the very young and the mentally ill) be held responsible for their actions is no more firmly rooted in reason than a sand castle is rooted in the beach. In stark contrast to the current paradigm, however, the emerging research on neuroplasticity, attention, and the causal efficacy of will supports the opposite view—one that demands the recognition of moral responsibility.

And it does something more. The implications of directed neuroplasticity combined with quantum physics cast new light on the question of humankind's place, and role, in nature. At its core, the new physics combined with the emerging neuroscience suggests that the natural world evolves through an interplay between two causal processes. The first includes the physical processes we are all familiar with—electricity streaming, gravity pulling. The second includes the contents of our consciousness, including volition. The importance of this second process cannot be overstated, for it allows human thoughts to make a difference in the evolution of physical events.

Because the question of mind—its existence and its causal efficacy—is central to our thesis, let us turn first to an exploration of a problem as ancient as philosophy and as modern as the latest discovery of genes that "cause" risk taking, or shyness, or happiness, or impulsivity—or any of the dozens of human behavioral traits that are now being correlated with the chemical messages encoded on our twisting strands of DNA.

Let us turn to the duality of mind and brain.

{ONE}

THE MATTER OF MIND

Nature in her unfathomable designs has mixed us
of clay and flame, of brain and mind, that the two things hang
indubitably together and determine each other's being, but
how or why, no mortal may ever know.

—*William James*

Principles of Psychology, *Chapter VI*

What is mind? No matter. What is matter? Never mind.

—*T. H. Key*

Of all the thousands of pages and millions of words devoted to the puzzle of the mind and the brain, to the mystery of how something as sublime and insubstantial as thought or consciousness can emerge from three pounds of gelatinous pudding inside the skull, my favorite statement of the problem is not that of one of the great philosophers of history, but of a science fiction writer. In a short story first published in the science and sci-fi magazine *Omni* in 1991, the Hugo-winning author Terry Bisson gets right to the heart of the utter absurdity of the situation: that an organ made from basically the same material ingredients (nucleated, carbon-based, mitochondria-filled cells) as, say, a kidney, is able to generate this ineffable thing called mind. Bisson's story begins with this conversation between an alien commander and a scout who has just returned from Earth to report the results of his reconnaissance:

"They're made out of meat."

"Meat?"

"There's no doubt about it. We picked several from different parts of the planet, took them aboard our recon vessels, probed them all the way through. They're completely meat."

"That's impossible. What about the radio signals? The messages to the stars?"

"They use the radio waves to talk, but the signals don't come from them. The signals come from machines."

"So who made the machines? That's who we want to contact."

"They made the machines. That's what I'm trying to tell you. Meat made the machines."

"That's ridiculous. How can meat make a machine? You're asking me to believe in sentient meat."

"I'm not asking you, I'm telling you. These creatures are the only sentient race in the sector and they're made of meat."

"Maybe they're like the Orfolei. You know, a carbon-based intelligence that goes through a meat stage."

"Nope. They're born meat and they die meat. We studied them for several of their lifespans, which didn't take too long. Do you have any idea of the lifespan of meat?"

"Spare me. Okay, maybe they're only part meat. You know, like the Weddilei. A meat head with an electron plasma brain inside."

"Nope, we thought of that, since they do have meat heads like the Weddilei. But I told you, we probed them. They're meat all the way through."

"No brain?"

"Oh, there is a brain all right. It's just that the brain is made out of meat."

"So . . . what does the thinking?"

"You're not understanding, are you? The brain does the thinking. The meat."

"Thinking meat! You're asking me to believe in thinking meat!"

"Yes, thinking meat! Conscious meat! Loving meat. Dreaming meat. The meat is the whole deal! Are you beginning to get the picture, or do I have to start all over?"

It was some 2,500 years ago that Alcmaeon of Croton, an associate of the Pythagorean school of philosophy who is regarded as the founder of empirical psychology, proposed that conscious experience originates in the stuff of the brain. A renowned medical and physiological researcher (he practiced systematic dissection), Alcmaeon further theorized that all sensory awareness is coordinated by the brain. Fifty years later, Hippocrates adopted this notion of the brain as the seat of sensation, writing in his treatise on seizures: "I consider that the brain has the most power for man. . . . The eyes and ears and tongue and hands and feet do whatsoever the brain determines . . . it is the brain that is the messenger to the understanding [and] the brain that interprets the understanding." Although Aristotle and the Stoics rejected this finding (seating thought in the heart instead), today scientists know, as much as they know anything, that all of mental life springs from neuronal processes in the brain. This belief has dominated studies of mind-brain relations since the early nineteenth century, when phrenologists attempted to correlate the various knobs and bumps on the skull with one or another facet of personality or mental ability. Today, of course, those correlations are a bit more precise, as scientists, going beyond the phrenologists' conclusion that thirty-seven mental faculties are represented on the surface of the skull, do their mapping with brain imaging technologies such as positron emission tomography (PET) and functional magnetic resonance imaging (fMRI), which pinpoint which brain neighborhoods are active during any given mental activity.

This has been one of the greatest triumphs of modern neuroscience, this mapping of whole worlds of conscious experience—from recognizing faces to feeling joy, from fingering a violin string to smelling a flower—onto a particular cluster of neurons in the brain. It began in the 1950s, when Wilder Penfield, a pioneer in the neurosurgery of epilepsy, electrically stimulated tiny spots on the surface of patients' brains (a painless procedure, since neurons have no feeling). The patients were flooded with long-forgotten memories of their grandmother or heard a tune so vividly that they asked the good doctor why a phonograph was playing in the operating theater. But it is not merely the precision of the mental maps that has increased with the introduction of electrodes—and later noninvasive brain imaging—to replace the skull-bump cartography beloved of phrenologists. So has neuroscientists' certainty that tracing different mental abilities to specific regions in the brain—verbal working memory to a spot beneath the left temple, just beside the region that encodes the unpleasantness of pain and just behind the spot that performs exact mathematical calculations—is a worthy end in itself. So powerful and enduring has been Alcmaeon's hypothesis about the seat of mental life, and his intellectual descendants' equating of brain and mind, that most neuroscientists today take for granted that once you have correlated activity in a cluster of neurons with a cognitive or emotional function—or, more generally, with any mental state—you have solved the problem of the origin of mental events. When you trace depression to activity in a circuit involving the frontal cortex and amygdala, you have—on the whole—explained it. When you link the formation of memories to electrochemical activities in the hippocampus, you have learned everything worth knowing about it. True, there are still plenty of details to work out. But the most deeply puzzling question—whether that vast panoply of phenomena encompassed by the word *mind* can actually arise from nothing but the brain—is not, in the view of most researchers, a legitimate subject for scientific inquiry. Call it the triumph of materialism.

To the mainstream materialist way of thinking, only the physical is real. Anything nonphysical is at best an artifact, at worst an illusion. In this school of philosophy, at least among those who don't dismiss the reality of mind entirely, the mind is the software running on the brain's hardware. Just as, if you got right down to the level of logic gates and speeding electrons, you could trace out how a computer told to calculate 7 × 7 can spit out 49, so you could, in principle, determine in advance the physical, neural correlates in the brain of any action the mind will ever carry out. In the process, every nuance of every mental event would be explained, with not even the smallest subtlety left as a possibly spontaneous (from the Latin *sponte*, meaning "of one's free will, voluntarily") occurrence.

A friend of mine, the neurosurgeon Joseph Bogen, recalled to me a remark that the Nobelist David Hubel made to him in 1984: "The word Mind is obsolete." Hubel was stating exactly the conclusion of researchers who equate their brain scans and neuronal circuit diagrams with a full understanding of mental processes. Now that we understand so much about the brain, this reasoning holds, there's no longer any need to appeal to such a naïve term, with its faint smack of folk psychology. As Hubel said to Bogen, the very word *mind* "is like the word *sky* for astronomers." It's only fair to note that this view has not been unanimous. In fact, no sooner had brain imaging technology produced its neat maps than neuroscientists began to question whether we will "understand the brain when the map . . . is completely filled with blobs," as the neurobiologists M. James Nichols and William Newsome asked in a 1999 paper. "Obviously not," they answered. Still, in many sophisticated quarters, *mind* was becoming not merely an obsolete word but almost an embarrassing one.

But if you equate the sequential activation of neurons in the visual pathway, say, with the perception of a color, you quickly encounter two mysteries. The first is the one that befuddled our alien commander. Just as the human brain is capable of differentiating light from dark, so is a photodiode. Just as the brain is capable of differentiating colors, so is a camera. It isn't hard to rig up a photodiode to emit a beep when it detects light, or a camera to chirp when it detects red. In both cases, a simple physical device is registering the same perception as a human brain and is announcing that perception. Yet neither device is conscious of light or color, and neither would become so no matter how sophisticated a computer we rigged it up to. There is a difference between a programmed, deterministic mechanical response and the mental process we call consciousness. Consciousness is more than perceiving and knowing; it is *knowing* that you know.

If it seems ridiculous even to consider why a handful of wires and transistors fails to generate subjective perceptions, then ask the same question about neurons outside the brain. Why is it that no neurons other than those in a brain are capable of giving the owner of that brain a qualitative, subjective sensation—an inner awareness? The activity of neurons in our fingertips that distinguish hot from cold, for example, is not associated in and of itself with conscious perception. But the activity of neurons in the brain, upstream of the fingertips' sensory neurons, is. If the connection linking the fingers to the brain through the spinal cord is severed, all sensation in those fingers is lost. What is it about the brain that has granted to its own neurons the almost magical power to create a felt, subjective experience from bursts of electrochemical activity little different from that transpiring downstream, back in the fingertips? This represents one of the central mysteries of how matter (meat?) generates mind.

The second mystery is that the ultimate result of a rain of photons falling on the retina is . . . well, a sense. A sense of crimson, or navy blue. Although we can say that *this* wavelength of light stimulates *this* photosensitive cone in the retina to produce *this* sense of color—650 nanometers makes people with normal color vision see red, for instance—science is silent on the genesis of the feeling of red, or cerulean, or other *qualia*. This is the term many philosophers have adopted for the qualitative, raw, personal, subjective feel that we get from an experience or sensation. Every conscious state has a certain feel to it, and possibly a unique one: when you bite into a hamburger, it feels different from the experience of chewing a steak. And any taste sensation feels different from the sound of a Chopin étude, or the sight of a lightning storm, or the smell of bourbon, or the memory of your first kiss. Identifying the locus where red is generated, in the visual cortex, is a far cry from explaining our sense of redness, or why seeing red *feels* different from tasting fettuccine Alfredo or hearing "Für Elise"—especially since all these experiences reflect neuronal firings in one or another sensory cortex. Not even the most detailed fMRI gives us more than the physical basis of perception or awareness; it doesn't come close to explaining what it feels like from the inside. It doesn't explain the first-person feeling of red. How do we know that it is the same for different people? And why would studying brain mechanisms, even down to the molecular level, ever provide an answer to those questions?

It is, when you think about it, a little peculiar to believe that when you have traced a clear causal chain between molecular events inside our skull and mental events, you have explained them sufficiently, let alone explained the mind in its entirety. If nothing else, there's a serious danger of falling into a category error here, ascribing to particular clusters of neurons properties that they do not possess—in this case, consciousness. The philosopher John Searle, who has probed the mysteries of mind and brain as deeply as any contemporary scholar, has described the problem this way: "As far as we know, the fundamental features of [the physical] world are as described by physics, chemistry and the other natural sciences. But the existence of phenomena that are not in any obvious way physical or chemical gives rise to puzzlement. . . . How does a mental reality, a world of consciousness, intentionality and other mental phenomena, fit into a world consisting entirely of physical particles in fields of force?" If the answer is that it doesn't—that mental phenomena are different in kind from the material world of particles—then what we have here is an *explana-*

tory gap, a term first used in this context by the philosopher Joseph Levine in his 1983 paper "Materialism and Qualia: The Explanatory Gap."

And so, although correlating physical brain activity with mental events is an unquestionable scientific triumph, it has left many students of the brain unsatisfied. For neither neuroscientist nor philosopher has adequately explained how the behavior of neurons can give rise to subjectively felt mental states. Rather, the puzzle of how patterns of neuronal activity become transformed into subjective awareness, the neurobiologist Robert Doty argued in 1998, "remains the cardinal mystery of human existence." Yet there is no faster way to discomfit a room of neuroscientists than to confront them with this mind-body problem, or mind-matter problem, as it is variously called. To avoid it, cellular neurophysiologists position their blinders so their vision falls on little but the particulars of nerve conduction—ions moving in and out, electrical pulses traveling along an axon, neurotransmitters flowing across a synapse. As the evolutionary biologist Richard Lewontin puts it, "One restricts one's questions to the domain where materialism is unchallenged."

Materialism, of course, is the belief that only the physical is ontologically valid and that, going even further, nothing that is not physical—of which mind and consciousness are the paramount examples—can even exist in the sense of being a measurable, real entity. (This approach runs into problems long before minds and consciousness enter the picture: time and space are only two of the seemingly real quantities that are difficult to subsume under the materialist umbrella.) For a sense of the inadequacy of equating what neurons do with what minds experience, consider this thought experiment, based on one first advanced by the Australian philosopher Frank Jackson. Imagine a color-blind neuroscientist who has chosen to study color vision. (Jackson called her Mary.) She maps, with great precision, exactly what happens when light of a wavelength of 650 nanometers falls on the eyes of a volunteer. She laboriously traces the pathway that analyzes color through the lateral geniculate body of the thalamus, along the sweeping fibers of the optic radiation, into the primary visual cortex. Then she carefully notes the activation of the relevant areas of the visual association cortex in the temporal lobe. The volunteer reports the outcome: he sees red! So far, so good. The neuroscientist has precisely described the stimulus—light of a precise wavelength. She has meticulously traced the brain circuits that are activated by this stimulus. And she has been told, by her volunteer, that the whole sequence adds up to a perception of red.

Can we now say that our neuroscientist knows, truly and deeply *knows*, the feeling of seeing red? She certainly knows the input, and she knows its neural correlates. But if she got out of bed the next morning to find that her color blindness had miraculously remitted, and her gaze fell on a field of crimson poppies, the experience of "red" at that instant would be dramatically and qualitatively different from the knowledge she had gained in the lab about how the brain registers the color red. Mary would now have the conscious, subjective, felt experience of color.

We needn't belabor the point that there is a very real difference between understanding the physiological mechanisms of perception and having a conscious perceptual experience. For now, let's say the latter has something to do with awareness of, and attention to, what is delivered for inspection by the perceptual machinery of the central nervous system. This conscious experience, this mental state called a sense of red, is not coherently described, much less entirely explained, by mapping corresponding neural activity. Neuroscientists have successfully identified the neural correlates of pain, of depression, of anxiety. None of those achievements, either, amounts to a full explanation of the mental experience that neural activity underlies. The explanatory gap has never been bridged. And the inescapable reason is this: a neural state is not a mental state. The mind is not the brain, though it depends on the material brain for its existence (as far as we know). As the philosopher Colin

McGinn says, "The problem with materialism is that it tries to construct the mind out of properties that refuse to add up to mentality."

This is not exactly the view you find expressed at the weekly tea of a university neuroscience department. For the most part, the inevitable corollary of materialism known as identity theory—which equates brain with mind and regards the sort of neuron-to-neuron firing pattern leading to the perception of color as a full explanation of our sense of red—has the field by the short hairs. The materialist position has become virtually synonymous with science, and anything nonmaterialist is imbued with a spooky sort of mysticism (cue the *Twilight Zone* theme). Yet it is a misreading of science and its history to conclude that our insights into nature have reduced everything to the material.

The advent of materialism is widely credited to Isaac Newton, who is considered the intellectual father of the view that the world is an elaborate windup clock that follows immutable laws. (Or, as Alexander Pope put it in his famous couplet, "Nature and Nature's laws lay hid in night:/God said, 'Let Newton be!' and all was light.") But that represents a misreading of Newtonian physics. It is true that, by discovering the law of gravity, and realizing that its manifestation on Earth (that famous, if apocryphal, falling apple) and its manifestation in space (tethering the Moon to Earth, and Earth and planets to the Sun) are simply different aspects of the same phenomenon, Newton in some sense largely eliminated the divine from the ongoing workings of the universe. But Newton himself did not believe in pure materialism. Although he rid his clockwork universe of the hand of God, Newton replaced it with something just as immaterial—fields of force. In contrast to the materialist doctrine, which holds that the world is a set of objects that interact through direct contact, Newton's theory of gravity posited action at a distance. Just how, exactly, does Earth keep the Moon from flying away into space? Through gravity. And what is gravity? An ineffable force that pervades all space and is felt over essentially infinite distances. There is no connective tissue, no intervening matter between the mutual gravitational pulls of objects separated by vast distances across a vacuum. You cannot touch a gravitational field (although you can, of course, feel its effects). Newton himself squirmed under the implications of this: "That one body may act upon another at a distance, through a vacuum, without the mediation of anything else . . . is to me so great an absurdity that I believe no man who has . . . any competent faculty of thinking can ever fall into it. Gravity must be caused by an agent . . . but whether this agent be material or immaterial is a question I have left to . . . my readers."

This is not the view that most people associate with classical Newtonian physics. Laypeople as well as most scientists believe that science regards the world as built out of tiny bits of matter. "Yet this view is wrong," argues Henry Stapp, a physicist at the Lawrence Berkeley National Laboratory high in the hills above Berkeley, California. At least one version of quantum theory, propounded by the Hungarian mathematician John von Neumann in the 1930s, "claims that the world is built not out of bits of matter but out of bits of knowledge—subjective, conscious knowings," Stapp says. These ideas, however, have fallen far short of toppling the materialist worldview, which has emerged so triumphant that to suggest humbly that there might be more to mental life than action potentials zipping along axons is to risk being branded a scientific naif. Even worse, it is to be branded nonscientific. When, in 1997, I made just this suggestion over dinner to a former president of the Society for Neuroscience, he exclaimed, "Well, then you are not a scientist." Questioning whether consciousness, emotions, thoughts, the subjective feeling of pain, and the spark of creativity arise from nothing but the electrochemical activity of large collections of neuronal circuits is a good way to get dismissed as a hopeless dualist.

Ah, that dreaded label.

The dualist position in the mind-matter debate dates back to the seventeenth-century French philosopher René Descartes (1596–1650). Although the problem of mind and matter is as old as the philosophy of the ancient Greeks, Descartes was the first modern scientific thinker to grapple seriously with the strangeness of mind, with the fact that the mental realm seems to be of an entirely different character from the material world. His solution was simplicity itself. He posited the existence of two parallel yet separate domains of reality: *res cogitans*, the thinking substance of the subjective mind whose essence is thought, and *res extensa*, or the extended substance of the material world. Mental stuff and material (including brain) stuff are absolutely distinct, he argued. Material substance occupies space (Descartes was big on space: he invented analytic, or Cartesian, geometry), and its behavior can be explained by one piece of matter's mechanically pushing around another piece of matter. Descartes believed that all living things, including all "brute animals," were just "automata or moving machines" that act "according to the disposition of their organs, just as a clock, which is only composed of wheels and weights, is able to tell the hours and measure the time more correctly than we can do with all our wisdom." In Descartes's mechanical clockwork cosmology, all bodies, including living bodies, were automatons, moving around like the mechanical puppets that were fashionable showpieces in the gardens of noblemen of the day. The human body was no exception. Descartes regarded the brain as a machine, subject to mechanistic, deterministic rules, and the body as an automaton. In his 1664 *Traite de l'homme*, Descartes included a charming illustration modeling reflexive behavior. He showed a man's foot edging into a fire; the message "hot!" is depicted traveling through sensory nerves to the head and then back down to a muscle in the leg. This path results in the foot's reflexively pulling out of the blaze. Descartes's careful tracing of the path is one of the earliest examples of those endless neural-correlates discoveries so beloved of twentieth-century neuroscientists.

Descartes defined mind, in contrast to matter, by what it lacks—namely, spatial extent and heft. And he recognized another difference. Reflexes and other attributes or expressions of matter, he argued, are subject to scientific inquiry. Conscious, subjective experience is not. Descartes's separation of nature into a physical realm and a mental/experiential realm, each dynamically independent of the other, thus gave an indirect benefit to science. The seventeenth century saw what threatened to be a to-the-death struggle between science and the Church, which perceived science as a threat. Descartes's declaration that reality divides neatly into two realms reassured the Church that the province of science would never overlap, and therefore never challenge, the world of theology and the spiritual. Science ceded the soul and the conscious mind to religion and kept the material world for itself. In return for this neat dividing up of turf, Descartes hoped, religious leaders would lay off scientists who were studying natural laws operating in the physical, nonmental realm. The ploy was only partly successful for Church-science relations. Descartes himself was forced to flee Paris for Holland in search of greater tolerance.

But this division of reality into mind and matter was also something of a scientific debacle. Separating the material and the mental into ontologically distinct realms raised the white flag early in the mind-body debate: science abandoned the challenge of explaining how the components of the physical world found expression in the mental world. And thus was Cartesian dualism born. Today, three and a half centuries later, his belief endures. If there is a single fundamental underpinning in the intellectual tradition of Western scientific thought, it is arguably that there exists an unbridgeable divide between the world of mind and the world of matter, between the realm of the material (which is definitely real) and the realm of the immaterial (which, according to the conventions of science, is likely illusory).

Yet Cartesian dualism ran into trouble almost immediately. Descartes's material automaton was, in its human form, an automaton with a difference: it was capable of voluntary, volitional, freely willed movement. By exerting its will, Descartes declared, the immaterial human mind could cause the material human machine to move. This bears repeating, for it is an idea that, more than any other, has thrown a stumbling block across the path of philosophers who have attempted to argue that the mind is immaterial: for how could something immaterial act efficaciously on something as fully tangible as a body? Immaterial mental substance is so ontologically different—that is, such a different sort of thing—from the body it affects that getting the twain to meet has been exceedingly difficult. To be sure, Descartes tried. He argued that the mental substance of the mind interacts with the matter of the brain through the pineal gland, the organ he believed was moved directly by the human soul. The interaction allowed the material brain to be physically directed by the immaterial mind through what Descartes called "animal spirits"—basically a kind of hydraulic fluid.

Even in his own time Descartes's dualism fell far short of carrying the day, and its principal antagonist, materialism, quickly reared its head. In the mid-1600s, with the advent of neuroscience, researchers began to piece together new theories of the relationship between mind and brain, discovering basic biological mechanisms underlying conscious feelings and thoughts. On the basis of these findings, the French physician Julien Offray de la Mettrie (1709–1751) asserted that mind and brain are merely two aspects of the same physical entity, and that this three-pound collection of cells sitting inside our skull either entirely determines, or is somehow identical with, mental experience. In his 1747 book *L'homme machine* (Man the Machine), La Mettrie gained notoriety by attempting to show that humans are in essence nothing but automatons. In this he was taking to its logical conclusion a chain of reasoning that had begun when Descartes proclaimed an entirely mechanical understanding of every living thing *save* humans. Even more than Descartes, La Mettrie applied the methods of experimental medical science to bolster his bold claim. He described the brain as the organ of thinking and maintained that brain size determines mental capacity. And he compared the workings of the brain to those of musical instruments. "As a violin string or a key of the clavichord vibrates and renders a sound," he wrote, "so the brain's chords struck by sound waves are stimulated to render or to repeat the words which touch it." Perhaps the most remarkable aspect of La Mettrie's perspective is how contemporary it sounds in this, the age of computer intelligence.

Thus were born the dueling ontologies, with partisans of matter like La Mettrie squaring off against those like Descartes who believed that mental events cannot all be reduced to physical ones. For more than three centuries after Descartes published his thesis, philosophers battled over which entity, mind or matter, was the basic stuff of the world. Philosophers including Leibniz, Berkeley, Hume, Kant, Mach, and James contended that matter is but a uniquely objective and substantial form of mind. This position is not far different from that held by many contemporary physicists, who believe that matter is merely a concentrated form of energy. It is this position that most closely mirrors my own. On the other side of the dualist divide, thinkers such as Hobbes, La Mettrie, Marx, Watson, B. F. Skinner, and Daniel Dennett have argued what has become the consensus position of mainstream science: that mind truly is, in essence, nothing but matter, and our subjective experience that mind is something special or different is just an illusion. The mind is entirely and completely derived from the brain's matter.

Within the scientific if not the philosophical community, the rise of scientific materialism in the midnineteenth century seemed to leave Cartesian dualism in the dust. Materialism not only became the reigning intellectual fashion; it emerged as virtually synonymous with science. In fields from biology to cosmology, science is portrayed as having vanquished the nonmaterial explanations that prescientific cultures advanced for natural phenomena. The mysterious forces once believed to trigger storms have been reduced to permutations of air pressure and temperature. The ghosts behind electric phenomena have been revealed as moving particles. The materialist view of mind holds that there is no more to all this than neurons doing their electrochemical thing. As Colin McGinn puts it: "This is not because neural processes merely cause conscious processes; it's because neural processes are conscious processes. Nor is it merely that conscious processes are an aspect of neural processes; it is rather that there is nothing more to a conscious state than its neural correlate."

The neural connections that form brain circuits are necessary for the existence of mind as we know it. To check this, simply imagine a skull emptied of its brain; when the brain is gone, so are the contents of the mind. Or consider that when the brain is damaged, so is the mind (usually). Or that those neural correlates we've been mentioning are undeniably real: when the mind changes, so does the brain (leaving aside which causes which). To some scientists this is all there is, and mind can thus be fully explained by brain, by matter. The materialist conceit holds that brain is all that's needed—there is no more to the feeling of fear than a tickling of the amygdala, no more to the sound of a whisper than an excitation in the auditory cortex. And there is no more to the sense of free will—the sense that we can choose whether to look left or right, to pick this flower or that, to slip in this CD or that one—than delusion and ignorance about what, exactly, the brain is doing in there. Mind, the theory goes, is just brain with more lyrical allusions. Neural correlates to every aspect of mind you can think of are not merely correlates; they are the essence of that aspect of mind. If introspection tells us otherwise—if it tells us that these "qualia" have an existence and a reality that transcend the crackling and dripping of neurons, and that the power to choose is not illusory—well, then introspection is leading us astray. If introspection tells us that there is more to mind than this, then introspection is flawed. What we think we know about our minds from the inside is wrong, as misguided as an aborigine's examining a television and concluding that there are living beings inside.

Materialism, it seems fair to say, has neuroscience in a chokehold and has had it there since the nineteenth century. Indeed, there are those in the neuroscience community whose reductionist bent is so extreme that they have made it their crusade "to eliminate mind language entirely," as the British neuroscientist Steven Rose bluntly puts it. In other words, notions such as feeling, and memory, and attention, and will—all crucial elements of mind—are to be replaced with neurochemical reactions. This materialist, reductionist camp holds that when we have mapped a mental process to a location in the brain, and when we've worked out the sequence of neurochemical releases and uptakes that is associated with it, we have indeed fully explained, and more important understood, the phenomenon in question. Mystery explained. Case closed.

Or is it? Some of the most eminent neuroscientists have questioned the materialist position on the mind-matter enigma. The Canadian neurosurgeon Wilder Penfield, after a long career dedicated to explaining the material basis of mind, in the end decided that brain-related explanations are intrinsically insufficient. Charles Sherrington, the founder of modern neurophysiology, contended in 1947 that brain processes alone cannot account for the full range of subjective mental phenomena, including conscious free will. "That our being should consist of two fundamental elements offers, I suppose, no greater inherent improbability than that it should rest on one only," he wrote. One of Sherrington's greatest pupils, Sir John Eccles, held similar views. Eccles won a Nobel Prize for his seminal contributions to our understanding of how nerve cells communicate across synapses, or nerve junctions. In his later years, he worked toward a deeper understanding of the mechanisms mediating the interaction of mind and brain—including the elusive notion of free will. Standard neurobiology tells us that tiny vesicles in the nerve endings contain chemicals called neurotransmitters; in response to an electrical impulse, some of the vesicles release their contents, which cross the synapse and transmit the impulse to the adjoining neuron. In 1986 Eccles proposed that the probability of neurotransmitter release depended on quantum mechanical processes, which can be influenced by the intervention of the mind. This, Eccles said, provided a basis for the action of a free will.

It is fair to say that the debate instigated by Descartes over the mind-body problem has not ended at all; it has instead become almost painfully sophisticated and complex. Among the warring theories in play today we have (in one contemporary rundown) "the identity theory, the central state theory, neutral monism, logical behaviorism, token physicalism and type physicalism, token epiphenomenalism and type epiphenomenalism, anomalous monism, emergent materialism, eliminative materialism, various brands of functionalism"—and, undoubtedly, enough additional isms to assign one to every working philosopher in the world. A few words on a small handful of these philosophies of mind and matter (listed from most to least materialistic) should capture the flavor of the debate and give a sense of the competing ideas.

- FUNCTIONALISM, or "Mentalistic Materialism" as the neurosurgeon Joe Bogen has termed it, denies that the mind is anything more than brain states; it is a mere by-product of the brain's physical activity. As the philosopher Owen Flanagan puts it, "Mental processes are just brain processes," and understanding what those brain processes are and how they work tells us all there is to know about what mind is. This view recognizes only material influences. Paul and Patricia Churchland and Daniel Dennett are leading advocates of such materialist views, which are closely akin to behaviorism. The materialist position goes so far as to deny the ultimate reality of mental "events" like our color-blind scientist's sudden experience of the redness of a peony, as well as the actual fact of consciousness itself. Other than the action potentials coursing through brain circuits, they insist, there is nothing more to the workings of the mind—at least, nothing that science needs to address. If we hold tenaciously to such quaint notions as experiential reality, consciousness, and the ontological validity of qualia, it is only out of ignorance: once science parses the actions of the brain in sufficient detail, qualia and consciousness will evaporate just as the "vital spark" did before biologists nailed down the nature of living things. Materialism certainly has one thing going for it. By denying the existence of consciousness and other mental phenomena, it neatly makes the mind-matter problem disappear. No mind, all matter—no mind-matter problem.

- EPIPHENOMENALISM acknowledges that mind is a real phenomenon but holds that it cannot have any effect on the physical world. This school acknowledges that mind and matter are two separate beasts, as are physical events and mental events, but only in the sense that qualia and consciousness are not strictly reducible to neuronal events, any more than the properties of water are reducible to the chemical characteristics of oxygen and hydrogen. From this perspective, consciousness is an epiphenomenon of neuronal processes. Epiphenomenalism views the brain as the cause of all aspects of the mind, but because it holds that the physical world is *causally closed*—that is, that physical events can have only physical causes—it holds that the mind itself doesn't actually cause anything to happen that the brain hasn't already taken care of. It thus leaves us with a rather withered sort of mind, one in which consciousness is, at least in scientific terms, reduced to an impotent shadow of its former self. As a nonphysical phenomenon, it cannot act on the physical world. It cannot make stuff happen. It cannot, say, make an arm move. Epiphenomenalism holds that the brain is the cause of all the mental events in the mind but that the mind itself is not the cause of anything. Because it maintains that the causal arrow points in only one direction, from material to mental, this school denies the causal efficacy of mental states. It therefore finds itself right at home with the fundamental assumption of materialist science, certainly as applied to psychology and now neuroscience, that "mind does not move matter," as the neurologist

C. J. Herrick wrote in 1956. Put another way, all physical action can be but the consequence of another physical action. The sense that will and other mental states can move matter—even the matter that makes up one's own body—is therefore, in the view of the epiphenomenalists, an illusion.

Although epiphenomenalism is often regarded these days as the only generally acceptable alternative to stark materialism, one problem with this position is that it contradicts our basic core experience that mental states really do affect our actions. To deny the causal efficacy of mental states altogether is to dismiss the experience of willed action as nothing but an illusion. Another critical problem with epiphenomenalism (and other schools that deny the causal efficacy of mind) was raised in 1890 by the psychologist and philosopher William James. The basic principles of evolutionary biology would seem to dictate that any natural phenomenon as prominent in our lives as our experience of consciousness must necessarily have some discernible and quantifiable effect in order for it to exist, and to persist, in nature at all. It must, in other words, confer some selective advantage. And that raises an obvious question: What possible selective advantage could consciousness offer if it is only a functionless phantasm? How could consciousness ever have evolved in the first place if, in and of itself, it does nothing? Why, in short, did nature bother to produce beings capable of self-awareness and subjective, inner experience? True, evolutionary biologists can trot out many examples of traits that have been carried along on the river of evolution although not specifically selected for (the evolutionary biologists Stephen Jay Gould and Richard Lewontin called such traits *spandrels,* the architectural term for the elements between the exterior curve of an arch and the right angle of the walls around it, which were not intentionally built but were instead formed by two architectural traits that were "selected for"). But consciousness seems like an awfully prominent trait not to have been the target of some selection pressure. As James put it, "The conclusion that [consciousness] is useful is . . . quite justifiable. But if it is useful, it must be so through its causal efficaciousness."

- EMERGENT MATERIALISM argues that mind arises from brain in a way that cannot be fully predicted from or reduced to brain processes. The attributes of mind, that is, cannot be explained solely by brain's physical activity. Further, according to this view, mind may have the power to effect both mental and physical change. Emergentists like Steen Rasmussen suggest that, sometimes, a high-order, emergent property like mind has the power to exert an effect on the lower-order processes that created it. In other words, what emerges can affect what it emerges from.

The Nobel-winning neuroscientist Roger Sperry taught at the California Institute of Technology from 1954 until his death in 1994. Best known for his study of "split brain" patients (many of whose surgeries severing the connections between the right and left cerebral hemispheres were actually performed by Joe Bogen), Sperry produced the most detailed and scientifically based version of emergent materialism. He variously called his own emergent theory "mentalism," "emergent mentalism," or just "the new mentalism." At first, he argued only that mind is not reducible to cerebral activity, echoing the mainstream emergent position that mind arises from brain as a unique entity whose attributes and power cannot be predicted, or even explained, from its material components alone. But later Sperry became uneasy with the triumph of materialism in neuroscience and what he called its "exclusive 'bottom-up' determination of the whole by the parts, in which the neuronal events determine the mental but not vice versa." As a result he later espoused a view that mental states can indeed have causal efficacy. In contrast to agnostic physicalism (discussed later), which allows mental states to influence other mental states only through the intermediary of the brain, emergent materialism grants to some mental states the power directly to change, shape, or bring into being other mental states, as well as to act back on cerebral states. In the years just before his death, Sperry hinted that mental forces could causally shape the electrochemical activity of neurons.

This represented a radical new vision of the causal relations between higher-order mental processes and neuronal events. What Sperry termed "mental forces" could, he argued, direct the electrochemical traffic between neurons at the cellular level. This view thus argues that emergent mental properties can exert top-down causal control over their component parts—"the downward control by mental events over the lower neuronal events." This, as we will see in Chapter 2, describes very well the control by an OCD patient's mind of his neuronal events, specifically the activity in the pathological circuits underlying the disease. Sperry was at pains to point out that his belief did not constitute dualism (that dreaded word!) in any Cartesian sense, but rather a radically revised form of materialism in which the mind is not only emergent but also causal. He maintained (as classical, non-science-based dualists do not) that the myriad conscious experiences cannot exist apart from the brain; he did not posit an unembodied mind or consciousness as, again, classical dualists do. The mental forces he considered causally efficacious were no spooky, nonmaterial, preternatural entities. As he put it in 1970, "The term [mental forces] . . . does not imply here any disembodied supernatural forces independent of the brain mechanism. The mental forces as here conceived are inseparably tied to the cerebral structure and its functional organization." They shape and direct the lower-level traffic of electrical impulses. The form of causal efficacy Sperry proposed was one that adherents of materialist, bottom-up determinism dismissed—namely, one in which "higher-level" mental properties exert causal control over the "lower level" of neurons and synapses. In this scheme, Sperry wrote in 1965, "the causal potency of an idea, or an ideal, becomes just as real as that of a molecule, a cell, or a nerve impulse." He fervently hoped that the new view of mind would integrate "traditionally conflicting positions on mind and matter, the mental versus the material," and that "science as a whole may be in the process of shifting away from its centuries-old microdeterminate materialist paradigm to a more valid macromental model for causal explanation and understanding."

Not even a Nobel Prize offered adequate shielding from the brickbats hurled at Sperry for this plunge into the mind-and-matter wars. When the English psychologist Oliver Zangwill visited Caltech in August 1970, as Joe Bogen recounts, he expressed to Sperry his concern that if "Sperry went on in this vein it is likely to diminish the impact of his many marvelous achievements." How, Bogen asked, did Sperry react? Very little, replied Zangwill. From about 1980, almost all of Sperry's writings were about consciousness and mental forces acting from the top down. When he was honored at Caltech in 1982 on the occasion of his Nobel, those who had come to know him only recently assumed, recalls Bogen, "that he's gone religious like so many old folks." By 1990, even Caltech professors who had known Sperry for four decades "had given up trying to defend or even to understand 'the philosophy of his later years,' as one of them put it."

Although Sperry put great stress on the reality of the mind in the causal chain, when pressed he seemed to fall back on classical materialist assumptions. He emphatically denied the importance of quantum mechanics for understanding mind-brain relations, insisting that Newtonian physics was entirely up to the task. "It remains true in the mentalist model that the parts . . . determine the properties of the whole, i.e. microdeterminism is not abandoned," he wrote in his last major paper. "The emergent process is . . . in principle, predictable." Thus the mental forces he was so fond of referring to were themselves determined from below. To those, like me, who were becoming committed to the genuine power of mental force and its integral role in a quantum-based mind-brain theory, Sperry's views seemed like a refined form of epiphenomenalism.

- AGNOSTIC PHYSICALISM also holds that mind derives exclusively from the matter of the brain. In contrast to the epiphenomenalists and functionalists, however, adherents of this school acknowledge that this may not be the whole story. That is what the "agnostic" part reflects: those who subscribe to this worldview do not deny the existence of nonmaterial forces, just as an agnostic does not actively deny the existence of God. Rather, they regard such influences, if they exist, as capable of affecting mental states only as they first influence observable cerebral states. William James falls into this camp. Joe Bogen is careful to distinguish physicalism from materialism. The former holds that the mental does not change without the physical's (that is, brain states) changing, too. This says nothing about the existence of nonmaterial influences on the mind. It simply asserts that any such influences must work through the brain in order to affect the mind. In contrast, materialism transcends physicalism in actively denying the existence of nonmaterial influences.

In explaining his own position, Bogen recounts an argument he once had with the philosopher Paul Churchland about the mystery of how brain produces mind, and the need some philosophers and neuroscientists perceive to invoke something immaterial and without spatial extent to affect the brain. Churchland burst out, "Throughout the history of this subject the mind has been considered to be between God and brain. But now you presume to put the brain between God and mind." To which Bogen replied, "Exactly so, which is how I can be a committed physicalist while remaining agnostic or even indifferent about the immaterial."

- Process philosophy, a school greatly influenced by Alfred North Whitehead, holds that mind and brain are manifestations of a single reality, one that is in constant flux. It thus is compatible with classical Buddhist philosophy, which views clear and penetrating awareness of change and impermanence (*anicca* in Pali) as the essence of insight. Thus, as Whitehead put it, "The reality is the process," and it is a process made up of vital transient "drops of experience, complex and interdependent." This view is strikingly consistent with recent developments in quantum physics.
- Dualistic interactionism holds that consciousness and other aspects of mind can occur independently of brain. In this view, mental states have the power to shape brain or cerebral states—and, going even further, the mind cannot in any sense be reduced to the brain. Although mind depends on brain for its expression, brain is by its very material nature not sufficient to explain mind completely, for consciousness and everything else lumped under this thing called mind are categorically different beasts from brain and everything else material. John Eccles, who along with the philosopher Karl Popper for many years gallantly championed this view, put it this way not long before his death: "The essential feature of dualist-interactionism is that the mind and brain are independent entities . . . and that they interact by quantum physics." Scientists and philosophers in this camp reject materialism to the point of actually positing a nonmaterial basis for the mind. Even worse, they seem to have a penchant for speaking about the possibility of life after death, something no self-respecting scientist is supposed to do in public (although both Eccles and Penfield did). Even scientists and philosophers who question whether simply mapping neural correlates can truly provide the ultimate answer have doubts about dualistic interactionism: neuroscientists may have worlds to go before they understand *how* brain gives rise to mind, but even in

a field not generally marked by certainty they are as sure as sure can be that it does, somehow, manage the trick.

Even this abbreviated rundown of mind-brain philosophies would not be complete without what the Australian philosopher David Chalmers calls "don't-have-a-clue materialism." This is the default position of those who have no idea about the origins of consciousness or the mind but assert that "it must be physical, as materialism must be true," as Chalmers puts it. "Such a view is held widely, but rarely in print." One might add that many working scientists hold this view without really reflecting on the implications of it.

Although none of these worldviews has ended the mind-matter debate, most philosophers who study consciousness nevertheless hew largely to some form of the reductive materialist creed. But there are notable exceptions. Dave Chalmers is one of those arguing for what he calls "a non-reductive ontology of consciousness"—in other words, an approach that does not reduce consciousness to a (mere) physical process. Chalmers says that he "started out life as a materialist, because materialism is a very attractive scientific and philosophical doctrine." But he became more and more dissatisfied with the dogmatic materialist ontology that posits that all aspects of consciousness are a logically entailed and perhaps metaphysically necessary outcome of materialist processes. He therefore began to focus on the explanatory gap between the material and the mental—between explanations of how neurons work, on the one hand, and our felt, inner awareness on the other. Even if we knew everything about every field and iota of matter in the universe, in other words, it is hard to see how that knowledge would produce that elusive "Aha!" moment, when we all say, *Oh, right, so that's how consciousness is done* (in the way we might react, say, to a materialist explanation of how the liver produces bile). Those neuronal mechanisms, Chalmers concluded, would never in and of themselves add up to consciousness. Physical form and function add up to more physical form and function. "To truly bridge the gap between the physical nature of brain physiology and the mental essence of consciousness, we have to satisfy two different conceptual demands," Chalmers told the public television series "Closer to the Truth" in 2000. "It's not yet looking very likely that we're going to reduce the mind to the brain. In fact, there may be systematic reasons to think there will always be a gulf between the physical and the mental."

If that gulf is unbridgeable, Chalmers therefore argues, consciousness might profitably be regarded as what he calls a "nonreductive primitive," a fundamental building block of reality, just as mass and electric charge, as well as space and time, are nonreductive primitives in theories of the physical world. Taking consciousness as a primitive rather than as an emergent property of the physical brain, Chalmers's search for a nonreductive ontology of consciousness led him to what he calls panprotopsychism. The *proto* reflects the possibility that the intrinsic properties of the basic entities of the physical world may be not quite mental, but that collectively they are able to constitute the mental (it is in this sense of *proto* that physics is "protochemical"). In this view, mind is much more fundamental to the universe than we ordinarily imagine. Panprotopsychism has the virtue of integrating mental events into the physical world. "We need psychophysical laws connecting physical processes to subjective experience," Chalmers says. "Certain aspects of quantum mechanics lend themselves very nicely to this."

In particular, if consciousness is an ontological fundamental—that is, a primary element of reality—then it may have the power to achieve what is both the best-documented and at the same time the spookiest effect of the mind on the material world: the ability of consciousness to transform the infinite possibilities for, say, the position of a subatomic particle as described by quantum mechanics into the single reality for that position as detected by an observer. If that sounds both mysterious and spooky, it is a spookiness that has been a part of science since almost the beginning of the twentieth century. It was physics that first felt the breath of this ghost, with the discoveries of quantum mechanics, and it is in the field of neuroscience and the problem of mind and matter that its ethereal presence is felt most markedly today. "Quantum theory casts the problem of the connection between our conscious thoughts and the activities of our brains in a completely new light," argues my physicist friend Henry Stapp. "The replacement of the ideas of classical physics by the ideas of quantum physics completely changes the complexion of the mind-brain dichotomy, of the connection between mind and brain."

As the millennium turned, a smattering of neuroscientists finally began to accept that consciousness and the mind (as opposed to mere brain) are legitimate areas of scientific inquiry. This is not to say that the puzzle of how mind is interconnected with brain is anywhere close to solution, but at least researchers are letting it into the lab, and that is a signal change from the recent past. "When I first got interested in [the problem of consciousness] seriously and tried to discuss it with brain scientists," recalled John Searle in 2000, "I found that most of them were not interested in the question. . . . Consciousness seems too airy-fairy and touchy-feely to be a real scientific subject." Yet today neuroscientists flock to conferences with *consciousness* in their names, write books with *consciousness* in their titles, and contribute to a journal that boldly proclaims itself the *Journal of Consciousness Studies*. Two Nobel laureates have moved on from the work that won them an invitation to Stockholm, to pursue the puzzle of consciousness: Francis Crick, who with James Watson shared the Nobel Prize in Physiology or Medicine for determining the structure of DNA, mused in his book *The Astonishing Hypothesis* that the seat of volition might be found in a deep crevasse in the cortex called the anterior cingulate sulcus; Gerald Edelman, who shared the 1972 Nobel for working out the molecular structure of antibodies, argued that consciousness arises from the resonating interplay of assemblies of neurons. More and more scholars are concluding that our deep inner sense of a mental life not fully encompassed by the electrochemical interactions of neuronal circuits is not delusional. As the German neuroscientist Wolf Singer puts it, these elements of consciousness "transcend the reach of reductionistic neurobiological explanations." More and more neuroscientists are admitting to doubts about the simplistic materialistic model, allowing, as Steven Rose does, that brain has an "ambiguous relationship to mind."

As a result, a whiff of dualism is once again rising, like the smoke from a long-extinguished campfire, within the scientific community. "Many people, including a past self of mine, have thought that they could simultaneously take consciousness seriously and remain a materialist," writes Dave Chalmers. "This is not possible. . . . Those who want to come to grips with the phenomenon must embrace a form of dualism. One might say: you can't have your materialist cake and eat your consciousness too." Instead, Chalmers argues that it is time, and it is necessary, to sacrifice the simple physicalist worldview that emerged from the scientific revolution and stood science in good stead for the last three centuries. Although philosophers and scientists both have been known to argue that materialism is the only worldview compatible with science—"dualism is inconsistent with evolutionary biology and modern physics and chemistry," Paul Churchland asserted in 1988—this is simply false. Nor is it justifiable to hew to materialism in the misguided belief that embracing dualism means embracing something supernatural, spiritualistic, nonscientific. To the contrary: scientists questioning reductive materialism believe that consciousness will turn out to be governed by natural law (even if they haven't a clue, yet, about what such a law might look like). As Chalmers says, "There is no a priori principle that says that all natural laws will be physical laws; to deny materialism is not to deny naturalism."

In a welcome irony, centuries of wrestling with the mind-matter problem that arose from the clash between dualism and materialism might come down to this: dualism, with its assertion that there are two irreconcilable kinds of stuff in the world, and materialism, with its insistence that there is only the material, should both be tossed on the proverbial trash heap of history. Dualism fails to explain the relationship between mind and matter, in particular how the former can be functionally conjoined with the latter; materialism denies the reality of subjective states of sentience. Dualism leads us to a dead end; materialism doesn't even let us begin the journey.

At a meeting on consciousness held in Tucson in the spring of 2000, I was delighted when the philosopher John Searle asserted his belief that volition and the will are real, able to influence how the material stuff of the brain expresses itself. After the session, Dave Chalmers needled me for enjoying Searle's talk so much. Well, I had: it seemed to me that Searle was the first mainstream philosopher to question whether the physical realm could account for all our mental experiences. Chalmers said that's not what he heard Searle say. In fact, Dave grinned, he would bet me twenty dollars that Searle had not at all denied "causal closure of the microphysical"—that is, the belief that only physical causes can bring about physical effects, which would preclude the nonmaterial mind's affecting the physical brain. I took the bet. Let's ask him, Dave said. No no, I objected: he'll deny it; let's get a copy of his paper and see for ourselves. During the break I went to the exhibit hall and found the *Journal of Consciousness Studies* booth, where Searle had dropped off a preliminary version of his paper. I wheedled a photocopy out of them. In it appeared this argument for the causal efficacy of the rational mind:

> *[Neuro]physiological determinism [coexisting] with psychological libertarianism is intellectually very unsatisfying because, in a word, it is a modified form of epiphenomenalism. It says that psychological processes of rational decision*

making do not really matter. The entire system is deterministic at the bottom level, and the idea that the top level has an element of freedom is simply a systematic illusion. . . . If [this] is true, then every muscle movement as well as every conscious thought is entirely fixed in advance, and the only thing we can say about psychological indeterminism at the higher level, is that it gives us a systematic illusion of free will.

[This] hypothesis seems to me to run against everything we know about evolution. It would have the consequence that the incredibly elaborate, complex, sensitive and—above all—biologically expensive system of conscious rational decision making would actually make no difference whatever to the life and survival of organisms. Epiphenomenalism is a possible thesis, but it is absolutely incredible, and if we seriously accepted it, it would make a change in our world view more radical than any previous change, including the Copernican Revolution, Einsteinian relativity theory and quantum mechanics.

We found Searle after the session, and Chalmers asked him point blank. "Of course I do not deny causal closure," Searle shot back. I wasn't surprised; most scientists seem to have a morbid fear of saying that anything nonphysical can have causal efficacy in the physical realm. The furthest most scientists and philosophers are willing to go is to acknowledge that what we think of as mental events act back on the brain only through the physical states that give rise to those mental events. That is, brain state A may give rise to mental state A' as well as brain state B—but the causal actor in this case was brain state A and not mental state A'.

Anyway, at an end-of-conference party that evening at Chalmers's house, I paid Dave the twenty dollars we'd bet. I did note, though, that Searle's insisting on causal closure of the physical world was logically inconsistent with his argument that volition is real and able to affect the physical matter of the brain. I hope the beginnings of an answer to this quandary will emerge from the data I will present showing the critical role of willed effort in generating self-directed cerebral change. In any event, as I said to Dave when I handed him the twenty dollars, "This story still has a long way to go."

Wrestling with the mystery of mind and matter is no mere academic parlor game. The rise of modern science in the seventeenth century—with the attendant attempt to analyze all observable phenomena in terms of mechanical chains of causation—was a knife in the heart of moral philosophy, for it reduced human beings to automatons. If all of the body and brain can be completely described without invoking anything so empyreal as a mind, let alone a consciousness, then the notion that a person is morally responsible for his actions appears quaint, if not scientifically naïve. A machine cannot be held responsible for its actions. If our minds are impotent to affect our behavior, then surely we are no more responsible for our actions than a robot is. It is an understatement to note that the triumph of materialism, as applied to questions of mind and brain, therefore makes many people squirm. For if the mysteries of the mind are reducible to physics and chemistry, then "mind is but the babbling of a robot, chained ineluctably to crude causality," as the neurobiologist Robert Doty put it in 1998.

In succeeding chapters we will explore the emerging evidence that matter alone does not suffice to generate mind, but that, to the contrary, there exists a "mental force" that is not reducible to the material. Mental force, which is closely related to the ancient Buddhist concepts of mindfulness and karma, provides a basis for the effects of mind on matter that clinical neuroscience finds. What is new here is that a question with deep philosophical roots, as well as profound philosophical and moral implications, can finally be addressed (if not yet fully solved) through science. If materialism can be challenged in the context of neuroscience, if stark physical reductionism can be replaced by an outlook in which the mind can exert causal control, then, for the first time since the scientific revolution, the scientific worldview will become compatible with such ideals as will—and, therefore, with morality and ethics. The emerging view of the mind, and of the mind-matter enigma, has the potential to imbue human thought and action with responsibility once again.

{ TWO }

BRAIN LOCK

To refrain from an act is no less an act than to commit one.

—Sir Charles Sherrington,

"The Brain and Its Mechanism," 1933

The important thing in science is not so much to obtain new facts as to discover new ways of thinking about them.

—Sir William Lawrence Bragg

Dottie was a middle-aged wife and mother by the time she walked into my office at the Obsessive Compulsive Disorder (OCD) Research Group at UCLA Medical Center in Westwood, but she had been in the grip of obsessive-compulsive disorder since she was a little girl of five. Early on, it was the numbers 5 and 6 that paralyzed her with fear, she told me with some distress. I soon learned why: her obsession with the "magical" powers of numbers still consumed large portions of her life. If, while driving, Dottie glimpsed a license plate containing either a 5 or a 6, she felt compelled to pull over immediately and sit at the side of the road until a car with a "lucky" number in its license plate passed by. Without a lucky number to counteract the digits of doom, Dottie was convinced, something terrible would befall her mother. She would sometimes sit in the car for hours, waiting for the fates to bestow permission to hit the road again. When Dottie had a son of her own, her obsession shifted. Now it was eyes: Dottie was certain that if she made the slightest misstep, her son would go blind. If she walked where someone with vision problems had walked, she would throw out her shoes; if she so much as heard the word *ophthalmologist* she would cringe in terror. As she spoke, I noticed the word *vision* written four times in the palm of her hand. Oh, that, she explained, eyes downcast: while she was watching television that afternoon, a terrifying thought about eyes had popped into her head. This was her way of exorcising it. If she hadn't, there was no telling what might have befallen her son's eyesight.

Obsessive-compulsive disorder is a neuropsychiatric disease marked by distressing, intrusive, unwanted thoughts (the obsession part) that trigger intense urges to perform ritualistic behaviors (the compulsion part). Together, obsessions and compulsions can quickly become all-consuming. In Dottie's case, the obsessive thoughts centered first on her mother's safety and then on her son's eyesight; her compulsions were the suite of "magical" behaviors she engaged in to ward off disaster to the people she loved. The unremitting thoughts of OCD intrude and lay siege to the sufferer's mind (*obsession* comes from the Latin verb that means "to besiege"), insisting that the doorframe you just brushed is contaminated with excrement, or that the bump in the road you just drove over was not an uneven patch of asphalt but a body sprawled on the pavement.

One of the most striking aspects of OCD urges is that, except in the most severe cases, they are what is called ego-dystonic: they seem apart from, and at odds with, one's intrinsic sense of self. They seem to arise from a part of the mind that is not you, as if a hijacker were taking over your brain's controls, or an impostor filling the rooms of your mind. Patients with obsessive-compulsive disorder experience an urge to wash their hands, for instance, while fully cognizant of the fact that their hands are not dirty. They ritualistically count the windows they pass, knowing full well—despite the contrary message from their gut—that failing to do so will not doom their child to immediate death. They return home to check that the front door is locked so often as to render them unable to hold a job, even though part of their brain knows full well that it is securely locked. They count the steps from their car to the door of the office where they have a job interview, hoping and praying that the number will turn out to be a perfect square, a prime, a member of the Fibonacci sequence, or something else magical, since if it is not, they must do an about-face and return to the car to try again. They do this time and again, knowing that the interview time is fast approaching and that—oh, God, they've lost out on another job because of this crazy disease. OCD has a lifetime prevalence of 2 to 3 percent; in round numbers, it affects an estimated one person in forty, or more than 6 million Americans, typically striking in adolescence or early adulthood and showing no marked preference for males or females.

Excessive and ritualized hand-washing may be the best known of the OCD compulsions, but there are scores of others. They include alphabetizing the contents of a pantry, repeatedly checking to see whether a door is locked or an appliance is turned off, checking over and over to see whether you have harmed someone (peeking in on a sleeping child every minute, for instance), following rituals to ward off evil (like scrupulously avoiding sidewalk cracks), touching or tapping certain objects continuously, being unable to resist counting (totting up, every day, the traffic lights you pass en route home), or even excessively making lists. OCD can manifest itself as obsessions about order or symmetry, as expressed in an irresistible need to line up the silverware just so, or as an obsession about hoarding, as expressed in never throwing out old magazines and newspapers. Paradoxically, perhaps, giving in to the urge to wash or check or count or sort, which the patient does in the vain hope of making the dreadful feeling recede, backfires. An OCD compulsion does not dissipate like a scratched itch. Instead, giving in to the urge exacerbates the sense that something is wrong. It's like chronic poison ivy of the mind: the more you scratch it, the worse it gets.

Someone with obsessive-compulsive disorder derives no joy from the actions she takes. This puts OCD in marked contrast to, for instance, compulsive gambling or compulsive shopping. Although both compulsive shoppers and compulsive gamblers lack the impulse control to resist another trip to the mall or another game of video poker, at least they find the irresistible activity, well, kind of fun. An OCD patient, in contrast, dreads the arrival of the obsessive thought and is ashamed and embarrassed by the compulsive behavior. She carries out behaviors whose grip she is desperate to escape, either because she hopes that doing so will prevent some imagined horror, or because resisting the impulse leaves her mind unbearably ridden with anxiety and tortured by insistent, intrusive urges. Since the obsessions cannot be silenced, the compulsions cannot be resisted. The sufferer feels like a marionette at the end of a string, manipulated and jerked around by a cruel puppeteer—her own brain.

Freud believed that OCD is a manifestation of deep emotional conflicts. As a result, patients who sought traditional psychiatric therapy for the illness were typically told that the rituals they performed or the thoughts they could not silence were rooted in sexual conflict and reflected, for instance, a repressed memory of childhood trauma. The content of the disease—why one patient can't stop thinking that she left the coffee maker on, while another is beset by a compulsion to wash doorknobs—may indeed reflect the individual's personal history. But as yet there is no biological explanation for why OCD expresses itself one way in one patient and a different way in another. Nor is it clear what the ultimate cause of obsessive-compulsive disorder is, though there is clearly a genetic contribution.

Until the mid-1960s, the psychiatric and psychological professions deemed OCD *treatment-intractable:* nothing could be done to release patients from its grip. "People didn't know what to do with OCD," says the clinical psychologist Michael Kozak, who spent nineteen years at MCP Hahnemann Hospital in Philadelphia studying the disease and its treatments. "They tried all sorts of things that didn't work very well, from electroshock and psychosurgery to any available drug and classical talk therapy." In the late 1960s and early 1970s, however, psychiatrists got a hand from serendipity: they noticed that when patients suffering from clinical depression were put on the tricyclic antidepressant clomipramine hydrochloride (Anafranil), some of them experienced relief from one or more of their OCD symptoms. Since clomipramine, among its many biochemical actions, strongly inhibits inactivation of the neurotransmitter serotonin (much as Prozac does), researchers suspected that amplifying the brain's serotonin levels might alleviate OCD.

There was at least one problem with this approach, however. Though clearly effective, clomipramine is a "dirty" drug, one with numerous pharmacological actions; as a result, it is associated with many unpleasant side effects. This problem led to the development of so-called selective serotonin reuptake inhibitors (SSRIs), such as Prozac, Paxil, Zoloft, Luvox, and Celexa, all of which specifically block the same mechanism that clomipramine acts on nonspecifically: the molecular pump that moves serotonin back into the neurons from which it was released, thus allowing more of the chemical to remain in the synapse. All of these SSRIs seem to be equally effective in treating OCD symptoms. For each of them, studies since the 1980s have shown, about 60 percent of patients respond at least somewhat, "and of those there's about a 30 to 40 percent reduction in symptoms," says Kozak. "So there's something real going on with the drugs. But when about half of the people aren't helped significantly and those who are helped are still left with 60 percent of their symptoms, we have a ways to go."

At about the same time that researchers stumbled onto clomipramine for OCD, Victor Meyer, a psychologist at Middlesex Hospital in London, began to develop what would emerge as the first effective behavioral therapy for the disease. In 1966 he tried out, on five patients in an inpatient psychiatric ward, what would become the most widely used psychological treatment for the next twenty-five years. Called exposure and response prevention (ERP), it consisted of exposing patients to the trigger that called forth obsessional thoughts and the compulsion to engage in a distress-relieving behavior. Meyer would tell a patient to leave her house, for instance, but prevent her from returning to check whether she had left the stove on. Or he would have her touch all the doorknobs in a public building but not allow her to wash her hands afterward. Or he would tell the patient to touch dried bird droppings but not allow her to wash (at least not right away). Meyer reported significant improvement in the patients he treated with exposure and response prevention. Edna Foa, who adopted the technique and added a detailed questionnaire to allow therapists to get at the patient's so-called fear structure—the layers of emotions that underlie the obsessions—introduced it into the United States.

Typically, the first exposure during therapy uses a trigger that the patient has assigned a low score on a scale of "subjective units of distress," or SUDs. The therapist (during in-office sessions) then prevents the patient from responding as he usually does—dashing to a sink to wash, for instance. *Prevents* can mean many things, from gentle coercion to physical restraint of the patient; from carefully explaining that if the patient complies he is likely to get better, to turning off the water in the bathroom of his room in a mental hospital. Exposures are also conducted at home; the patient works at stopping himself from acting on his compulsive urges. The patient, needless to say, can become extremely anxious during this phase, which often lasts an hour or more. Ideally, however, as therapy continues, he begins to master his responses to triggers further up the distress scale, the anxiety ignited by the triggers lessens, and he gains control of his thoughts and actions.

Controversy swirls around exposure and response prevention therapy, however. The most common claim for the treatment is that three out of four patients who complete therapy do well, experiencing a 65 percent reduction in their OCD symptoms. But that little phrase "who complete" hides a land mine. "The trouble is, a lot of people won't do it at all, they're so afraid to confront their obsessions and not be allowed to carry out their compulsions," says Kozak. During his work with Edna Foa in Philadelphia, where they developed one of the best programs in the United States, some 25 percent of patients refused to undergo a single session once they learned what was involved. With less adept clinicians, refusal rates can run even higher. Some clinicians manipulate their dropout rates by fiddling with the entry criteria: by doing a little exposure and response prevention and rejecting patients who can't take it, researchers make their results look better. Even then, 10 to 30 percent of patients who agree to start therapy drop out. And not every clinician practicing exposure and response prevention has done it well, or wisely. "There have been quite some mistakes, with therapists abusing the method or going faster than patients would have liked," says Dr. Iver Hand of the University of Hamburg in Germany, a pioneer in the field who developed a variant of ERP. "It is easy for a badly-trained therapist to abuse the method." Compared to drugs, behavioral therapy seemed to produce better results for patients who could tolerate it. But the hidden statistics made it clear: for millions of OCD patients, exposure and response prevention was not the answer.

This was the state of play when I entered the field in the mid-1980s. It wasn't so much psychology, or even physiology, that attracted me to the study of obsessive-compulsive disorder. It was philosophy. OCD, I thought, offered a wedge into the mind-brain enigma. Because symptoms are usually so clear-cut that patients can describe precisely what they feel, I realized that there should be no problem establishing the mental, experiential aspect of the disease. And since it was becoming clear even in the 1980s that psychiatric illness was rooted in the functional neuroanatomy of the brain, I was also optimistic that it would be possible to establish what was happening in the brain of people with the disease. Finally, the disease's ego-dystonic nature suggested that although the brain was acting up, an intact mind was struggling to overcome it: the events of the brain and the state of the mind were, at least partly, separable. Obsessive-compulsive disorder thus seemed to be the perfect vehicle for pursuing such profound questions as the schism between mind and brain and, in particular, the distinction between active and passive mental activity: the symptoms of OCD are no more than the products of passive brain mechanisms, but patients' attempts to resist the compulsions represent active, mental effort.

What attracted me most to the psychological treatment of OCD, however, was a tantalizing possibility. Cognitive therapy—a form of structured introspection—was already widely used for treating depression. The idea is to help patients more clearly assess the contents of their thought stream, teaching them to note and correct the conceptual errors termed "cognitive distortions" that characterize psychopathological thinking. Someone in the grips of such thinking would, for instance, regard a half-full glass not merely as half-empty but also as fatally flawed, forever useless, constitutionally incapable of ever being full, and fit only to be discarded. By the mid-1980s, cognitive therapy was being used more and more in combination with behavioral therapy for OCD, and it seemed naturally compatible with a mindfulness-based perspective. If I could show that a cognitive-behavioral approach, infused with mindful awareness, could be marshaled against the disease, and if successful therapy were accompanied by changes in brain activity, then it would represent a significant step toward demonstrating the causal efficacy of mental activity on neural circuits.

So in February 1987 I launched a group therapy session for OCD patients, meeting every Thursday afternoon, in conjunction with an ongoing study of the disease's underlying brain abnormalities that my colleagues and I at the UCLA School of Medicine had begun in 1985. One of the first patients in the group was a man who could not stop washing. His wife, driven to distraction by his compulsions, was on the verge of leaving him. Although the man felt incapable of resisting the urge to wash, at the same time he had clear insight into how pathological his behavior was. After nearly a year of group therapy, with winter approaching, he said, "That's it. I've had it. This winter I'm not doing washing compulsions. I'm not going through another winter with raw, red, cracked, chapped, painful hands. I'd rather be dead." This was a level of resolve neither I nor anyone in the group had seen before. Over the next few weeks, he actually managed to pull it off. He held his washing to normal levels and made it through the winter without chapped hands.

That case was uppermost in my mind as we delved ever deeper into the study of OCD's underlying neuroanatomy. Two years earlier we had studied depression, observing (as many other groups would later) that the brains of depressed patients are often marked by changes in cortical activity as detected on *positron emission tomography* (PET), the noninvasive imaging technique that measures metabolic activity in the brain. We had begun studying obsessions after observing that many patients with depression had intrusive, obsessional thoughts. The obvious question arose: What brain changes mark OCD itself? An advertisement that we placed in the local paper, asking, "Do you have repetitive thoughts, rituals you can't control?" brought an overwhelming response. Over the next several years, we invited about fifty of those respondents to the UCLA Neuropsychiatric Institute to undergo a full assessment for possible OCD.

In an analysis of the PET scans of twenty-four patients, published in a series of papers in the late 1980s, we pinpointed several brain structures that seemed to be consistently involved in obsessive-compulsive disorder. Compared to the brains of normal controls, the brains of our OCD volunteers showed hypermetabolic activity in the orbital frontal cortex, which is tucked into the underside of the front of the brain above and behind the eyes (hence its name) as shown in Figure 1 on page 63. The scans showed, too, a trend toward hyperactivity in the caudate nucleus. Another group had found that a closely related structure, the anterior cingulate gyrus, was also pumped up in the brains of OCD patients.

By 1990, five different studies by three different research teams

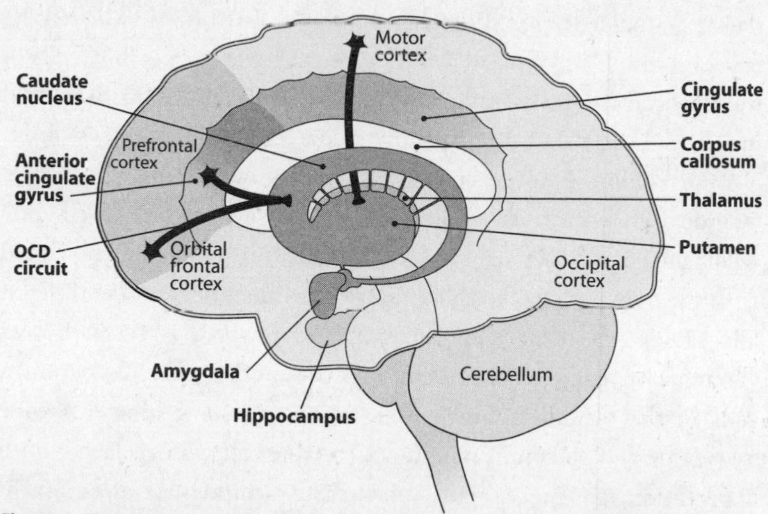

Figure 1: This side view of the brain shows some of its key structures, including those involved in OCD. In the "OCD circuit," neurons that project from the orbital frontal cortex and the anterior cingulate gyrus to the caudate nucleus are overactive, generating the persistent sense that something is amiss.

had all shown elevated metabolism in the orbital frontal cortex in patients with OCD, so this structure consumed our attention. We combed the literature for clues to what the orbital frontal cortex (OFC) does for a living in the normal human brain. Our first major clue came from studies by the behavioral physiologist E. T. Rolls at Oxford University in the late 1970s and early 1980s (studies whose results were later echoed by researchers elsewhere). In one key set of experiments, Rolls and his colleagues taught rhesus monkeys that every time they saw a blue signal on a monitor, licking a little tube in their cage would get them a sip of black currant juice, one of their favorite beverages. Licking the tube in the absence of the blue light would do nothing. Good Pavlovians all, the monkeys learned quickly to lick the tube when they saw blue. Through electrodes implanted in the brains of these alert animals, Rolls observed that the orbital frontal cortex now became active as soon as blue appeared.

Then Rolls switched signals on his little furry subjects: now green meant juice and blue meant salt water, which monkeys (being no fools) despise. When the monkeys saw blue and licked the tube, but got brine instead of the juice they were expecting, cells in their orbital frontal cortex went ballistic, firing more intensely and in longer bursts than the cells did when the tube contained juice. Yet these cells did not respond when the monkeys sipped salt water outside the test situation. Instead, this group of cells fired only when the color previously associated with juice became associated with the delivery of something nonrewarding, or even of nothing at all. The mere absence of an expected reward, it seemed, was enough to trigger intense activity in these OFC cells. Apparently, cells in the orbital frontal cortex fire when something has gone awry, as when an actual experience (getting salt water) clashes with expectation (getting currant juice). The orbital frontal cortex, it seems, functions as an error detector, alerting you when something is amiss—if you're a rhesus monkey, getting a mouthful of brine when you're expecting currant nectar is the essence of amiss. Expectations and emotions join together here to produce a sort of neurological spellcheck.

If cells of the orbital frontal cortex do indeed function as rudimentary error detectors, then they should quiet down when expectation and reality are back in harmony. And that is what the Oxford group found. Once the monkeys learned to associate green with juice, OFC cells quieted down, firing in shorter and less intense bursts than they did when they detected an error in the world around them. From these experiments, it seemed clear to me that error-detection responses originating in the orbital frontal cortex could generate an internal sense that something is wrong, and that something needs to be corrected by a change in behavior. They could generate, that is, the very feeling that besets OCD patients. With this realization, I got a sense of real excitement, for this was our first solid clue about the physiological meaning of the PET data showing that OCD patients have a hyperactive orbital frontal cortex: their error-detection circuitry seems to be inappropriately stimulated. As a result, they are bombarded with signals that something is amiss—if not brine subbing for fruit juice, then an iron left plugged in, a germ unscrubbed. If the error-detection system is spotting things out of whack everywhere in a person's environment, the result is like a computer's spellcheck run amok, highlighting every word in a document. Intense and persistent firing in the orbital frontal cortex, it seemed, would cause an intense visceral sensation that something is wrong, and that action of some kind—be it alphabetizing cans or checking whether appliances are turned on—is needed to make things right again. In fact, the reason for the visceral sense of dread that OCD patients suffer is that the OFC (and related structures like the anterior cingulate gyrus) is wired directly into gut-control centers of the brain. Small wonder, then, that the ERROR! ERROR! message triggers such a stomach-churning panic. Monkeys quiet their error messages by correcting their responses: they stop sipping in response to that deceptive blue signal and try other options. What about OCD patients, I wondered? How can they quiet their faulty error-detection circuit?

In 1997, some clever studies expanded the job description of the orbital frontal cortex and its neighbor, the anterior cingulate, to account even more fully for this inchoate sense of dread. Researchers led by Antoine Bechara and Antonio Damasio at the University of Iowa had volunteers play a sort of gambling game, using four decks of cards and $2,000 in play money. On each card was written a dollar amount either won or lost. All the cards in the first and second decks brought either a large payoff or a large loss, $100 either way, simulating the state of affairs that any savvy investor understands: the greater the risk, the greater the reward. Cards in decks 3 and 4 produced losses and wins of $50—small risk, small reward. But the decks were stacked: the cards were arranged so that those in decks 3 and 4 yielded, on balance, a positive payoff. That is, players who chose from decks 3 and 4 would, over time, come out ahead. The losses in decks 1 and 2 were not only twice as large, moreover, but more common, so that after a few rounds players found themselves deep in the hole. A player who chose from the first two decks more than the second two would lose his (virtual) shirt.

Normal volunteers start the game by sampling from each of the four decks. After playing a while, they began to generate what are called anticipatory skin conductance responses when they are about to select a card from the losing decks. (Skin conductance responses, assessed by a simple electronic device on the surface of the skin, gauge when sweat glands are active. Sweat glands are controlled by the autonomic nervous system, whose level of activity is a standard measure of arousal and anxiety—and thus the basis for lie detectors.) This skin response occurred even when the player could not verbalize why decks 1 and 2 made him nervous; nevertheless, he began to avoid those decks. Patients with damage to the inferior (or *ventral*, meaning the "underside of") prefrontal cortex, however, played the game differently. They neither generated skin conductance responses in anticipation of drawing from the risky decks, nor shied away from these decks. They were instead drawn to the high-risk decks like high-rollers to the $500 baccarat table and never learned to avoid them.

Bechara and Damasio suggest that, since normal volunteers avoided the bad decks even before they had conceptualized the reason but after their skin response showed anxiety about those decks, something in the brain was acting as a sort of intuition generator. Remarkably, the normal players who were never able to figure out, or at least articulate, why two of the decks were chronic losers still began to avoid them. Intuition, or gut feeling, turned out to be a more dependable guide than reason. It was also more potent than reason: half the subjects with damage to the inferior prefrontal cortex (which includes the orbital frontal cortex) eventually figured out why, in the long run, decks 1 and 2 led to net losses and 3 and 4 led to net wins. Even so, amazingly, they still kept choosing from the bad decks.

Decision making, then, clearly has not just a rational but also an emotional component. Damage to the inferior prefrontal cortex seems to rob patients of key equipment for accessing intuition. This finding is particularly important, I realized, because it mirrors the situation in OCD patients, who have the opposite malfunction of the very same brain area. OCD patients, who have an *overactive* inferior prefrontal cortex, get an excessive, intrusive feeling that something is wrong, even when they know that nothing is. In patients in the gambling study, these areas were damaged and therefore *underactive;* these patients failed to sense that something was wrong even when they knew, rationally, that something was. The normal subjects in the gambling study felt something was wrong when it was wrong, even if they didn't know why. This all constitutes powerful evidence that the orbital frontal cortex is involved in generating the intuitive feeling "Something is wrong here."

A second overactive region we detected on the PET scans of the brains of OCD patients was the *striatum.* This structure is composed of two major information-receiving structures, the *caudate nucleus* and the *putamen,* which nestle beside each other deep in the core of the brain just in front of the ears. The entire striatum acts as a sort of automatic transmission: the putamen acts as the gear shift for motor activity, and the caudate nucleus serves a similar function for thought and emotion. The striatum as a whole receives neuronal inputs from so many other regions of the brain that it rivals, for sheer complexity, the central switching station for the busiest telecom center imaginable, with signals arriving and departing in a buzz of perpetual activity. All areas of the cortex send neural projections to the striatum; so do parts of the thalamus and the brainstem, as shown in Figure 2 on page 69.

But what particularly intrigued me was the fascinating traffic pattern connecting the striatum and the cortex. One set of neuronal projections into the striatum originates in the prefrontal cortex, especially in regions associated with planning and executing such complex behaviors as the manipulation of mental images. Small clusters of projections formed by these prefrontal arrivals are called *matrisomes*. The matrisomes are typically found near distinct microscopic patches that stipple the striatum; they are called *striosomes*. The striosomes, too, receive some input from the prefrontal cortex, in particular the areas most intimately associated with emotional expression: the orbital frontal cortex and anterior cingulate cortex. These are the very cortical structures that PET scans have shown to be overactive in OCD. But the primary inputs to these striosomes are the polar opposites of the thoughtful, rational prefrontal cortex: the striosomes are also bombarded with messages from the *limbic system*. The limbic system comprises the structures that play a critical role in the brain's emotional responses, particularly fear and dread. It is the limbic system's core structure, the amygdala, that seems to generate fear and dread. And it is the amygdala that projects most robustly into the striosomes' distinctive patches.

The striatum, and especially the caudate, can thus be thought of as a neuronal mosaic of reason and passion. It sits smack dab at the confluence of messages bearing cognitive content (courtesy of the matrisomes, where inputs arrive from the rational prefrontal cortex) and messages shot through with emotion (thanks to the striosomes, landing zones for inputs from the limbic system). The juxtaposition of striosomes and matrisomes therefore seems highly conducive to interactions between emotion and thought. Since the striosomes receive projections primarily from the emotional centers of the limbic system and the matrisomes receive projections from the higher cognitive centers of the prefrontal cortex, together they provide the perfect mechanism of integrating the messages of the heart with those of the mind.

In the mid-1990s researchers discovered a subset of highly specialized nerve cells that provide a key to understanding how the brain integrates reason and emotion. Called *tonically active neurons* (TANs), these cells often sit where striosomes and matrisomes meet, as Ann Graybiel and her colleagues at the Massachusetts Institute of Technology discovered. The TANs are thus perfectly

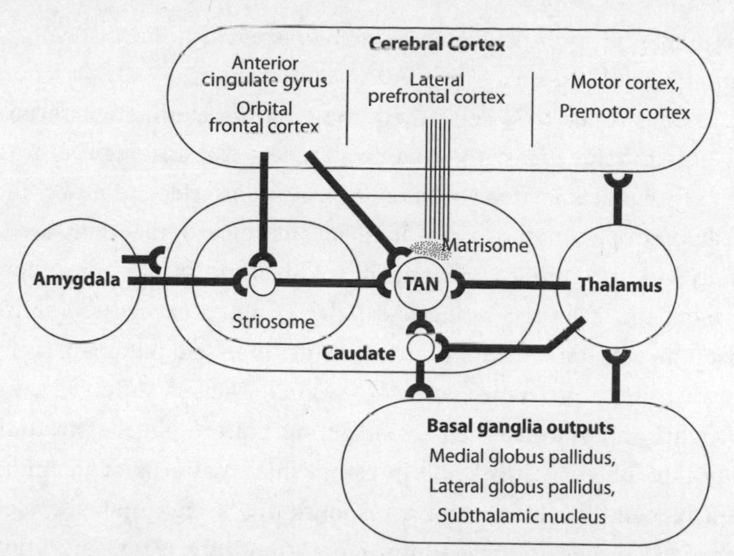

Figure 2: Cells in the caudate known as tonically active neurons (TANs) tend to be found between striosomes and matrisomes. Striosomes are areas where information from an emotion-processing part of the brain, the amygdala, reaches the caudate; matrisomes are clumps of axon terminals where information from the thinking, reasoning cerebral cortex reaches the caudate. By virtue of their position, TANs can integrate emotion and thought. They fire in a characteristic pattern when the brain senses something with positive or negative emotional meaning. Cognitive-behavioral therapy may change how TANs respond to OCD triggers.

positioned to integrate information from both structures and, by implication, from the intensely passionate limbic system and the eminently reasonable prefrontal cortex.

TANs respond dramatically to visual or auditory stimuli that are linked, through behavioral conditioning, to a reward. As a result of this finding, Graybiel's team began to suspect that TANs play a central role in behavioral responses to hints of an upcoming reward. In a series of experiments on macaque monkeys, the MIT scientists found that TAN firing rates changed when a once-neutral cue became associated with a reward. Let's say, for instance, that a visual cue such as a flashing light means that the monkey will get a reward (juice) if it performs a simple behavioral response (licking a spoon). When TANs detect a potential reward, that is, they pause at first and then fire faster. But TANs do not respond to the light cue if the monkey has not learned to associate it with a reward. As the monkey's brain learns to recognize a reward, TANs fire in a characteristic pattern.

Thanks to its TAN cells, then, the striatum is able to associate rewarded behavior with particular cues. Because TANs can quickly signal a switch in behavioral response depending on the meaning of a stimulus ("That light means juice!"), they may serve as a sort of gating mechanism, redirecting information flow through the striatum during learning. As noted earlier, the entire striatum acts as an automatic transmission: the putamen shifts between motor activities, and the caudate nucleus shifts between thoughts and emotions. Different gating patterns in the striatum may thus play a critical role in establishing patterns of motor as well as cognitive and emotional responses to the environment. Such patterned responses are nothing more than habits. Indeed, Graybiel has shown that the striatum can play a fundamental role in the development of habits. Our best guess is that the tonically active neurons underpin the gating of information through the striatum and thus its role in the formation of habits. What seems to happen is that distinct environmental cues, associated with differing emotional meanings, elicit different behavioral and cognitive responses as TANs shift the output flow of the striatum. In this way TANs may serve as the foundation for the development of new patterns of activity in the striatum.

Most important, TANs could be crucial to the acquisition of new behavioral skills in cognitive-behavioral therapy. In neurological terms, we could say that cognitive-behavioral therapy teaches people purposefully to alter the response contingencies of their own TANs. This is a crucial point. Such therapy teaches people to alter, by force of will, the response habits wired into their brains through TANs. In the case of OCD, therapy teaches patients to reinterpret their environment and apply their will to alter what had been an automatic behavioral response to disturbing feelings. If that happens often enough, then the new response—the new behavioral output—should itself become habitual. The key to a successful behavioral intervention in OCD, it seemed to me, would be to teach the striatum new gating patterns.

The gating image turns out to be particularly apt in light of what we have learned about the striatum's two output pathways: one direct and one indirect. The indirect pathway takes the scenic route, running from the striatum through the globus pallidus, to the subthalamic nucleus, back to the globus pallidus, and finally to the thalamus and cortex. The direct pathway runs through the globus pallidus, then straight to the thalamus and back to the cortex. The crucial difference is that the direct pathway provides activating input to the thalamus, but the indirect pathway provides inhibitory input. Thus the direct and indirect output pathways from the striatum have opposite effects. The direct pathway tends to activate the cortex, the indirect pathway tends to quiet the cortex.

The striatal gate determines which pathway nerve impulses will follow. Recall that the striatum receives input from the entire cortex, with the caudate specifically receiving strong input from the prefrontal areas. Prefrontal inputs include those from the orbital frontal cortex and anterior cingulate error-detection circuitry. In 1992 Lew Baxter, my longtime colleague at UCLA, dubbed the circuit containing the orbital frontal cortex and its connections to the basal ganglia the "worry circuit." It is now often called "the OCD circuit." When this circuit is working properly, the result is a finely tuned mechanism that can precisely modulate the orbital frontal cortex and anterior cingulate by adjusting the degree to which the thalamus drives these areas. When that modulation is faulty, as it is when OCD acts up, the error detector centered in the orbital frontal cortex and anterior cingulate can be overactivated and thus locked into a pattern of repetitive firing. This triggers an overpowering feeling that something is wrong, accompanied by compulsive attempts somehow to make it right. The malfunction of the OCD circuit that our UCLA group found in OCD patients therefore makes sense. If the exquisite balance of the direct and indirect pathway outputs from the basal ganglia is impaired, it can cause the orbital frontal cortex to become stuck in the "ERROR! ERROR!" mode.

When the striatum is working normally, it gates the vast array of information about the environment that flows into it from the cortex and initiates what Ann Graybiel has termed "chunks of action repertoires." These chunks help form "coordinated, sequential motor actions" and develop "streams of thought and motivation." Thus a single bit of information, such as the feel of a stick shift in your hand, can initiate a complex behavior, for instance, a series of foot movements on the clutch and hand movements on the stick. But in OCD patients the striatum, our PET scans showed, is not even close to functioning normally. It does not gate properly, leading to serious overactivity in the orbital frontal cortex. The intrusive, persistent sense in OCD that something is wrong seems to be the result of orbital frontal cortex neurons' becoming chronically activated (or inadequately inactivated) as a result of a gating problem, which causes the direct-output pathway to overwhelm the indirect one.

In OCD, the striatum—in particular, the caudate nucleus—appears to be failing to perform its gating function properly. It has become like an automobile transmission that fails to shift. Most people have brains that shift gears automatically, but OCD patients have a sticky manual transmission. As a result, the direct pathway seems stuck in the "on" position. This is what I came to call Brain Lock: the brain can't move on to the next thought and its related behavior. Instead, such evolutionarily ancient drives as washing and checking for danger keep breaking through, creating a sense of being overwhelmed by these feelings and urges. The feeling of being "stuck in gear," which often manifests itself as the feeling of needing to get things just right, also explains why an OCD patient finds it so hard to change the compulsive behavior, and why doing so requires such focused and even heroic effort. Medications that block the neuronal reuptake of serotonin can help by at least partially decreasing the intensity of OCD urges, probably by helping to rebalance the direct and indirect pathways.

A third brain region implicated in OCD is the anterior cingulate gyrus, which also sends projections to the striosomes of the caudate nucleus. Located behind and above the orbital cortex, the cingulate also has connections to the vital brain centers that control the gut and the heart. This structure is probably responsible for generating the gut-churning sense among OCD sufferers that some cataclysm will befall them if they fail to act on their compulsion, say, to tap the steering wheel ten times (or one hundred!) before turning the ignition. The anterior cingulate seems to amplify the gut-level feeling of anxiety and dread.

Even as our UCLA group was working out the OCD circuit, a study from the other side of the country confirmed what we were finding. Researchers at a Massachusetts General Hospital (MGH) group led by Scott Rauch used both PET and functional magnetic resonance imaging (fMRI) scans to measure cerebral blood flow in the brains of eighteen OCD patients. The scientists deliberately created an environment designed to agitate: when a patient settled into the PET scanner, the researchers placed beside him a dirty glove or other OCD-triggering object. The patient's anxiety level soared. At the same time, the MGH group reported in 1994 and 1996, the PET and fMRI scans consistently picked up significant increases in cerebral activity in the orbital frontal cortex, the anterior cingulate gyrus, and the caudate nucleus—exactly the structures found to be hypermetabolic in our PET scans at UCLA. Looking at the scans, you could almost see the brain desperately emitting "TILT! TILT" messages, signaling that something was dreadfully wrong. The conclusion was clear: when OCD urges become more intense as a result of exposure to a trigger such as a dirty object, circuitry involving these three brain structures becomes more active.

A picture of the brain abnormalities underlying OCD was emerging. The malfunctions center on circuitry within the orbital frontal cortex, containing "error alarm" circuits, and the basal ganglia, which acts as an automatic transmission or switching station. The circuit responsible for detecting when something is amiss in the environment, centered on the orbital frontal cortex, becomes inappropriately and chronically activated in OCD, probably because a malfunction in the gating function of the caudate nucleus allows the prefrontal cortex to be stimulated continuously. The result is a pattern of intrusive, persistent thoughts and feelings that something is wrong or excessively risky. Interconnections among the orbital prefrontal cortex, anterior cingulate, and caudate may allow this circuit to become self-sustaining and thus extremely difficult to break out of—as any OCD patient can attest. The result is a perseveration of the thoughts and urges that OCD creates. That these abnormalities leave the superior prefrontal regions, and thus higher cognitive function, essentially untouched seems consistent with the ego-dystonic nature of OCD's intrusive thoughts and urges—that is, with the fact that they are experienced as alien to the patient's sense of who she is, and apart from the natural flow of her stream of consciousness.

This neurobiologically based view of OCD did not exactly take psychiatry by storm. At a meeting on anxiety disorders in the early 1990s, I was presenting some recent findings, on a poster, about the brain mechanisms underlying OCD. A leading behavioral therapist strolled up to me, stopped, and glanced at the poster. Looking me up and down, she spit out, "The whole notion you have of the brain causing OCD is ridiculous!" Well, I asked, what else might be the cause, if not the brain? "I'll tell you what causes OCD. Try not to think of a pink elephant. Can you do it?" Without waiting for an answer, she continued, "There! That is what causes OCD"—and she walked away.

Fortunately, responses like this were the exception rather than the norm, and the overheated brain circuitry underlying OCD, as we and others were working out, offered a glimmer of hope for patients. Since the most evolutionarily recent (and thus sophisticated) prefrontal parts of the brain are almost entirely spared in OCD, the patient's core reasoning power and sense of identity remain largely intact. They might thus be relied on, I reasoned, to play a part in therapy. Now that science had pretty much nailed down the brain abnormalities at the root of OCD, I was ready for the next step. I set out to find a treatment that would alter metabolic activity in the guilty triad: the orbital prefrontal cortex, the anterior cingulate gyrus, and the caudate nucleus. In particular, I had a hunch that any successful treatment would probably have to enhance the gating function of the caudate so that the worry circuit could be quieted and the patient enabled to resist OCD urges.

By mid-1987 I was leading my OCD therapy group of eight to ten patients every Thursday afternoon from 4:30 to 6:00 on the second floor of the Neuropsychiatric Institute. Though the behaviorists had pioneered an effective approach to treating what had, before 1966, been regarded as an untreatable condition (largely because psychoanalytic attempts to treat it generally yielded abysmal results), I was reluctant to adopt exposure and response prevention wholesale. I recoiled at the intense distress it caused many patients; I just couldn't see myself hauling patients to a public restroom, forcing them to wipe their hands all over the toilet seats, and then preventing them from washing. And since experienced clinicians were already estimating (at least in their more candid conversations) that at least 25 to 30 percent of OCD patients refuse, or are unable, to comply with exposure response prevention, I already knew that such therapy alone could never be the final answer to OCD. But most of all, I hated the way exposure and response prevention rendered patients almost completely passive during therapy.

The culture of behaviorism and its approach to therapy disturbed me at a philosophical level, too. Behaviorists adopted only treatment techniques that could also be used to train an animal. They were enormously proud of this, believing it somehow rendered their approach more "scientific." But I balked at it. I didn't doubt that behaviorist approaches could play a useful and even necessary role for patients who suffer from severe cases of OCD, especially in the early stages of treatment. But for patients capable of willful self-directed activity either at the outset or once the most crippling symptoms lifted, I was convinced that it was time to try a different approach. Throughout 1987, as the OCD therapy group blossomed, I decided to extract the useful parts of behaviorist practice—those that could be applied as part of self-directed treatment—and integrate them with the uniquely human characteristics that a willful, conscious person could use in treatment.

A good part of my decision reflected a change in my own life the summer before. In August 1986, I had resumed the daily practice of Buddhist mindfulness meditation, which I had begun in 1975 but fell away from in 1979. In those years, I had been deeply influenced by *The Heart of Buddhist Meditation*, by the German-born Buddhist monk Nyanaponika Thera. In this book, Nyanaponika coined the term *Bare Attention*. As I noted in the Introduction, this mental activity, he wrote, is

> *the clear and single-minded awareness of what actually happens* to *us and* in *us, at the successive moments of perception. It is called "Bare" because it attends just to the bare facts of a perception as presented either through the five physical senses or through the mind . . . without reacting to them by deed, speech or by mental comment which may be one of self-reference (like, dislike, etc.), judgment or reflection. If during the time, short or long, given to the practice of Bare Attention, any such comments arise in one's mind, they themselves are made objects of Bare Attention, and are neither repudi-*

ated nor pursued, but are dismissed, after a brief mental note has been made of them.

Bare Attention, the key to Buddhist meditation, is the act of viewing one's experience as a calm, clear-minded outsider. "Mindfulness is kept to a bare registering of the facts observed," wrote Nyanaponika. A method for doing this, developed by Nyanaponika's meditation teacher, the Burmese master Mahasi Sayadaw, is called "making mental notes." This involves mindfully noting the facts as they are observed in order to enhance the mental act of bare registration.

Thus mindfulness, or mindful awareness, was very much on my mind when I began the OCD group the following February with several UCLA colleagues. We kept, at first, well within the tradition of cognitive-behavioral therapy, teaching patients to correct the cognitive distortions described earlier. But I wasn't content with that. The cognitive approach might be fine for depression, in which there are genuine cognitive distortions that need to be corrected ("It's not true that everyone hates me; those who like me include . . .") and the correction actually helps. But this doesn't work with OCD. To teach a patient to say "My hands are not dirty" is just to repeat something she already knows. The problem in OCD is not failure to realize that your hands are clean; it's the fact that the obsession with dirt keeps bothering you and bothering you until you capitulate and wash—yet again. Cognitive distortion is just not an intrinsic part of the disease; a patient basically knows that failing to count the cans in the pantry today won't really cause her mother to die a horrible death tonight. The problem is, she doesn't feel that way.

Because cognitive therapy alone seemed to lack what OCD patients needed, I cast about for something else. My return to meditation now convinced me that the best way to treat OCD would involve an approach informed by the concept of mindfulness. I felt that if I could help patients to experience the OCD symptom without reacting emotionally to the discomfort it caused, realizing instead that even the most visceral OCD urge is actually no more than the manifestation of a brain wiring defect, it might be tremendously therapeutic. The more patients could experience the feeling impersonally, as it were, the less they would react emotionally or take it at face value. They would not be overwhelmed by the sense that the obsession had to be acted on and could better actualize what they knew intellectually: that the obsession makes no sense. The appropriate cognitive awareness was already on board: patients generally know, with the rational, thinking part of their mind, that it makes no sense to check the oven a dozen times before leaving the house. It was now necessary for patients to engage their healthy emotions to strengthen that insight and act on it. To do this would require persistent effort, and habitual practice would be crucial.

Might the use of mindful awareness, I wondered, help an OCD patient achieve that goal? Might mindfulness practice, and systematic mental note taking (as people do during meditation), allow OCD patients to become instantly aware of the intrusion of symptoms into conscious awareness and then to redirect attention away from these persistent thoughts and feelings and onto more adaptive behaviors? It seemed worth investigating whether learning to observe your sensations and thoughts with the calm clarity of an external witness could strengthen the capacity to resist the insistent thoughts of OCD. Not that I had any illusions about how easy it would be. To register mentally the arrival of each and every OCD obsession and compulsion, and to identify each as a thought or urge with little or no basis in reality, would require significant, willful effort. It would not be sufficient just to acknowledge superficially the arrival of such a symptom. Such superficial awareness is essentially automatic, even (almost) unconscious. Mindful awareness, in contrast, comes about only with conscious effort. It is the difference between an off-handed "Ah, here's that feeling that I have to count cans again," and the insight "My brain is generating another obsessive thought. What does it feel like? How am I responding? Does the feeling make sense? Don't I in fact know it to be false?"

I began showing patients in the treatment group their PET scans, to drive home the point that an imbalance in their brains was causing their obsessive thoughts and compulsive behaviors. Initially, some were dismayed that their brain was abnormal. But generally it dawned on them, especially with therapy, that they are more than their gray matter. When one patient, Dottie (the woman with the 5s and 6s obsession), exclaimed, "It's not me; it's my OCD!" a light went off in my head: what if I could convince patients that the way they responded to the thoughts of OCD could actually change their brains? I developed a working hypothesis that making mental notes could be clinically effective and decided to introduce mindfulness into the OCD clinic. Making mental notes became, in my own mind, "Relabeling" the feeling that accompanies an OCD obsession. This would be the first step in what came to be called the Four Step method.

Having rejected standard exposure and response prevention, I instead told patients to describe their symptoms and the situations in which they arose. I then explained that the feeling that the door is unlocked, for instance, is the disorder itself, and that our brain imaging research had shown that the cause was a biochemical imbalance in the brain. I never told patients just to resist the urge and it would go away. Instead, I emphasized the importance of identifying as clearly and quickly as possible the onset of an OCD symptom—not just recognizing that an obsessive thought was intruding or a compulsive urge was demanding to be carried out, but recognizing exactly what each of these feelings was. As soon as the thought that your hands are dirty seizes your attention, I counseled them, use mindfulness to enhance awareness of the fact that you do not truly think your hands need washing; rather, tell yourself that you are merely experiencing the onslaught of an obsessive thought. The patient would start saying to herself, That's not an urge to wash; that's a bothersome thought and an unpleasant feeling caused by a brain wiring problem. Or, if the compulsion to check a door lock intruded, the patient was to regard it as the result of a nasty compulsive urge, and not of any real need to check the lock. The feeling of doubt, I told patients repeatedly, is a false message, due to a jammed transmission in the brain. To enhance the recognition that the thoughts and urges are symptoms of OCD, I taught patients to make real-time mental notes, in effect creating a running stream of mindful commentary on what they were experiencing. This enabled them to keep a rational perspective on the intrusive thoughts and urges and not get caught up in automatic compulsive responses and thus a destructive run of compulsive rituals.

By refusing to accept obsessive thoughts and compulsive urges at face value, and instead recognizing that they are inherently false and misleading, the patients took the first step toward recovery. Done regularly, Relabeling stops the unpleasant feelings of OCD from being unpleasant in the same way: understanding their true nature gives a feeling of control, even of freedom. By Relabeling their thoughts and urges as manifestations of a medical disorder, patients make a purposeful cognitive shift away from self-identification with the experience intruding into the stream of consciousness.

The week after patients started relabeling their symptoms as manifestations of pathological brain processes, they reported that they were getting better, that the disease was no longer controlling them, and that they felt they could do something about it. I knew I was on the right track. By this time, the PET data had clearly shown that the orbital frontal cortex of OCD patients is hypermetabolic. One day, just a few months after starting the group, I happened to be carrying around some plain black-and-white PET scans. One patient asked me, "Doc, can you just tell me why the damn thing keeps bothering me—why it doesn't go away?" I looked at him. "You want to know why it doesn't go away?" I asked. "I'll show you why it doesn't go away. You see this dark spot in the brain on this scan? That is why: it means this region of the brain is hugely overactive in people with OCD. That's why the bad feeling doesn't go away."

It was as if a lightbulb went off in his head—indeed in all the patients' heads. At that moment what was to become the second of the Four Steps, Reattribute, was born. Whenever a patient told me an obsession was bothering her, I responded, This is why: I printed slides of color PET scans and showed patients the neuroanatomical basis of their symptoms. *This* is why you feel you have to wash, or check, or count, I said, photographic evidence in hand. This reattribution of OCD feelings to a brain glitch was the breakthrough that pushed us beyond simple Relabeling. Cognitive techniques that merely teach the patient to recognize OCD symptoms as false and distorted—something called *cognitive restructuring*—do not make much of a dent in OCD. Relabeling was essentially just a form of cognitive restructuring. Reattributing went further: having Relabeled an intrusive thought or insistent urge as a symptom of OCD, the patient then attributes it to aberrant messages generated by a brain disease and thus fortifies the awareness that it is not his true "self." By first making mental notes of the arrival of an OCD obsession, and immediately attributing it to faulty brain wiring, I hoped, patients could resist that false message. "The brain's gonna do what the brain's gonna do," I told them, "but you don't have to let it push you around."

Two patients in particular picked up on this idea. One was Anna, then twenty-four, a graduate student in philosophy. She asked her boyfriend about every detail of his daily life because she was obsessed with the (baseless) suspicion that he was unfaithful. Although she never truly believed he was cheating on her, she was unable to stop obsessing about it. What had he eaten for lunch? Who were his girlfriends when he was a teenager? Did he ever look at pornographic magazines? Had he had butter or margarine on his toast? The slightest discrepancy in his accounts set Anna off, making her whole world crumble under the suspicion that he had betrayed her. The other patient was Dottie, then fifty-two, whose obsession with "magical numbers" is described at the beginning of this chapter. Both women realized that the reason they were experiencing these false thoughts was an abnormality in their brain's metabolism. "Once I learned to identify my OCD symptoms as OCD rather than as 'important' content-laden thoughts that had to be deciphered for their deep meaning," Anna explained later, as described in my 1996 book *Brain Lock*, "I was partially freed from OCD."

As I worked with Dottie and Anna as well as the other group members throughout 1988 and 1989, I began using Relabeling and Reattributing via mindfulness as a core part of their treatment. Accentuating Relabeling by Reattributing the condition to a rogue neurological circuit deepens patients' cognitive insight into the true nature of their symptoms, which in turn strengthens their belief that the thoughts and urges of OCD are separate from their will and their self. By Reattributing their symptoms to a brain glitch, the patients recognize that an obsessive thought is, in a sense, not "real" but, rather, mental noise, a barrage of false signals. This improves patients' ability not to take the OCD thoughts at face value. Reattributing is particularly effective at directing the patient's attention away from demoralizing and stressful attempts to squash the bothersome OCD feeling by engaging in compulsive behaviors. Realizing that brain biochemistry is responsible for the intensity and intrusiveness of the symptoms helps patients realize that their habitual frantic attempts to wash (or count or check) away the symptoms are futile.

Relabeling and Reattributing reinforce each other. Together, they put the difficult experience of an OCD symptom into a workable context: Relabeling clarifies what is happening, and Reattributing affirms why it's happening, with the result that patients more accurately assess their pathological thoughts and urges. The accentuation of Relabeling by Reattributing also tends to amplify mindfulness. Through mindfulness, the patient distances himself (that is, his locus of conscious awareness) from his OCD (an intrusive experience entirely determined by material forces). This puts mental space between his will and the unwanted urges that would otherwise overpower the will.

Besides Relabeling and Reattributing their OCD symptoms, I realized, patients needed to turn their attention to something else, performing an activity other than the one being urged on them by their stuck-in-gear brain. It seemed a good idea to make it a systematic part of the treatment, akin to the practice of methodically directing attention "back to the breath" when the mind wanders during meditation. So I gave it a name: Refocusing. It evolved to become the core step of the whole therapy, because this is where patients actually implement the willful change of behavior. The essence of applying mindful awareness during a bout of OCD is thus to recognize obsessive thoughts and urges as soon as they arise and willfully Refocus attention onto some adaptive behavior.

Directed mental focusing of attention becomes the mind's key action during treatment. The goal of this step is not to banish or obliterate the thought, but rather to initiate an adaptive behavior unrelated to the disturbing feeling even while the feeling is very much present. Refocusing on such a behavior, and thus resisting the false message to carry out the OCD compulsion, requires significant willpower, for the feeling that something must be washed or checked is still very much a part of the inner experience. Although the patient has Relabeled and Reattributed the obsessions and compulsions to brain pathology, the anxiety and dread still feel frighteningly real. Early in treatment, I therefore suggested to patients that they Refocus on a pleasant, familiar "good habit" kind of behavior. This is when biological reasoning became crucial: I specifically wanted patients to substitute a "good" circuit for a "bad" one. The diversion can be almost anything, although patients began telling me that physical activity—gardening, needlepoint, shooting baskets, playing computer or video games, cooking, walking—was especially effective. That is not to say it was easy. To the contrary: Refocusing attention away from the intrusive thought rather than waiting passively for the feeling to go away is the hardest aspect of treatment, requiring will and courage.

Soon after I explained the Refocus step to one patient, Jeremy, he began carrying around a small notebook in which he wrote ways to Refocus whenever a compulsive urge arose. On its cover, he had written "caudate nucleus." In what he called his "refocus diary," Jeremy told me, he recorded how he prevented himself from responding to an OCD urge and which alternative behavior he used. The diary, it turned out, not only increases a patient's repertoire of Refocus behaviors, but also boosts confidence by highlighting achievements: see, yesterday when I had a seemingly irresistible urge to count cans I did some needlepoint instead. Many patients were helped by selecting one Refocus task daily as the "play of the day," to remember and review as a form of positive feedback and self-empowerment. Over the course of treatment, patients slowly developed the sense that they could control their response to the OCD intrusions and that well-directed effort really does make a difference.

Early on, I developed a "fifteen-minute rule." The patient had to use an "active delay" of at least fifteen minutes before performing any compulsive act. Setting a finite length of time to resist giving in seems to help patients (for the same reason, probably, that devout Catholics find it easier to give up drinking or smoking for the forty days of Lent than they do to give up bad habits for an open-ended period). The fifteen minutes should not be just a passive waiting period, however. Rather, it must be a period of mindful adaptive activity intended to activate new brain circuitry, with the goal of pursuing the alternative activity for a minimum of another fifteen minutes. This seems to be the length of time generally needed for most patients' OCD urges to diminish noticeably. When a patient's mind is invaded by obsessive thoughts, even brief periods of Refocusing help, for they demonstrate that it is not essential to squelch intrusive thoughts entirely in order to engage in healthier behaviors.

Refocusing also alleviates the overwhelming sense of being "stuck in gear." This is where Relabeling and Reattributing come in: both help keep patients' minds clear about who they are and what the disease process is. This mental clarity has tremendous therapeutic power, for it keeps the Refocusing process moving forward. It also reinforces the insight that active will is separable from passive brain processes—an awareness that forms the core of the quantum perspective on the mind-brain interface, as we shall explore later.

At the neurological level, the rationale for Refocusing is straightforward. Our PET scans had shown that the orbital frontal cortex, the caudate nucleus, and the thalamus operate in lockstep in the brain of an OCD sufferer. This brain lock in the OCD circuit is undoubtedly the source of a persistent error-detection signal that makes the patient feel that something is dreadfully wrong. By actively changing behaviors, Refocusing changes which brain circuits become activated, and thus also changes the gating through the striatum. The striatum has two output pathways, as noted earlier: direct and indirect. The direct pathway tends to activate the thalamus, increasing cortical activity. The indirect pathway inhibits cortical activity. Refocusing, I hoped, would change the balance of gating through the striatum so that the indirect, inhibitory pathway would become more traveled, and the direct, excitatory pathway would lose traffic. The result would be to damp down activity in this OCD circuit.

When patients changed the focus of their attention, in other words, the brain might change, too. I thought that if I could somehow induce the patient to initiate virtually any adaptive behavior other than whatever the compulsion was, this process would activate neuronal circuitry different from the pathways that were pathologically overactive. Then I could exploit the brain's tendency to pick up on repetitive behaviors and make them automatic—that is, to form new habits. Ideally, this alternative behavior would be one the patient already knows so well that it is almost automatic. When patients change their focus from "I have to wash again" to "I'm going to garden," I suspected, the circuit in the brain that underlies gardening becomes activated. If done regularly, that would produce a habitual association: the urge to wash would be followed automatically by the impulse to go work in the garden. I therefore began encouraging patients to plan sequences of

Refocusing behaviors that they could call on, in order to make them as automatic as possible. Refocusing is the step that, more than any other, produces changes in the brain.

In the fall of 1988 a UCLA medical student was working as my cotherapist in the OCD group sessions. We had recently begun using the group as part of a major research study on the effect of psychological interventions for OCD on brain function. Robert Liberman, who was supervising this student, asked me one day how I was conducting the group therapy. When I explained the Relabeling and Reattributing steps, and how I was teaching patients to recognize that their brain is sending them a false message, Liberman was intrigued. I had to meet a friend of his, he said: Dr. Iver Hand of the University of Hamburg in Germany. Hand had developed a technique called *exposure response management*, which is based on the insight that there is no need to make an OCD patient wait passively in an angst-ridden state for her compulsive urge to dissipate. If you instead help her to manage the anxiety caused by the exposure, Hand found, she will tolerate more exposures and improve more quickly. When I dug up some of his published papers, I saw that Hand had found that when patients acquired specific cognitive skills, they were better able to tolerate the presence of, say, a dirty washrag, and therefore more exposures. They even began to do some of the treatment on their own. I recognized a kindred spirit: Hand was finding that patients could learn to exploit their understanding of OCD to manage their anxiety.

Iver and I met in San Francisco at the American Psychiatric Association conference in 1989 and hit it off immediately. The following spring, Liberman suggested that Iver and I write the OCD chapter for a textbook on biobehavioral treatments for psychiatric disorders. We holed up at the Veterans Administration Hospital in Brentwood, a few blocks from my office. The chapter would never be written (partly because Iver and I could never quite reconcile our beliefs about whether biology or psychosocial factors caused

OCD; he was convinced that OCD symptoms are the product of a patient's need to distance himself from intimate relations). But we agreed strongly on approaches to treatment. We spent hours in the coffeehouses of West L.A., debating whether exposure and response prevention was mechanistic and inhumane. Iver argued that his version of ERP was nothing of the sort: because he varied the exposures and, critically, motivated patients to resist the compulsion, he very much involved patients in their own treatment rather than treating them as a behaviorist's pet pigeons. As Iver talked, it hit me: up to that point I was explaining treatment in a sort of shoot-from-the-hip style. If I could explain things to patients more methodically, perhaps by breaking mindfulness into discrete, straightforward, teachable steps, it could become the basis for self-treatment.

I was sitting at the keyboard, typing out a case history to describe the treatment, with Iver beside me. How to explain what I was doing with patients? Okay, Relabel, Reattribute, Refocus—but what else was going on? It suddenly hit me. In 1989 I had begun reading the Austrian economist Ludwig von Mises, who defined *valuing* as "man's emotional reaction to the various states of his environment, both that of the external world and that of the physiological conditions of his own body." This was exactly what the OCD therapy was changing. Combining Buddhist philosophy with Austrian economics, I had a name for the last of the Four Steps: Revalue. "This might actually be important," I thought, for I now had, in a simple and usable form, a strategy for treating OCD: Relabel, Reattribute, Refocus, Revalue.

Revaluing is a deep form of Relabeling. Anyone whose grasp of reality is reasonably intact can learn to blame OCD symptoms on a medical condition. But such Relabeling is superficial, leading to no diminution of symptoms or improved ability to cope. This is why classical cognitive therapy (which aims primarily to correct cognitive distortions) seldom helps OCD patients. Revaluing went deeper. Like Relabeling, Reattributing, and Refocusing, Revaluing was intended to enhance patients' use of mindful awareness, the foundation of Theravada Buddhist philosophy. I therefore began teaching Revaluing by reference to what Buddhist philosophy calls wise (as opposed to unwise) attention. Wise attention means seeing matters as they really are or, literally, "in accordance with the truth." In the case of OCD, wise attention means quickly recognizing the disturbing thoughts as senseless, as false, as errant brain signals not even worth the gray matter they rode in on, let alone worth acting on. By refusing to take the symptoms at face value, patients come to view them "as toxic waste from my brain," as the man with chapped hands put it.

In both my individual and my group practice, I was getting encouraging results with the Four Steps by the early 1990s. With continued self-treatment—for I always intended that patients be able to follow the treatment regimen on their own—the intensity of their OCD symptoms kept falling. As it did, the patients found they needed to expend less effort to dismiss OCD symptoms through Relabeling, and less effort to Refocus on another behavior.

Some of the OCD patients, especially those willing to be treated without drugs, were recruited into the brain imaging study that Lew Baxter and I were starting, with the goal of measuring whether the positive behavioral changes we were seeing in patients were accompanied by brain changes. Our UCLA group therefore performed PET scans on eighteen drug-free OCD patients before and after they underwent ten weeks of the Four Steps, with individual sessions once or twice a week in addition to regular group attendance. The patients who signed on exhibited moderate to quite severe symptoms. What they all had in common was a willingness to be PET-scanned twice and to try a largely self-directed, drug-free treatment. Twelve of the patients improved significantly during the ten-week study period. In these, PET scans after treatment showed significantly diminished metabolic activity in both the right and the left caudate, with the right-side decrease particularly striking. (See Figure 3.) There was also a significant decrease in the

Figure 3. PET scan showing decreased energy use in the right caudate nucleus (which appears on the left side in a PET scan) in a person with OCD after successful treatment with the Four-Step Method. PRE shows the brain before and POST ten weeks after behavioral therapy with no medication. Note the decrease in "size," which signifies decrease in energy use, in the right caudate (rCd) after doing the Four-Step Method. The drawings show where the caudate nucleus is located inside the head.

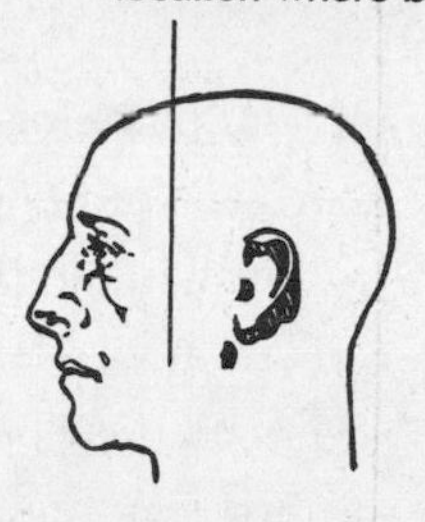

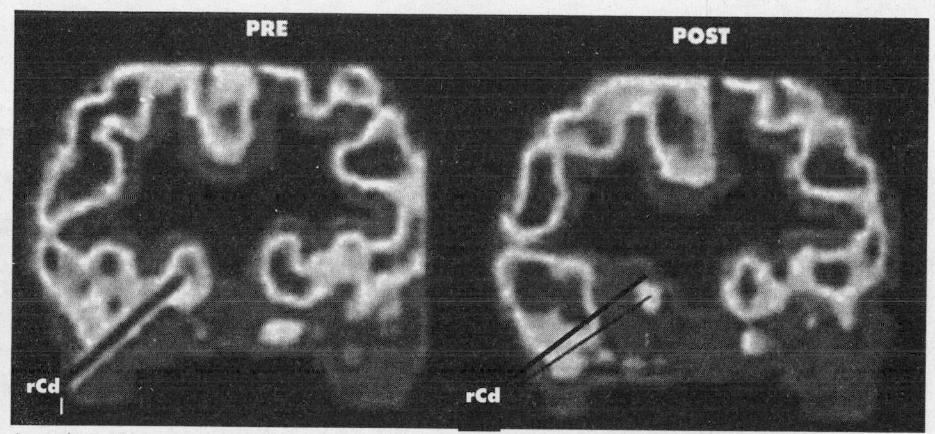

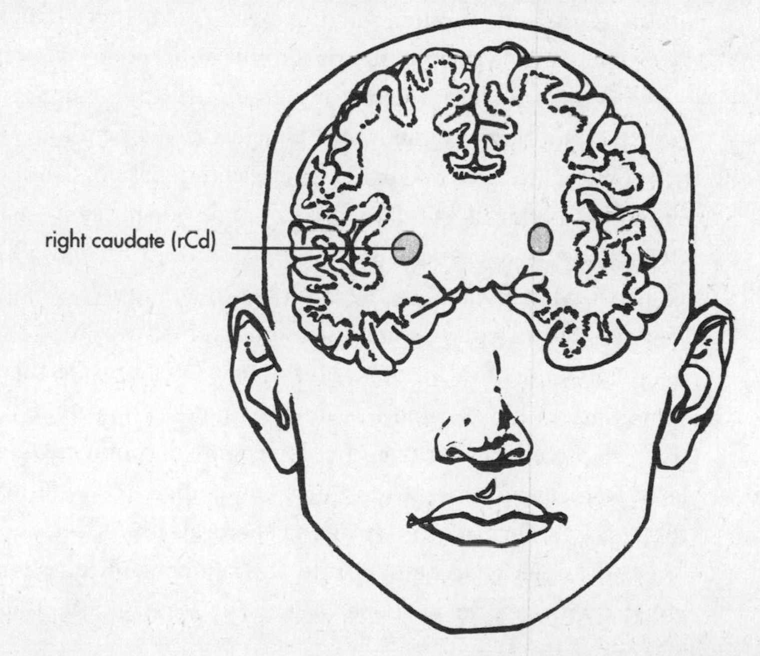

abnormally high, and pathological, correlations among activities in the caudate, the orbital frontal cortex, and the thalamus in the right hemisphere. No longer were these structures functioning in lock-step. The interpretation was clear: therapy had altered the metabolism of the OCD circuit. Our patients' brain lock had been broken.

This was the first study ever to show that cognitive-behavior therapy—or, indeed, any psychiatric treatment that did not rely on drugs—has the power to change faulty brain chemistry in a well-identified brain circuit. What's more, the therapy had been self-directed, something that was and to a great extent remains anathema to psychology and psychiatry. The changes we detected on PET scans were the kind that neuropsychiatrists might see in patients being treated with powerful mind-altering drugs. We had demonstrated such changes in patients who had, not to put too fine a point on it, changed the way they thought about their thoughts. Self-directed therapy had dramatically and significantly altered brain function. There are now a wealth of brain imaging data supporting the notion that the sort of willful cognitive shift achieved during Refocusing through mindful awareness brings about important changes in brain circuitry as we will see in later chapters.

With this evidence in hand, my group therapy sessions increasingly took on the air of an informal neuroscience seminar. In addition to showing PET scans, I began to lecture patients on the OCD circuit. If the basal ganglia is like a car's transmission—which in OCD patients can stick like the gear shift in an old Plymouth Valiant—what I was showing them was that simply by practicing, they could learn how to shift behavioral gears themselves, changing the functioning of the brain's transmissions. As a result, their OCD symptoms would become less intense, and shifting to an alternative, adaptive behavior would become more automatic. Done regularly, Refocusing strengthens a new automatic circuit and weakens the old, pathological one—training the brain, in effect, to replace old bad habits programmed into the caudate nucleus and basal ganglia with healthy new ones. When the focus of attention shifts, so do patterns of brain activity. (Quantum physics, as we'll see later, is consistent with this.) Regular Refocusing helps patients resist giving in to OCD thoughts and urges because engaging in intentional rather than automatic behavior—gardening rather than counting cans—puts in play different brain circuitry. Just as the more one performs a compulsive behavior, the more the urge to do it intensifies, so if a patient resists the urge and substitutes an adaptive behavior, the metabolic activity of the caudate, anterior cingulate, orbital frontal cortex, and thalamus changes in beneficial ways. The bottom line, I told my patients, is that Refocusing holds out the tantalizing promise of systematically substituting healthy circuitry for pathological circuitry—of literally reprogramming your brain.

In the winter of 1995–1996, Eda Gorbis began work as cotherapist in my OCD group. Gorbis came by her interest in fears honestly: born and raised in what was then the Soviet Union, she grew up in an atmosphere poisoned by very real threats, yet one where imagined threats thrived, too. Even as a child Gorbis was acutely aware that some of her parents' friends were riddled with anxiety over the dangers inherent in their world, while others seemed immune to them. The question of why—what enabled one person to shrug off real and ever-present threats while another became psychologically crippled by them?—lingered in her mind for years, even after she and her family fled the Soviet Union when Gorbis was a young teenager. She hopscotched to five different countries, before she arrived in the United States and earned the degree in clinical psychology that would let her pursue an answer to the question of her childhood. Baxter and I had just opened the OCD treatment center, and Gorbis signed on as a volunteer.

As had virtually everyone else in the field, she had pored over *Stop Obsessing*, by Edna Foa and Reid Wilson, which laid out the standard behaviorist approach to OCD: expose patients to the "triggers" that cause them distress (have them touch a doorknob if they are obsessed with germs, for instance), but prevent them from engaging in the ritualistic behavior that ordinarily dissipates that distress (prohibit them from running to a sink to wash, in this case). "I had the book like a Bible on my night table," Gorbis said. But even as she practiced what Foa preached, Gorbis had her doubts. "The strict behaviorist approach seemed, to me, a bit too mechanical," she recalls. "It was treating patients as if they had no humanity; it was not recognizing that they had a thinking, feeling mind inside." Despite her doubts, in the autumn of 1995 Gorbis left her family and spent several months with a group of leading behaviorists, including Foa. When she returned to UCLA that winter, she began coleading my OCD group, integrating the Four Steps with her own approach to behavior therapy.

As its reputation grew, the UCLA Four Steps approach began to draw intractable OCD cases from around the United States, people so enslaved by their obsessions and compulsions that they could hardly get through a day, much less hold a job. Yet by blending the Four Steps with standard behavioral therapy methods, Gorbis was achieving a success rate of over 80 percent with no relapse to anything close to pretreatment severity. That compares to Foa's 65 to 75 percent initial success rate (that is, excluding relapse rates, which are significant), and 60 percent or less at other centers toeing the strict behaviorist line. "We were changing the lives of people who before had been almost totally paralyzed by their OCD," Gorbis says. And she was not removing the rearview mirrors from patients' cars. "Mindfulness became an empowering tool for the patients, giving them—finally—control over their lives," she says. By the late 1990s, the UCLA group was treating hundreds of patients a year, and the Four Steps was at the center of the group practice. It was gratifying to get independent confirmation of the power of this approach in 2002, when Dr. Nili Benazon of Wayne State University published a major study showing that a mindfulness-based cognitive-behavioral method closely related to ours is very effective at treating children with OCD.

As I thought about the therapy sessions, and of how the

patients' mental effort and acts of will had the power to regate the circuitry of their brain, a simple but deeply important question arose. What happens at the instant a person decides not to wash her hands, after decades of doing so in response to the false signals from the orbital cortex and despite her anterior cingulate's making her heart race and her gut churn? Why and how does this person switch gears, activating circuits in the dorsal prefrontal cortex connecting to adaptive basal ganglia circuits, rather than the OCD circuits connecting the orbital frontal cortex to the anterior cingulate and caudate? (See Figure 4.) At the instant of activation, both circuits—one encoding your walk to the garden to prune roses, the other a rush to the sink to wash—are ready to go. Yet something in the mind is choosing one brain circuit over another. Something is causing one circuit to become activated and one to remain quiescent. What is that something? William James posed the question this way: "We reach the heart of our inquiry into volition when we ask, by what process is it that the thought of any given action comes to prevail stably in the mind?"

The demonstration that OCD patients can systematically alter their brain chemistry through cognitive-behavioral therapy such as the Four Steps regimen has inescapable implications for theories trying to explain the relationship between mind and brain. As I began to consider how best to make the OCD work relevant to questions of how the mind can change the brain, I became more and more intrigued by the idea that there must be a force to account for the observed brain changes. The willful effort OCD patients generate during treatment, I suspected, was the most reasonable way to account for the generation of this force. The results achieved with OCD supported the notion that the conscious and willful mind differs from the brain and cannot be explained solely and completely by the matter, by the material substance, of the brain. For the first time, hard science—for what could be "harder" than the metabolic activity measured by PET scans?—had weighed in on the side of mind-matter theories that, as explained in

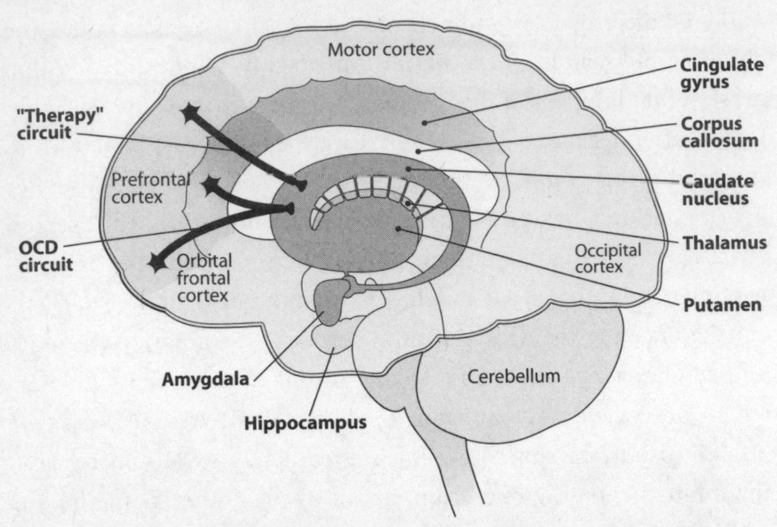

Figure 4: The exertion of willful effort during cognitive-behavioral therapy can activate a "therapy" circuit in the dorsal prefrontal cortex. This can help to override the effects of the OCD circuit.

the previous chapter, question whether mind is nothing but matter. The changes the Four Steps can produce in the brain offered strong evidence that willful, mindful effort can alter brain function, and that such self-directed brain changes—*neuroplasticity*—are a genuine reality. Let me repeat this: the Four Steps is not merely a self-directed therapy; it is also an avenue to self-directed neuroplasticity.

I anticipated the objections that materialist reductionists would raise. Surely what is happening here, they would say, is that one part of the brain is changing another. The brain is changing itself; there is no need to invoke a separate, nonmaterial entity called mind to account for the changes documented by the PET scans. But a materialist explanation simply cannot account for these findings. To train people suffering from OCD requires tapping into their natural belief in the efficacy of their own willful actions. Explanations based exclusively on materialist causation are both infeasible and inappropriate for conveying to OCD patients the steps they must follow to change their own brain circuitry systematically. In order to work, behavioral medicine (of which the Four Steps is an example) absolutely requires the use of the patient's inner experience, including the directly perceived reality of the causal efficacy of volition. The clinical and physiological results achieved with OCD support the notion that the conscious and willful mind cannot be explained solely and completely by matter, by the material substance of the brain. In other words, the arrow of causation relating brain and mind must be bidirectional. Conscious, volitional decisions and changes in behavior alter the brain. And as we will see, modern quantum physics provides an empirically validated mathematical formalism that can account for the effects of mental processes on brain function.

The demonstrated success of mindfulness-based cognitive-behavioral therapy for OCD led me to posit a new kind of studyable force. I called it directed mental force. It would arise, I thought, from willful effort. What mental force does is activate a neuronal circuit. Once that new circuit begins to fire regularly, an OCD patient does not need as much effort to activate it subsequently; the basal ganglia, responsible for habitual behaviors, take care of that. My still-nascent thesis held that directed mental force is the physical aspect of the willful effort to bring healthy circuitry on line. With regular use of the frontal cortex, changes occur in the gating function of the caudate, and mental function improves. Relabeling and Refocusing attention begin to be automatic. In this way, frontal cortex thought processes begin to be wired directly into the caudate. As the brain takes over, less mental force is needed.

Mental force needs the brain to express itself. But it is more than brain, and not reducible to brainstuff. In the fractions of a second when the brain might activate either the pathological circuit underlying a dash to the sink to wash or the healthy circuit underlying a stroll to the garden to prune, mental force amplifies activity in the healthy circuit. You can generate a mental force that activates one circuit rather than another. In a more innocent age, we called that will. But the very idea that the brain can change at all, much less that it can change in response to mind, first had to overcome a century-old dogma.

{THREE}

BIRTH OF A BRAIN

Martha Curtis was, if not a musical prodigy, then certainly musically gifted. She was playing the piano at age five and at nine took up the violin, eventually coaxing from the instrument passionate and even heartbreaking concertos. But something else made Martha stand out: she had begun suffering convulsions at age three and a half. Her doctors diagnosed her condition as epilepsy and started her on the standard medication prescribed to control the seizures. But the seizures only continued, and by the time she was eleven, they were sometimes leaving the little girl unconscious on the floor, terrifying her parents. Martha soldiered on, however, and won a place in the junior orchestra at Michigan's Interlochen Arts Camp, from whose academy she graduated as her class's salutatorian. But by the time she entered the Eastman School of Music in the mid-1970s, she was seizing on stage. As a twenty-something, while performing with various orchestras, Martha had seizures that punched through the pharmaceutical overlay of the drugs frequently and relentlessly.

In April 1990, she suffered four grand mal seizures, three while performing. Knowing that no orchestra would let her back on stage if she kept seizing, she sought help at the Cleveland Clinic. There, the neurologist Hans Luders took Martha off drugs and admitted her to an inpatient epilepsy unit, where electrodes could monitor her temporal lobes twenty-four hours a day. The electroencephalogram showed a constant storm of abnormal electrical activity emanating from Martha's right temporal lobe and spreading over her entire brain like a fast-moving squall—the hallmark of epilepsy. Surgery to remove the spawning ground of the storms, Luders told his patient, was the only option: the quantity of carbamazepine (Tegretol) needed to quiet the pathological electrical activity, Luders said, was already toxic. There was one problem, however. The right temporal lobe seems to be where the brain stores musical memories. Removing it might well eliminate Martha's epilepsy; it might also leave her unable to play the violin ever again. That was something she could hardly face. It was only because she had music in her life than she had been able to bear her illness. "I am alive today," she said in 2000, "because I had a violin."

Martha had surgery in January 1991. As soon as she left intensive care, she took up her violin and, fearing the worst, tried to play a Bach composition. She chose it because, before her surgery, she had found it one of the hardest pieces to play from memory. She nailed it. But although her musical ability seemed intact, her brain seemed to have been left too intact: the surgery had apparently not removed enough of her right temporal lobe (specifically, it had left behind too much of the hippocampus): Martha's seizures persisted. She returned to Cleveland for a second operation. This surgery removed all the hippocampus and much of the amygdala, but the seizures continued, for they were originating from a specific tiny spot in the amygdala. But still Martha could play. When she asked for a third surgery, her doctors warned that taking away so much of her right temporal lobe could prove catastrophic, leaving her paralyzed or even dead. But Martha had decided that she simply could not go on living with the unpredictable and debilitating seizures.

By the time she emerged from the third surgery, close to 50 percent of her right temporal lobe, including the entire hippocampus, was gone. So were her seizures. Her musical memory, however, was very much intact, allowing her to memorize complex pieces even better than before her surgeries, when the anticonvulsants left her in a mental fog. Her brain, doctors concluded, must have been damaged when she was still a toddler, probably by the measles she contracted at age three. Because Martha had begun learning music at such a young age, her brain, it seems, adapted to the damage, with the result that regions other than the abnormal right temporal lobe were drafted during childhood to support musical memory. Because the real estate that the brain usually zones for musical memory was essentially unusable, the brain—exploiting its youthful plasticity—picked up and moved its musical operations across the neural road.

At Johns Hopkins University Medical Center, surgeons challenged the adaptability of a child's brain even more. In 2000 a three-and-a-half-year-old girl named Alexandria Moody had arrived at the Baltimore hospital from her home in Indiana with her mother and stepfather, suffering from chronic seizures. Her physicians back home suspected the little girl was suffering from a brain aneurysm, but an MRI revealed something completely unexpected: the entire left hemisphere of Alex's brain had suffered severe developmental abnormalities. The seizures seemed to be emanating from there, concluded John Freeman, a specialist in pediatric epilepsy. He recommended a complete hemispherectomy—removal of the entire left side of the brain. The operation sounds radical, and it is. But starting in the mid-1980s it became the treatment of choice for children suffering from uncontrollable and often life-threatening seizures due to developmental abnormalities, stroke, or Rasmussen syndrome that do not respond to drugs. Although the brain's deep structures (the brainstem, thalamus, and basal ganglia) are left intact, patients almost always suffer some paralysis on the side of the body opposite the lost hemisphere. But the reward is generally worth the risk: in June 2001, Hopkins surgeons performed their one hundredth hemispherectomy.

The pediatric neurosurgeon Ben Carson performed the operation on Alexandria. Having done more than eighty hemispherectomies since 1985, he was optimistic. "If you see the patients who have had hemispherectomies, you're always amazed," he said. "Here they are, running, jumping, talking, doing well in school. They're able to live a normal life"—despite losing half their brain. What saves these children is their youth. "You can take out the right half of the brain or the left half," Carson said. "Plasticity works in both directions. The reason it works so well in very young patients is that their neurons haven't decided what they want to do when they grow up. They're not committed. So they can be recruited to do other things. Whereas if I had a hemispherectomy it would be devastating." The worst a child suffers from losing half her brain, however, is some impairment of the peripheral vision and fine motor skills on one side of the body.

The brain of a child is almost miraculously resilient, or plastic: surgeons can remove the entire left hemisphere, and thus (supposedly) all of the brain's language regions, and the child still learns to talk, read, and write as long as the surgery is performed before age four or five. Although in most people the left cerebral hemisphere supports language, the brain, it seems, can roll with the punches (or the surgery) well enough to reassign language function to the right cerebral hemisphere, all the way over on the other side of the head. Therefore, if the brain suffers damage before age two, and loses the areas originally designated as language regions, it usually reorganizes itself to reassign those language functions to another area. By age four to six, a brain injury that wipes out the original language areas usually leaves the child with a profound learning deficit, although she will typically retain the language she had learned up to then. After six or seven, however, the brain is already becoming set in its ways, and loss of its language regions can leave a severe and lasting language deficit. If an adult suffers damage to the left perisylvian, the site of language areas in the brain, the result is typically (though as recent stroke research shows, not always) permanent *aphasia,* the inability to use or understand words. A preschooler can recover from the loss of half her brain, but a little lesion in the same hemisphere leaves an elderly stroke patient mute. So although the brain of a young child retains impressive plasticity, that malleability yields, within a few short years, to something like neural obstinacy: the brain balks at rearranging itself in the face of changed circumstances.

As far as scientists can tell, then, a young brain can usually compensate for injury to a particular region by shifting the function of the damaged region to an unaffected region. But this comes at a cost. The area to which an otherwise-lost function is shifted becomes neurologically crowded, says Jordan Grafman of the National Institute of Neurological Disorders and Stroke, part of the National Institutes of Health. As a result, when the brain tries to execute two tasks in adjacent regions it can cause a sort of traffic jam. Grafman offers the example of an adolescent boy whose brain had been injured years before in a freak childhood accident. His right parietal lobe, a structure that supports visual and spatial skills, suffered a lesion. Yet despite the injury, the boy developed normal visual and spatial skills. Oddly, however, he had great difficulty with math, which is normally a function of the left parietal lobe. Through brain imaging, researchers learned that functions ordinarily controlled by the (injured) right side of the brain had moved over to the left hemisphere. Spatial skills typically develop before math skills do. As a result, when it came time for the child to learn math, the region of his brain that would ordinarily be responsible for that function had already been taken, and there was little neural real estate left to support mathematical reasoning.

Young brains are also relatively nimble at a form of neuroplasticity called *cross-modal reassignment.* This occurs when a brain region that ordinarily handles one form of sensory input does not receive the expected input. Rather than sit around unused, it seems, that region becomes more receptive to receiving a different input, as a satellite dish receiving no signal when pointed in one direction shifts to catch signals from another direction. Such reassignment within the brain seems to explain the changes that occur in children who become blind at a very young age. The visual cortex no longer receives sensory input from the retina through the optic nerve, and as a result the somatosensory cortex, which receives tactile input, invades areas normally dedicated to processing visual input. People who have been blind from birth often have an exquisitely sensitive sense of touch, particularly if they learn to read Braille when still young. The feel of the raised dots is processed in the visual cortex.

Similarly, in children who are congenitally deaf, the brain seems to reassign its auditory cortex (which is receiving no auditory information) to process visual information instead. In one clear demonstration of this, scientists exposed subjects who had been deaf since birth to a flash of light in their peripheral vision: the auditory cortex experienced a wave of electrical activity, showing that it had been rewired for sight rather than sound. What seems to have happened is that, during gestation, a visual neuron from the retina took a wrong turn and found itself in the auditory cortex. Under normal circumstances, the connections that neuron formed with other neurons would wither away; retinal neurons just don't make connections in the auditory cortex. But in a deaf child auditory neurons are silent and so offer no competition for the wayward retinal neuron. Synapses made by the wayward neuron survive and actually come in handy: congenitally deaf people typically perform better on tests of peripheral vision than people with normal hearing, probably thanks to these extra visual synapses. And the deaf often use their auditory cortex to register sign language; people with normal hearing typically use the visual cortex.

Since the early 1990s, MIT researchers led by Mriganka Sur had been probing the limits of neuroplasticity in somewhat unheralded lab animals: newborn ferrets. In these animals as well as in humans, the optic and auditory nerves grow from the eye and the ear, respectively, through the brainstem and thalamus before reaching the visual or auditory cortex. In humans, as we'll discuss later, this basic wiring plan is present at birth; in ferrets, however, these connections reach the thalamic way station to the cortex only after birth. In their breakthrough experiment, the MIT scientists took advantage of this delay. They lesioned the normal inputs to the auditory thalamus on one side of the brain. With the competition out of the way, as it were, projections from the retina, arriving at the thalamus, grew into the auditory cortex. Now the auditory cortex on that side was receiving signals from the retina and only the retina.

The result: When the ferrets were shown flashes of light on the rewired side of their brain, they responded not as if they saw the light but as if they heard it. The retinal nerve had carried the signal to the auditory cortex. This part of the brain, normally dedicated to sensing sounds, had been rewired to respond to sight.

Whatever the zoning law that originally destined this patch of cortex to bloom into primary auditory cortex, on receiving input from the retina it was transformed into the animal's primary visual cortex. The result: when the ferrets were shown a flash of light, they saw it with their auditory cortex. And there was more. Just as in the visual cortex of normal ferrets, the "auditory" cortex of rewired ferrets contained neurons that specialized in inputs of different spatial orientations—vertical, horizontal, and everything in between. The ferrets consistently responded to a light stimulus presented to the rewired auditory cortex as if it were indeed a light signal, even though the retinal neurons carrying the input fed in to what is normally the turf of "auditory" cortex. This bears emphasizing. Whether the nerves run from the retina or from the cochlea, they carry signals in the same way, through electrical impulses that I'll discuss later. There is nothing inherently "sightlike" or "soundlike" in the signals. It was once considered a fundamental principle of brain organization that the way signals are perceived depends on which part of the brain processes them. In the rewired ferrets, retinal nerves carry signals to what had been auditory cortex. Yet the rewiring had given the auditory cortex a new identity, turning it into a de facto visual cortex. As Michael Merzenich of the University of California, San Francisco, commented, "The animals 'see' with what was their auditory cortex. . . . [I]n these rewired animals, the experience of sight appears to arise from visual inputs to the auditory cortex area." The findings reminded Merzenich of a comment William James once made: if we could tinker with the nerves so that exciting the ear activated the brain center concerned with vision, and vice versa, then we would "hear the lightning and see the thunder."

Before exploring further the neuroplasticity of the developing brain, let's review some basic neurobiology. First, some elementary anatomy: a neuron in the brain consists, typically, of a cell body called the *soma*—Greek for "body"—which measures 10 to 100 micrometers across (100 micrometers equals 0.1 millimeter). The soma contains all the little goodies that keep the cell metabolizing and synthesizing proteins and performing all the other housekeeping functions that cells of all kinds carry out. From the soma sprout numerous multibranched tentacles, called *dendrites,* like snakes from Medusa's head. The dendrites' chief function in life is to receive incoming electrochemical messages from other neurons and carry the messages to the cell they're part of. Dendrites are relatively thick where they emerge from the cell body but divide at dozens if not hundreds of branch points, becoming thinner and wispier each time. The number of dendrites varies tremendously, depending on the function of the cell.

Neurons also sprout from their soma a single *axon,* a long fiber that extends away from the cell body like a string from a balloon and whose job is to carry information to another neuron. This information takes the form of electrical currents. Where an axon from a transmitting neuron terminates on a receiving neuron, it develops special structures, including little holding tanks for neurochemicals. These vesicles release chemicals that transmit messages to the next cell in the circuit. In this way neurons transmit informa-

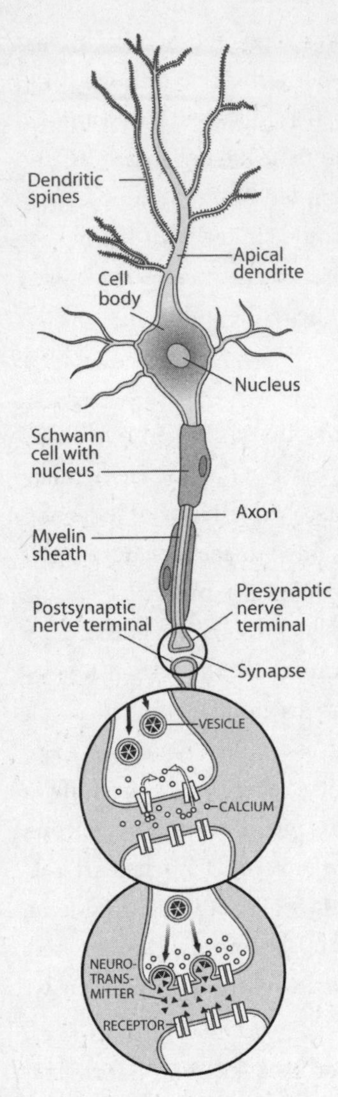

Figure 5: This neuron is typical of those that project from the cortex to the striatum. The inset shows the critical role that calcium ions play in triggering the release of neurotransmitter from vesicles in the presynaptic neuron into the synapse.

tion along their axons and on to the next neuron. So, again, dendrites receive, axons send. Axons can be as long as a meter, or as short as a few tenths of a millimeter, as are those that carry signals within a single part of the brain. Because these short-axon neurons link neurons locally, they are the most important players in the game of information processing: highly evolved animals have relatively more short-axon neurons than long-axon ones, reflecting their role in integrating and processing information. Exactly how many different types of neurons fill the brain remains an open question, although fifty seems to be a reasonable guess.

Despite their diversity of shape, size, types of connections, and neurochemical content, all neurons carry information in much the same way: they chatter away in electrochemical language. Information transmitted by one neuron and received by another takes the form of electrical signals generated by charged atoms, or *ions*—in particular, positively charged sodium and potassium ions or negatively charged chloride ions. The flux of ions across the cell membrane of a neuron is meticulously regulated by membrane pumps to give the inside of the cell a net negative charge relative to its surroundings. Rapid changes in the flux of ions can generate a moving pulse of electric charge called an action potential. Like a bulge zipping down a jump rope, the action potential speeds down the axon at up to 200 miles an hour in vertebrates (though only 30 to 40 miles an hour in invertebrates). It is the physical embodiment of the information sent from one neuron to another.

At the end of the axon lies the synapse, which is actually just—well, almost nothing, actually. To be more precise, the *synapse* consists of the axon of a transmitting neuron (called the presynaptic neuron), the dendrite or soma of a receiving neuron (the postsynaptic neuron), and the gap one-millionth of a centimeter wide between them. The synaptic gap, first named by the physiologist Sir Charles Sherrington a century ago, is reminiscent of the almost-touch between the finger of God and the finger of Adam that Michelangelo painted on the ceiling of the Sistine Chapel. For in that achingly close encounter lies a world of potential—in the case of neurons, the potential to hand off the signals that find expression as thoughts, emotions, and sensory perceptions.

Neurons take E. M. Forster's dictum "Only connect" to extremes. The average brain neuron (insofar as there is such a beast) forms about 1,000 synaptic connections and receives even more. Many neurons receive as many as 10,000 inputs, and some cells of the cerebellum receive up to 100,000. When the pulse of charge arrives at the synapse, it stimulates the entry of calcium ions, which triggers the process by which those tiny vesicles in the presynaptic neuron release neurotransmitters. Since the discovery of the first neurotransmitter in 1921, their number has, as of this writing, topped sixty. Neurotransmitters come in a range of molecular types, from amino acid derivatives to gases like nitric oxide (NO). Because neurotransmitters are the language of neuronal communication in the brain, drugs for mental disorders ranging from depression to anxiety to OCD target them. Valium, for instance, facilitates the effects of the neurotransmitter gamma-aminobutyric acid (GABA).

Molecules of neurotransmitter diffuse across the synapse to the postsynaptic neuron. There, the molecules act as a little armada of space vehicles, docking with tailor-made receptors on the postsynaptic neuron as rovers dock with the mother ship. And, not to belabor the analogy, when the neurotransmitters dock, they unleash a flurry of activity inside the neuron not unlike that unleashed when space pods dock: cascades of very complex molecular interactions including ion fluxes that eventually make the postsynaptic neuron more electrically positive. Once the postsynaptic neuron crosses an electrical threshold, it fires an action potential of its own, shooting it off to the next neuron in the circuit. And the electrochemical activity that underlies the thoughts, emotions, and sensory processing within the brain keeps going.

Although this hurly-burly of electrochemical activity is often thought of as turning on activity in the brain (of being *excitatory,* in neuroparlance), in fact synaptic transmission can also be inhibitory. The preceding example describes an excitatory neuron, in which the released neurotransmitters bind to receptors on the postsynaptic neuron and cause it to become more positive. If it is sufficiently more positive, it fires its own action potential. Inhibitory neurons have an opposite effect. In this case, the flux of ions increases the negative charge across the membrane, thus decreasing the possibility that an action potential will be triggered. Synapses between such neurons are therefore called inhibitory.

One additional concept is necessary for any discussion of neuroplasticity, and this is the notion of altering the strength of synapses. At first blush it seems nonsensical to talk about changing the strength of what is, after all, only a gap. But by "altering synaptic strength," we mean making the postsynaptic cell more likely to initiate an action potential, and keep the information transmission going, than it was before. This, as far as neuroscientists can tell, is the basis not only of memory but also of the wiring together of millions of neurons into functional circuits. How might such functional circuits form? The electrical impulses that shoot down an axon cannot vary in amplitude; neurons either fire or don't fire (this is known as the all-or-none property of neurons). So if the incoming electrical signal is invariant, then the only plausible suspect for the focus of change induced by neural activity is the synapse.

In 1949, the Canadian psychologist Donald Hebb proposed that learning and memory are based on the strengthening of synapses that occurs when pre- and postsynaptic neurons are simultaneously active. Somehow, he suggested, either the presynaptic neuron or the postsynaptic neuron (or both) changes in such a way that the activation of one cell becomes more likely to cause the other to fire. Although the notion was plausible from the moment Hebb first advanced it, there was not exactly a rush to the lab bench to test it. Hebb, after all, was a mere psychologist, not a neuroscientist. (Hebb was also the first to float the concept, in the late 1940s, of an "enriched environment" as a cause of behavioral improvements—an idea that, in its 1990s incarnation, would launch a thousand *Baby Einstein* videos.) Eventually, however, neuroscientists amassed data showing that Hebb was on to something: electrically stimulating cortical cells to fire simultaneously strengthened their synaptic connections.

As you might guess, this kind of increased synaptic strength is a key to the formation of enduring neuronal circuits and has become known by the maxim "Cells that fire together, wire together." As best neuroscientists can determine, Hebbian plasticity begins with the release from presynaptic neurons of the neurotransmitter glutamate. The glutamate binds to two kinds of receptors on the postsynaptic neuron. One receptor notes that its own neuron, the postsynaptic one, is active; the other notes which presynaptic neurons are simultaneously active. The postsynaptic neuron therefore detects the simultaneous occurrence of presynaptic and postsynaptic activity. The ultimate result is that a particular action potential whizzing down the axon of a presynaptic neuron becomes more efficient at causing the postsynaptic neuron to fire. When that happens, we say that there has been an increase in synaptic strength. The two neurons thus become locked in a physiological embrace, allowing the formation of functional circuits during gestation and childhood. The process is analogous to the way that traveling the same dirt road over and over leaves ruts that make it easier to stay in the track on subsequent trips. Similarly, stimulating the same chain of neurons over and over—as when a child memorizes what a cardinal looks like—increases the chances that the circuit will fire all the way through to completion, until the final action potential stimulates the neuron in the language centers and allows the kid to blurt out, "Cardinal!" As a result of Hebbian plasticity, the brain has learned that a crimson bird is called a cardinal. This same pathway crackles with electrical activity whenever you recall a cardinal, and the more you replay this memory, the greater the efficiency with which you can call it up. Changes in synaptic strength thus seem to underlie long-term memory, which must, by its very nature, reflect enduring (if not permanent) changes in the brain regions where memories are stored.

Altering connections in a way that strengthens the efficiency of a neuronal circuit over the long term was the first kind of neuroplasticity to be discovered. Plasticity must be a response to experience; after all, the only thing the brain can know and register about some perception is the pattern of neural activity it induces. This neural representation of the event somehow induces physical changes in the brain at the level of neurons and their synapses. These physical changes "allow the representation of the event to be stored and subsequently recalled," says Tim Bliss of the National Institute for Medical Research in Mill Hill, England. In a very real sense, these physical changes *are* the memory.

As much as any other neuroscientist, Dr. Eric Kandel of Columbia University has worked out the molecular changes that accompany Hebbian learning and the formation of memories. Kandel works with the unprepossessing creature called *Aplysia californica*, otherwise known as a sea snail, which resembles nothing so much as a crawling, bruise-colored blob with ears. *Aplysia*'s nerve cells are the largest (as far as scientists know) of any animal's; actually being able to see what you're investigating, without having to resort to stains and microscopes, makes the task of working out circuitry a lot simpler. So does having to keep track of a mere 20,000 nerve cells (compared to the 100 billion of the human brain).

Kandel and his colleagues made their first breakthrough when they played Pavlov, and instead of using dogs used *Aplysia*. They sprayed one of the sea snail's sensitive spots with water—this stimulus makes the creature snap back inside its mantle—and simultaneously gave it an electric shock. The result was sensitization: *Aplysia* jerked back inside its mantle whenever the scientists jolted it ever so slightly. This, in the world of the sea snail, counts as learning: *Aplysia* is remembering that a touch is followed by a nasty shock and so scoots back inside its protective mantle when it experiences the touch. In much the same way, Pavlov's dogs learned to salivate at the sound of a bell because, during training, food had been paired with that sound.

After identifying the neural circuits underlying this and other simple behaviors, Kandel and a series of collaborators were able to determine how the circuits change as *Aplysia* learns to respond to the different stimuli. They found, for instance, that the sensitized neurons had undergone a long-lasting change: when excited (by the touch), they discharge more neurotransmitter than do neurons of *Aplysia* that have not undergone sensitization. They also found that after a period of stimulation, certain reflex actions can be enhanced for significant periods of time—hours or even days. These stimuli give rise to increased levels of a so-called secondary messenger molecule, called cyclic AMP (or cAMP to its friends). The rise in cAMP levels results in the activation of certain genes in the nucleus of the nerve cell; the gene activation leads to the synthesis of new proteins, some of which appear to play a role in establishing new synaptic connections. It is these connections, neuroscientists now agree, that are the basis for long-term memory. Experience, then, produces stable, observable changes in what passes for *Aplysia*'s brain, changes that mammals also undergo, as Kandel showed in the 1990s when he added mice to his menagerie of lab animals and replicated the work in rodents.

The molecular basis of memory and learning, the discovery of which earned Kandel a share of the 2000 Nobel Prize in physiology or medicine, stands as one of the best understood of the changes the brain undergoes. It is one of the mechanisms that underlie the plasticity of the developing brain. Changes in how an organism interacts with its environment result in changes in connectivity.

We've spent some time on synaptic efficiency, and the "cells that fire together, wire together" mantra, because similar phenomena seem to underlie the plasticity of the developing brain, the diminution of plasticity in the mature brain, and the possibility of directed or induced neuroplasticity in every brain. At the beginning of the 1990s, neuroscientists had only a general idea of how a few embryonic cells transform themselves into a brain, a spinal cord, and a skein of peripheral nerves, all hooked up in such a way as to form a sensing, thinking, feeling human being. Wiring up the circuits of neurons that support those tricks is, to put it mildly, a daunting task. Neurons must appear in the right place at the right time and in the right quantity, to be sure. But contrary to Woody Allen's conclusion that 90 percent of life is just showing up, for a neuron, showing up is just the start. The axons that shoot out of the neurons must also find their way to the correct target cells and make functional connections, and in the last few years researchers started to glimpse how the brain does it. The key finding is that the brain wires itself in response to signals it receives from its environment, a process very similar to that underlying neuroplasticity in the adult brain, too.

It has become a cliché to note that the human brain is, as far as we're aware, the most sophisticated and complex structure in the known universe. A newborn brain contains something on the order of 100,000,000,000—that's 100 billion—nerve cells. That is most of the neurons a brain will ever have. Although 100 billion is an impressive number, it alone cannot explain the complexity, or the power, of the brain that lets us see and hear, learn and remember, feel and think; after all, a human liver probably contains 100 million liver cells, but if you collect 1,000 livers, you fall quite a few synapses short of a brain. The complexity of a brain, as distinct from a liver, derives chiefly from the connections that its neurons make. Neurons consist of that cell body we described, of course, but it is the neuron's accessories—axons and dendrites—that truly set a neuron apart from a liver cell.

Axons and dendrites enable neurons to wire up with a connectivity that computer designers can only fantasize about. Each of the 100 billion neurons connects to, typically, anywhere from about a few thousand to 100,000 other neurons. The best guess is that, at birth, each neuron makes an average of 2,500 of these specialized junctions, or synapses; reaches a connectivity peak of 15,000 synapses at age two or three; and then starts losing synapses in a process called pruning. If we take a conservative mean for the number of connections (1,000), then the adult brain boasts an estimated 100,000,000,000,000—100 trillion—synapses. Other estimates of the number of synapses in the adult brain go as high as 1,000 trillion.

Although it would be perfectly reasonable to posit that genes determine the brain's connections, just as a wiring diagram determines the connections on a silicon computer chip, that is a mathematical impossibility. As the Human Genome Project drew to a close in the early years of the new millennium, it became clear that humans have something like 35,000 different genes. About half of them seem to be active in the brain, where they are responsible for such tasks as synthesizing a neurotransmitter or a receptor. The brain, remember, has billions of nerve cells that make, altogether, trillions of connections. If each gene carried an instruction for a particular connection, we'd run out of instructions long before our brain reached the sophistication of, oh, a banana slug's. Call it the genetic shortfall: too many synapses, too few genes. Our DNA is simply too paltry to spell out the wiring diagram for the human brain.

Before we explore what makes up the shortfall, it's only fair to give genes their due by describing some aspects of brain development that they do deserve credit for. Since fetal brains follow almost identical developmental sequences and reach the same developmental milestones at about the same time, it's safe to say that the overall pattern of brain development is surely under genetic control (which is not to say that developmental neuroscientists have figured out how the genes do it). The brain starts down the developmental highway soon after a sperm fertilizes an egg. By the fourteenth day, what is now a ball of hundreds of cells folds in on itself, resembling a cleft in a plump chin: cells on the outer surface infold, until they arrive in the interior of the ball. As the ball of cells continues folding in, it simultaneously lengthens, forming a tube. One end will become the spinal cord; the other will develop into the brain. At about three weeks the embryo begins to produce neurons, reaching a peak of production in the seventh week and largely finishing by the eighteenth. Running the numbers shows what a prodigious feat neurogenesis is: since a newborn enters the world with 100 billion or so neurons in its brain, and since the lion's share of neurogenesis is completed just short of halfway through gestation, the fetal brain is producing something on the order of 500,000 neurons every minute during the high-production phase, or 250,000 per minute averaged over the entire nine months in utero. More than 90 percent have formed midway through gestation. After nine months, the newborn's brain is a jungle of that estimated 100 billion nerve cells.

From stem cells on the walls of the brain-to-be's ventricles, immature neurons are born. Starting in the second trimester of pregnancy, these protoneurons immediately begin to migrate outward in a journey so challenging that it has been likened to a baby's crawling from New York to Seattle and winding up in the precise neighborhood, on the right street, at the correct house that he was destined for from the moment he left Manhattan. These baby neurons follow a sort of cerebral interstate, a structure of highways laid down by cells called *radial glia*. These cells form an intracranial road network complete with rest stops (for the glial cells also nourish the traveling neurons). Protoneurons destined for the cortex observe a first-to-go, first-to-stop rule: those that first leave the ventricular walls stop in the closest cortical layer. The second wave of émigrés continues on to the second-closest layer, and the subsequent waves migrate past each formative layer before stopping at an ever-more-distant layer, until all six cortical layers are populated. Once the immature neurons are in place, the radial glia vanish. How neurons realize that they have reached their final destination remains a mystery, too. But we do know that it is only when the immature neurons are in place that they become full-fledged neurons and put down roots, blossoming with dendrites and sprouting an axon by which they will communicate with, and form a circuit with, other neurons near and far.

Timing also seems to be under clear genetic control. For instance, the *sulci*—invaginations, or fissures—that divide one lobe of the brain from another emerge at what seem to be genetically programmed times: the central sulcus, dividing the frontal lobe from the parietal, appears around the twentieth week of gestation and is almost fully formed in the seventh month. At about the fifteenth week after conception a body map appears in the brainstem and then in the thalamus (a sort of relay station for incoming sensory input), whose neurons begin forming synapses in what will be the brain's somatosensory cortex. Only several weeks into the final trimester do thalamic axons begin to form synapses on cortical neurons that will be their (normally) permanent partners. In fact, it is the arrival of these thalamic axons that turns this strip of cortex into the somatosensory region. By the third trimester, if all is going as it should, most of the neurons have found their place, and, although the circuits are only rough approximations of what they will ultimately become, at least the brain's major structures have taken shape.

At birth, the spinal cord and brainstem are just about fully formed and functional. That works out well, since it is these structures that carry out such vital tasks as thermoregulation, heartbeat, and reflexes such as grasping, sucking, and startling. Soon after birth the cerebellum and midbrain become *myelinated* (encased in the fatty coating of myelin that enables them to carry electrical impulses efficiently); the thalamus, basal ganglia, and parts of the limbic system follow suit in the first and second years after birth. Finally, the cerebral cortex, led by sensory areas, comes fully on line. At birth the somatosensory cortex, which processes the sense of touch, is a mess, with neurons from different points on the body converging in cortical regions that overlap so much that a newborn cannot tell where she is being touched. But through the experience of touch the somatosensory cortex develops into a precise "map," which means that one spot receives tactile stimuli from the lips and only the lips, and another receives tactile stimuli from the right knee and only the right knee, until every speck of skin is represented. This maturation proceeds from head to toe, with the mouth the first part of the body to become touch-sensitive. The rest of the cortex follows on the somatosensory toward maturity: motor regions first, followed by the parietal, temporal, and frontal association cortices (the seats of judgment, reason, attention, planning, and language, among other higher-order function), which are still forming in the late teens.

It is not merely gross anatomic structures of the brain that form during gestation and early childhood. Moving a few billion neurons to a particular site doesn't give you a working brain any more than throwing a few million integrated circuits into a plastic box gives you an iMac. All of those nerve cells need not only to find their way to the right location in the nascent brain but, crucially, to make the right connections once there, so that a taste detected on the tongue can make its way to the brainstem and from there to cortical regions that will identify it as, say, vanilla, or a tickle on the right cheek will be transformed into electrochemical signals that reach the part of the somatosensory cortex responsible for processing tactile sensations on that cheek.

Forming the right connections is no simple matter; that baby crawling from New York to a particular address in Seattle has it easy compared to what neurons face. Consider the challenge faced by neurons of the nascent visual system. Their goal is to form a functional pathway that goes like this: neural signals from the rods and cones of the eye's retina travel to retinal interneurons, which hand them off to retinal ganglion cells (which constitute the optic nerve) and then to a relay station called the *lateral geniculate* nucleus, where axons from the left eye alternate with axons from the right eye to form eye-specific layers. From there, the signal travels to cells in the primary visual cortex all the way in the back of the brain, where clusters of neurons receiving signals from the left eye form separate, alternating layers with clusters of neurons receiving input from the right eye. In order to effect this signal transfer properly, then, axons from the eye must grow to the correct spot in the lateral geniculate nucleus (LGN). Axons growing from the LGN must resist the urge to terminate in synapses in the auditory or sensory cortex (where they arrive first) and instead continue growing until they reach the appropriate target all the way back in the primary visual cortex. More than that, cells lying beside each other in the retina must extend their axons to neighboring neurons in the lateral geniculate, which must in turn extend their axons to neighboring cells in the visual cortex: for the ultimate goal is for each clump of a few hundred neighboring neurons in the visual cortex to form synapses only with neurons responding to the same little region of the baby's visual field. How in the world do they do it?

The first part of the axon's journey—the extension of its tip in the right direction—is the easy part. That's because target neurons emit come-hither signals called *trophic factors*. These provide general directions that tell each growing axon from the retina to form a synapse with a particular target neuron in the visual cortex; which neurons emit trophic factors and which traveling axons respond to them seems to be under genetic control. Using a specialized tip called a *growth cone,* the axon is programmed to sniff out and grow toward molecules of specific trophic factors scattered along the predestined route, like Hansel and Gretel following the trail of crumbs through the forest path back to their cottage. Growth cones have what seem to be genetically encoded receptor molecules that make the axon elongate and migrate in the direction of attractant molecules. In this sense, target selection is preprogrammed, with genes directing axons and dendrites to grow to their approximate final destinations. At the same time neurons with which the axon is not destined to form a synapse release chemical signals that repel the axons. Between the attractants and repellents—which have been given names like *netrin* (from the Sanskrit *netra*, meaning "leader" or "guide") and *semaphorin* (*semaphor* is Greek for "signal")—the axon stays on track to its destination.

Once the axon has arrived in Seattle, to continue the analogy, it still has to find the right neighborhood and the right house that will be its ultimate cellular address. And this is anything but preprogrammed, for as we've seen humans simply don't have enough genes to specify every connection that the neurons in the brain must forge. To make up the genetic shortfall, it slowly dawned on neuroscientists in the 1980s, the brain must assemble itself into functional circuits through experience. Once the genes have run out, experience directs the axons to complete their journey.

The factor that provides the developing brain with the right connections is, ironically, an early profusion of wrong connections. The fetal brain is profligate in its overproduction of both neurons and synapses. Not all the axons trying to reach out and touch another will manage it; the failures die. About half the neurons that form in the fetal brain die before the baby is born: 200 billion neurons, give or take, are reduced to the 100 billion of a newborn as neurons that fail to form functional synapses vanish.

Synapses are pruned even more ruthlessly. Spinal cord synapses begin forming by the fifth week of embryogenesis, cortical synapses are forming at seven weeks, and *synaptogenesis* (synapse formation) continues through gestation and well into childhood. By one count, each cortical neuron forms, at the height of synaptogenesis, 15,000 connections: that works out to 1.8 million synapses per second from the second month in utero to the child's second birthday. Which synapses remain, and which wither away, depends on whether they carry any traffic. If they do not, then like bus routes that attract no customers, they go out of business. By one estimate, approximately 20 billion synapses are pruned every day between childhood and early adolescence. It's survival of the busiest. Like a cable TV subscription canceled because nobody's watching, synaptic connections that aren't used weaken and vanish. Here is where the power of genes falls off rapidly: genes may lead neurons to make their initial, tentative connections and control the order in which different regions of the brain (and thus physical and mental capacities) come on line, but it's the environmental inputs acting on the plasticity of the young nervous system that truly determine the circuits that will power the brain. Thus, from the earliest stages of development, laying down brain circuits is an active rather than a passive process, directed by the interaction between experience and the environment. The basic principle is this: genetic signals play a large role in the initial structuring of the brain. The ultimate shape of the brain, however, is the outcome of an ongoing active process that occurs where lived experience meets both the inner and the outer environment. As we will see, as the prefrontal circuitry matures, volitional choice can become a critical element in shaping the architecture bequeathed by both genetic factors and environmental happenstance.

Although the gross appearance and morphology of the brain change little after birth, neuroplasticity allows it to undergo immense changes at the cellular level, changes that underlie the unfolding cognitive and other capacities of a child. One of the starkest demonstrations of this has come from studies of speech perception. Newborns can hear all the sounds of the global Babel: the French *u* in *du*, the Spanish *n*, the English *th*. When one of the sounds stimulates hairs in the cochlea, the sound becomes translated into an electrical impulse that finds its way to the brain's auditory cortex. As a racer hands off the baton in a relay, each neuron between ear and cortex passes the electrical impulse to the neuron beyond. After enough repetitions of a sound, the synapses connecting all those neurons have been strengthened just as Hebb described. The result is that this precise neuronal pathway responds to *th* every time, culminating in the stimulation of a dedicated cluster of neurons in the auditory cortex that produces the subjective sense that you have heard the sound *th*. The result is that networks of cells become tuned to particular sounds in the language the newborn constantly hears.

Space in the auditory cortex is limited, of course. Once the Hebbian process has claimed circuits, they are hard-wired for that sound; so far, neuroscientists have not observed any situations in which the Hebbian process is reversed so that someone born into a sea of, say, Finnish loses the ability to hear Finnish's unique sounds. Although a growing appreciation of the plasticity of the adult brain has now overturned the idea that it is impossible to learn a second language and speak it without an accent after the age of twelve, without special interventions the auditory cortex is like development in a close-in suburb: it's all built up, and there are no empty lots that can be dedicated to hearing new sounds.

Patricia Kuhl, a leading authority in speech development, got a dramatic demonstration of this. In order to test Japanese adults and children on their ability to distinguish various phonemes, she made an audio disk with the various sounds and took it with her during a visit to the language lab of colleagues in Tokyo. Before testing any volunteers, she first wanted to demonstrate the disk to the Japanese scientists. As "rake, rake, rake" intoned through the high-end Yamaha speaker, her colleagues leaned forward expectantly for the sound change she had told them was coming. The disk segued into "lake, lake, lake," Kuhl had said—but the Japanese, all proficient at English, still leaned in expectantly. They literally could not hear any difference between the sound of *lake* and the sound of *rake*.

The difference lay in their brains. Children who grow up hearing the sounds of a language form dedicated circuits in their auditory cortex: the brains of the children Kuhl left behind in Seattle, where she is a professor at the University of Washington, had been molded by their auditory experience to discriminate *r* from *l*. When? The seven-month-old Japanese babies whom Kuhl tested had no trouble discriminating *r* from *l*. But ten-month-olds were as deaf to the difference as adults. When Kuhl did a similar test of Canadian babies raised in English-speaking homes, she got the same results: six-month-olds could distinguish Hindi speech sounds even though those sounds were not part of their auditory world; by twelve months they could not. Between six and twelve months, Kuhl concludes, babies' brains begin the "use it or lose it" process of pruning unused synapses. The auditory cortex loses its sensitivity to phonemes that it does not hear every day. This may be why children who do not learn a second language before puberty rarely learn to speak it like natives.

The reverse is true, too: connections that are used become stronger, even permanent elements of the neural circuitry. A newborn forms millions of connections every day. Everything he sees, hears, feels, tastes, and smells has the potential to shape the nascent circuits of his brain. The brain is literally wired by experience, with sights, sounds, feelings, and thoughts leaving a sort of neural trace on the circuits of the cortex so that future sights, sounds, feeling, thoughts, and other inputs and mental activity are experienced differently than they would otherwise be. In the case of Kuhl's young subjects, she speculates, the phonemes a child hears strengthen the auditory synapses dedicated to that sound; repeatedly hearing the sound broadens a child's perceptual category for that sound, crowding out sounds with a similar auditory profile, until the number of auditory neurons dedicated to those neighboring sounds eventually dwindles to literally zero.

Clearly, the hardware of the brain is far from fixed at birth. Instead, it is dynamic and malleable. As far back as the 1960s and 1970s, researchers were documenting that rats raised in a lab cage with wheels to run on and ladders to scamper up, as well as other rats to interact with, grew denser synaptic connections and thicker cortices than rats raised with neither playmates nor toys. The "enriched" environment was closer to the world a rat would experience in the wilds of New York City, for example. The cortical differences translated into functional differences: rats with the thicker, more synaptically dense cortices mastered mazes and found hidden food more quickly than did rats from the poorer environments, who had thinner cortices.

From the moment a baby's rudimentary sensory systems are operational (which for hearing and tactile stimulation occurs before birth), experiences from the world beyond begin to impinge on the brain and cause brain neurons to fire. Let's return to the example of the visual system, which, when we left it, had axons from retinal neurons projecting into the nascent visual cortex. Alone among the senses, the visual system receives no stimulation until after birth. In the fourth week of gestation the eye begins to form. Synapses form first in the retina, then in subcortical visual areas, followed by the primary visual cortex, and, finally, higher visual centers in the temporal and parietal lobes. In the second trimester, the visual system experiences its growth spurt: between fourteen and twenty-eight weeks, all of the 100 million neurons of the primary visual cortex form. They start to make synapses in the fifth month and continue making them for another year at the staggering rate of 10 billion per day, until the visual cortex reaches its maximal density at eight months of age. At birth, a baby sees the world through a glass darkly. More synapses responsible for processing motion than for perceiving form have come on line, with the result that babies detect movement better than they do shape. The entire visual field, in fact, consists of no more than a narrow tunnel centered on the line of sight, and the baby's visual resolution is about one-fortieth that of a normal adult. Making the world look even odder (not that the baby has much to compare it to), newborns lack depth perception. They can see clearly only eight or so inches out. Normally, however, vision improves by leaps and bounds in the first few weeks, and by four months the baby can perceive stereoscopic depth. By six months, visual acuity has improved fivefold, and color vision, acuity, and volitional control of eye movements have all emerged. The tunnel of view expands, until by twelve months it is equivalent to an adult's. Around a child's first birthday, he sees the world almost as well as a normal adult.

The accurate wiring of the visual cortex that underlies the gradual improvement in vision occurs only if the baby receives normal visual stimuli. In other words, it is electrical activity generated by the very act of seeing that completes the wiring of the visual cortex. Again, although genes have played matchmaker between axons and neurons so that webs of neurons take shape, the number of human genes falls well short of the number needed to specify each synapse even within the visual cortex. Genes get the neurons to the right city and dispatch their axons to the general vicinity of neurons with which they will form synapses, but the baby's unique interaction with the environment has to take it from there. Scientists concluded in the 1990s that one of the main ways axons and dendrites make connections—and neurons therefore form circuits—is by firing electrical signals, almost at random, and then precisely honing the crude pattern to fit the demands of experience. "Even before the brain is exposed to the world, it is working to wire itself," says the neuroscientist Carla Shatz, one of the pioneers in the field of brain development. She became a neuroscientist because, while she was in high school, her grandmother suffered a debilitating stroke that paralyzed one side of her body; Shatz vowed to join the search for how the nervous system works and settled on developmental neuroscience. "The adult pattern of connections emerges as axons remodel by the selective withdrawal and growth of different branches," she says. "Axons apparently grow to many different addresses within their target structures and then somehow eliminate addressing errors."

That "somehow" remains a holy grail of developmental neuroscience. Axons seem to be fairly promiscuous in their choice of target neurons: any will do. But then competition among inputs sorts out the axons, leading to areas with specific functions. An early clue to how they manage this feat came in the 1970s. David Hubel and Torsten Wiesel, working at Harvard, hit on the simple experiment of depriving newborn cats of visual input by sewing shut one of their eyes. After even a week of sightlessness, axons from the lateral geniculate nucleus representing the closed eye occupied a much smaller area of the cortex than did axons representing the normal eye. Then the scientists opened the eye that had been sewed shut, so that both eyes would deliver (it was thought) equal stimuli to the brain. They recorded from neurons in the primary visual cortex of the kittens, who by this time were at least four months old. Despite the fact that both eyes were open, virtually all of the visual cortex received input only from the eye that had been open since the kitten's birth. Neurons from the eye that had been sewed shut formed few functional connections; it was as if the synapses had melted away from disuse. If kittens do not receive visual input between thirty and eighty days after birth (a window of time now known as the critical period), it is too late: the unused eye is blind forever. The development of visual function in the cortex depends on visual inputs; visual deprivation early in life changes the cortex, Hubel and Wiesel found. Their work won them the 1981 Nobel Prize in physiology or medicine.

Extra activity—the opposite of deprivation—also changes the cortex, found the Harvard team. Brain regions stimulated by a kitten's normal, open eye invaded and colonized regions that should have handled input from its closed eye. As a result, input from the normal eye reached a disproportionately large number of neurons in the visual cortex. Rather than the eyes' having equal representation, most of the cortical cells received input only from the normal eye. This eye was now driving cells usually devoted to the shut eye. Yet this was not the case when the scientists recorded from cells in the retina or the lateral geniculate nucleus. That is, areas that should respond to stimulus of the normal eye did so, and areas that should respond to stimulus of the once-closed eye did not. Apparently, the majority of the changes induced by depriving the brain of visual input occurred in the cortex, not earlier in the visual pathway. This discovery suggested that if axons are carrying abnormally high electrical traffic, they take over more space in the visual cortex.

Other experiences produce even curiouser changes in the cortex. Neurons in the visual cortex turn out to be specialists. Some respond best to the sight of an edge oriented vertically, others to an edge oriented horizontally. There are roughly equal numbers of cells that respond to each orientation. But if the world of a newborn cat is skewed so that it sees only vertical lines, then in time most of its neurons will respond only to vertical edges. When a kitten is reared in a room painted with vertical stripes, very few of its direction-sensitive neurons turn out to specialize in horizontal stripes, but there is a profusion of vertical specialists. The animal literally cannot see horizontal lines but perceives vertical ones with great acuity. Human brains bear similar marks of early visual experience: a 1973 study found that Canadian Indians, growing up in teepees, had better visual acuity for diagonal orientations than people reared in the modern Western standard of horizontal ceiling joints and vertical wall joints.

The key role that experience plays in wiring the human brain shows up most clearly when sensory stimuli are completely absent. In a human version of the kitten's sewn-shut eyelids, babies are sometimes born with a cataract—an opaque lens—in one eye or both. This congenital defect strikes about 1 baby in 10,000 in the United States. If both eyes suffer cataracts, the brain never develops the ability to see normally. If only one eye has a cataract, vision in the affected eye never develops, as a result of active suppression by the normal eye, explains Ruxandra Sireteanu of the Max-Planck Institute for Brain Research in Frankfurt: "The two eyes compete for a glimpse of the outer world. If left untreated, vision in the affected eye, and the ability to use both eyes for binocular vision, will be permanently lost. It is not actually the eye that suffers, but the brain." In this case acuity, the size of the field of vision, and the ability (or, rather, inability) to see stereoscopically remain stuck at the level of a newborn. Without normal visual input, the developing brain fails to establish the amazing network of connections that allow it to receive, process, and interpret electrical signals triggered by the fall of light on the retina.

Nowadays, doctors realize that removing the affected lens and replacing it with a clear, artificial one (the same sort of cataract surgery performed on elderly people) allows the brain to receive clear, sharp signals. What had remained unknown until recently, however, was how much visual input was required for the repaired eye to begin seeing. Was it hundreds of hours? Or mere seconds? In a 1999 study, researchers at the Hospital for Sick Children in Toronto went a long way toward finding an answer. They studied twenty-eight babies who had been born with dense cataracts: sixteen of the babies suffered cataracts in one eye, twelve had them in both. Between one week and nine months of age the babies underwent surgery to remove the cataracts; they had therefore been deprived of normal visual input for that length of time. Within days or weeks of surgery, the babies were fitted with contact lenses so that, for the first time in their lives, "visual input was focused on the retina." To test the babies' eyesight, the researchers showed them black-and-white vertical stripes on a gray background. By varying the spacing between the stripes, and using infants' well-established propensity to pay attention to novel objects, the researchers were able to measure the babies' visual resolution. Although the initial acuity of the treated eye or eyes was significantly worse than that of babies born with normal vision, it improved after as little as a single hour of this exposure. "It appears that lack of visual exposure maintains the visual cortex of infants at the newborn level," says Sireteanu. "But even a one-hour exposure to the visual world sets the stage for rapid visual improvements."

Other sensory systems are similarly dependent on environmental input to finish the neuronal wiring that remains to be done once genetic instructions are exhausted. Auditory neurons first appear at three weeks after conception, and auditory centers in the brainstem emerge by thirteen weeks. In experiments with lab animals, removing the cochlea of one ear soon after birth reduces the number as well as the size of auditory neurons in the brainstem. Since these neurons receive no input, they essentially fold their tents and go out of business. But that kind of brain change can apparently be reversed, again through sensory input. In congenitally deaf children given cochlear implants, which bypass the damaged sensory hair cells of the inner ear and carry acoustical signals directly to the cortex, the sudden onset of sound after weeks or months of silence leads to nearly (depending on the age at which the implants are given) perfect speaking and hearing, as well as normal language development.

A 1999 experiment with kittens revealed the neurological basis for the rapid improvement. Researchers led by Rainer Klinke at the Physiologisches Institut of Frankfurt, Germany, tested the effects of the cochlear implants on a group of three- to four-month-old kittens that were deaf at birth. Brain imaging had already shown that the unstimulated auditory nervous system in the deaf kittens was not developing as it does in normal cats. But after the implants, the cats began to respond to sounds in the same way as cats born with normal hearing. Their auditory cortex changed, too: within a short time the size of the region of auditory cortex that responded to sound increased, the amplitude of the electrical signals in the auditory cortex soared, and measures of information processing within the cortex (called *long-latency neural responses*) increased. The experiment offers some explanation for the fact that cochlear implants in children born deaf prove quite successful. "It is quite likely that a similar 'awakening' of the visual cortex takes place in congenitally blind infants newly exposed to visual information after cataract removal and the fitting of an artificial lens," says Sireteanu.

A child need not suffer a congenital defect to demonstrate the power of sensory experience to wire the brain. Consider the somatosensory cortex. As mentioned, during fetal development it forms a rudimentary "map" of the body, so that the area receiving signals from the right hand abuts the area that receives signals from the right arm, which in turn abuts the area that receives signals from the right shoulder—you get the idea. But even at birth the somatosensory cortex has a long way to go. It is only through the experience of sensation that the map becomes precise. What seems to happen is that a baby is touched on, say, the back of her right hand. The neurons that run from that spot into the brain therefore fire together. But a neuron from the right wrist that took a wrong turn and wound up in the hand area of the somatosensory cortex is out of step. It does not fire in synchrony with the others (since touching the back of a hand does not stimulate the wrist). The tentative synapse it formed in the somatosensory cortex therefore weakens and eventually disappears. The region of the somatosensory cortex that receives inputs from the back of the right hand now receives inputs only from the back of the right hand. As similar mismappings are corrected through experience and the application of the "neurons that fire together wire together" rule, a once-ambiguous map becomes sharp and precise.

That the newborn brain needs stimulation in order to form the correct wiring pattern is uncontested. A bitter and somewhat politicized dispute has arisen, however, over the meaning of *stimulation*. To many neuroscientists, it means nothing more than what a baby with functioning senses would receive from being alive and awake in the everyday world: sights, sounds, tastes, touches, and smells. Numerous observations have documented that babies who are severely neglected, and who do little but lie in their crib for the first year or more of life, develop abnormally: few can walk by the age of three, and some cannot sit up at twenty-one months. Whether more stimulation, especially cognitive stimulation, would produce even better wiring is hotly contested. Selling desperate parents video tapes that promise to turn their not-yet-crawling baby into a junior Einstein, persuading them to fill the house with the strains of Mozart, admonishing them that a family meal without a math lesson using forks and spoons is a missed opportunity—such gambits have given "stimulation" a bad name. It is clear that some amount of stimulation is crucial to the development of the human brain. But in all likelihood, it's enough for the baby to explore the environment actively and interact with parents, to play peekaboo and hide-and-seek, to listen to and participate in conversations.

The great wave of synaptic sprouting and pruning was supposed to wash over the brain in infancy and toddlerhood, finishing in the first few years of life. But in the late 1990s scientists at several institutions, including here at UCLA, rocked the world of neuroscience with the discovery that a second wave of synaptic sprouting occurs just before puberty. In 1999, neuroscientists led by Elizabeth Sowell of UCLA's Lab of Neuro Imaging MRI compared the brains of twelve- to sixteen-year-olds to those of twenty-somethings. They found that the frontal lobes, responsible for such "executive" functions as self-control, judgment, emotional regulation, organization, and planning, undergo noticeable change during late adolescence: they start growing at ten to twelve years (with girls' growth spurt generally occuring a little earlier than boys'), much as they did during fetal development. And in another surprising echo of infancy, the frontal lobes then shrink in people's twenties as extraneous branchings are pruned back into efficient, well-organized circuitry. No sooner have teens made their peace (sort of) with the changes that puberty inflicts on the body than their brain changes on them, too, reprising a dance of the neurons very much like the one that restructured the brain during infancy. Contrary to the notion that the brain has fully matured by the age of eight or twelve, with the truly crucial wiring complete as early as three, it turns out that the brain is an ongoing construction site. "Maturation does not stop at age 10, but continues into the teen years and even the 20s," says Jay Giedd of the National Institute of Mental Health, whose MRI scans of 145 healthy four- to twenty-one-year-olds also found that the gray matter in the frontal lobes increased through age eleven or twelve. "What is most surprising is that you get a second wave of overproduction of gray matter, something that was thought to happen only in the first 18 months of life. Then there is a noticeable decline. It looks like there is a second wave of creation of gray matter at puberty, probably related to new connections and branches, followed by pruning."

Again as during fetal development, synapses that underlie cognitive and other abilities stick around if they're used but wither if they're not. The systematic elimination of unused synapses, and thus unused circuits, presumably results in greater efficiency for the neural networks that are stimulated—the networks that support, in other words, behaviors in which the adolescent is actively engaged. Just as early childhood seems to be a time of exquisite sensitivity to the environment (remember the babies who dedicate auditory circuits only to the sounds of their native language, eliminating those for phonemes that they do not hear), so may adolescence. The teen years are, then, a second chance to consolidate circuits that are used and prune back those that are not—to hard-wire an ability to hit a curve ball, juggle numbers mentally, or turn musical notation into finger movements almost unconsciously. Says Giedd, "Teens have the power to determine their own brain development, to determine which connections survive and which don't, [by] whether they do art, or music, or sports, or videogames."

This second wave of synaptogenesis is not confined to the frontal lobes. When the UCLA team scanned the brains of nineteen normal children and adolescents, ages seven and sixteen, they found that the parietal lobes (which integrate information from far-flung neighborhoods of the brain, such as auditory, tactile, and visual signals) are still maturing through the midteens. The long nerve fibers called white matter are probably still being sheathed in myelin, the fatty substance that lets nerves transmit signals faster and more efficiently. As a result, circuits that make sense of disparate information are works in progress through age sixteen or so. The parietal lobes reach their gray matter peak at age ten (in girls) or twelve (in boys) and are then pruned. But the temporal lobes, seats of language as well as emotional control, do not reach their gray matter maximum until age sixteen, Giedd finds. Only then do they undergo pruning. The teen brain, it seems, reprises one of the most momentous acts of infancy, the overproduction and then pruning of neuronal branches. "The brain," says Sowell, "undergoes dynamic changes much later than we originally thought."

The wiring up of the brain during gestation, infancy, and childhood—and, we now know, adolescence—is almost as wondrous as the formation of a living, breathing, sensing, moving, behaving organism from a single fertilized ovum. The plasticity of the young brain is based on the overabundance of synapses, which allows only those that are used to become part of enduring circuits that underlie thinking, feeling, responding, and behaving. But does the dance of the neurons eventually end?

The great Spanish neuroanatomist Santiago Ramon y Cajal concluded his 1913 treatise "Degeneration and Regeneration of the Nervous System" with this declaration: "In adult centres the nerve paths are something fixed, ended, immutable. Everything may die, nothing may be regenerated." Cajal based his pessimistic conclusion on his meticulous studies of brain anatomy after injury, and his gloomy sentiment remained neuroscience dogma for almost a century. "We are still taught that the fully mature brain lacks the intrinsic mechanisms needed to replenish neurons and reestablish neuronal networks after acute injury or in response to the insidious loss of neurons seen in neurodegenerative diseases," noted the neurologists Daniel Lowenstein and Jack Parent in 1999.

This doctrine of the adult hard-wired brain, of the loss of neuroplasticity with the end of childhood, had profound ramifications. It implied that rehabilitation for adults who had suffered brain damage was useless. It suggested that cognitive rehab in psychiatry was a misbegotten dream. But the doctrine, as was becoming apparent even as Lowenstein and Parent wrote, was wrong. Years after birth, even well into adolescence, the human brain is still forming the circuits that will determine how we react to stress, how we think, even how we see and hear. The fact that (even) adults are able to learn and that learning reflects changes in synapses tells us that the brain retains some of its early dynamism and malleability throughout life. The adult has the ability not only to repair damaged regions but also to grow new neurons. Even the adult brain is surprisingly plastic. Thus the power of willful activity to shape the brain remains the working principle not only of early brain development, but also of brain function as an ongoing, living process.

Even as I recorded changes in the brains of my OCD patients after mindfulness-based cognitive-behavioral therapy, a decades-long dogma was beginning to fall. Contrary to Cajal and virtually every neuroscientist since, the adult brain can change. It can grow new cells. It can change the function of old ones. It can rezone an area that originally executed one function and assign it another. It can, in short, change the circuitry that weaves neurons into the networks that allow us to see and hear, into the networks that remember, feel, suffer, think, imagine, and dream.

{FOUR}

THE SILVER SPRING MONKEYS

We surgically abolish pain.
—*Edward Taub*

When night fell, Ingrid Newkirk would stand outside, walkie-talkie in hand, playing lookout as Alex Pacheco slipped through the darkened labs of the Institute for Behavioral Research (IBR) in Silver Spring, Maryland, taking photos and scribbling in his notebook. Over the course of several weeks in late August and early September 1981, Pacheco surreptitiously escorted a number of sympathetic veterinarians and primatologists who supported animal rights through the facility, showing them the rusty cages encrusted with monkey feces, cages whose bent and broken wires poked up from the floor like stakes that threatened to impale the monkeys. He showed them the signs of rodent infestation and of insects. But mostly he showed them the animals: sixteen crab-eating macaques, all adult males, and one adult female rhesus. Among the seventeen, thirty-nine fingers had been gnawed off, and arms were covered with oozing, unbandaged lesions. Was this standard for a primate lab, Pacheco asked each expert, or was something wrong—very wrong—here?

What Pacheco didn't know was that experiments conducted in this lab, grisly though they may have been, would overturn the dogma that the adult brain cannot be rewired.

In May 1981 Pacheco, then a twenty-two-year-old political science student at George Washington University, had applied for a job at the privately owned lab, four miles east of the nation's premier biomedical campus, the National Institutes of Health (NIH), in Bethesda. He was trying to decide whether he wanted to make a career of biomedical research, Pacheco had told Edward Taub, IBR's chief scientist. He was so fascinated by animal research, in fact, that he would gladly accept the unpaid, volunteer position that Taub offered him. As Taub told the *Washington Post* ten years later, he went home that night and raved to his wife, Mildred, an opera singer, about the "marvelous student" he had just met: "I told him there was no position, but he volunteered to work, out of pure interest," Taub marveled. But Pacheco's "pure interest" was not what Taub thought it was.

As a student at Ohio State University, Pacheco had burned with the passion of a true believer, and his true belief was that animals were needlessly subjected to cruelty, even torture. He had organized protests against the local farmers' practice of castrating their pigs and cattle without anesthetic; angry ag majors threatened to do the same to Pacheco one night. Pacheco had hoped to study for the priesthood, but instead he moved east and joined Newkirk, nine years his senior and already an experienced animal rights leader, to form People for the Ethical Treatment of Animals (PETA). Newkirk, who had worked undercover at a Maryland animal shelter and exposed the appalling conditions there, suggested that Pacheco do the same in a lab. He therefore obtained a Department of Agriculture list of federally funded facilities where scientists used animals for biomedical research; of them, IBR was closest to his apartment in Takoma Park, Maryland. Exactly what IBR did, Pacheco had no idea. As it turned out, Taub would be conducting experiments on the very species of animal that Pacheco, who had divided his childhood between Mexico and the United States, once had as a pet. Pacheco's Chi-Chi, like most of Taub's animals, was a crab-eating macaque (that's the species' common name, not its dietary preference) or cynomolgus monkey, *Macaca fascicularis*.

What we do here, Taub explained to Pacheco, is deafferent the monkeys' limbs. Afferent, or sensory, input from the body enters the spinal cord over the dorsal (back) routes to the spinal nerves. If nerves innervating some part of the body—an arm, for instance—are cut where they enter the cord, then that part of the body loses all sensation. The animal no longer feels its arm, or leg, or whatever limb has been deafferented. Taub was particularly interested in depriving single limbs, usually one arm, of afferent input and observing how that affected the animal's use of the arm. Because only the sensory nerve from the arm, and not the motor nerve, would be severed, it seemed logical that the adult animal would be able to continue using the arm. But a leading theory in behavioral psychology held that sensory input was crucial to motor function. This is what Taub planned to test. To do so, he performed the surgery on nine of his sixteen crab-eating macaques: Billy, one of the monkeys, had the sensory nerves to both arms severed, and the eight others had the nerve to a single arm severed. Seven other macaques, and Sarah, the lone rhesus and the only female, served as the controls. Taub had just received an $180,000 grant from NIH to continue this investigation.

Taub found that the deafferented monkeys, consistent with a long-standing theory in neurophysiology, did not naturally use the deafferented limb; without feeling, the arm seemed to hang like a dead weight, and the monkeys in many cases appeared to forget it was even there. So Taub tested whether the monkeys' reluctance, or refusal, to use the impaired limb could be overcome. Using a straitjacket to restrain the good arm, which the monkeys favored for tasks from ambulating to eating, he left them no option (if they wanted to get around and to eat) but to use the damaged one. He strapped the monkeys into chairs and, if they failed to flex the "affected arm," as Taub delicately called the seemingly useless one, administered a strong electric shock to them. Subjected to these massively stressful "negative reinforcements," the animals did indeed move the senseless limb. Eventually, Taub would examine the monkeys' nerves to see whether they had undergone any change—to see whether, for instance, any had grown back as a result of the animals' being forced to use the deafferented arm. That would mean killing the monkeys. But if the experiments showed what Taub expected and hoped, the results just might lead to new treatments for victims of stroke and brain trauma.

Pacheco wasn't interested in looking that far ahead. All he saw was what lay before him, and what he saw sickened him. The animals were housed in rusty old cages whose filth was surprising, for a lab; Pacheco reeked every night, recalls Newkirk, with whom he was living. Inspectors from the U.S. Department of Agriculture, which enforces lab animal laws, visited Taub's lab during Pacheco's time there; they reported no serious deficiencies. Given that, Pacheco wasn't sure what, exactly, would qualify as a deficiency (apparently not the monkey corpse he had found floating in a vat of formaldehyde). The living ones might well have envied the dead: they would spin around constantly, bouncing off the cage walls, masturbating compulsively—as macaques caged alone are wont to do. But these animals had an additional habit that horrified Pacheco. They chewed their deafferented limbs raw and gnawed off their fingers. From the shoulder down, of course, the affected limbs had no feeling.

Pacheco began documenting what he saw, shooting photographs and taking notes. When he told Taub he wanted to work nights and weekends, a grateful (and unsuspecting) Taub gave him the keys to the lab. That's how Pacheco came to be inside on those late summer nights in 1981, with Newkirk acting as lookout, each equipped with a walkie-talkie they had bought at a toy store. From August 24 to September 4, Pacheco sneaked five veterinarians and primatologists into the lab. Animal rights sympathizers, they provided the PETA pair with affidavits testifying to the conditions of the animals and of the lab. The next month, Newkirk and Pacheco took the affidavits and photos to the local police department.

The Montgomery County, Maryland, police raided the Institute for Behavioral Research on September 11, seizing all seventeen monkeys: Adidas, Allen, Augustus, Big Boy, Billy, Brooks, Charlie, Chester, Domitian, Hard Times, Hayden, Montaigne, Nero, Paul, Sarah, Sisyphus, and Titus. (Some of the monkeys had been named by Taub's students and assistants; the classical names came from Taub himself, who had long felt that some of the Roman emperors had not received their due from historians.) Taub had not been working that Friday, but when he rushed to the lab after an assistant phoned to tell him about the raid, he couldn't believe what was happening. He told a reporter, "I'm surprised, distressed and shocked by this. There is no pain in these experiments. We surgically abolish pain." Although neither Taub's experimental methods, nor the conditions in his lab, were grossly out of line with then-common practice, on September 28 the prosecutor charged Taub with seventeen counts of animal cruelty. The saga of the Silver Spring monkeys—one that would drag through the courts for ten years, embroil powerful congressmen in the fate of seventeen monkeys, and do more than any other single incident to launch the animal rights movement in the United States—had begun.

Edward Taub has mellowed in his later years. Yet you can still detect, in the self-assurance, shadows of the arrogance that once so infuriated other researchers that their faces would turn crimson and their lips would be reduced to sputtering as, in one case that he delights in telling, they lost their adopted English and fell back on their native Finnish to denounce him at a scientific meeting. And you can still detect the figurative thumbing-of-the-nose at scientific paradigms that provoked his bosses, even when they grudgingly granted him permission to carry out studies that threatened (or promised?) to topple those paradigms, to demand that he allow "observers" in, to keep an eye on him. And you can still detect the almost naïve view that scientific truth would vanquish ignorance and sentimentality. But although Taub had no trouble questioning the received wisdom in neuroscience and harbored no doubts that he, an outsider from the lowly field of behavioral psychology, had the right to question neuroscience "facts" dating back a century, it never dawned on him that using what were then (regrettably) not-uncommon laboratory procedures would earn him a singular distinction: the first scientist ever charged with animal cruelty.

Taub was born in New York City in 1931. After receiving his undergraduate degree in psychology from Brooklyn College in 1953, he began working toward his doctorate at Columbia University. As a grad student, he worked with monkeys at the Jewish Chronic Disease Hospital in Brooklyn. It was there that he was introduced to the experimental procedure that would first break him and then make him: limb deafferentation. "This was my advantage: that I was a psychologist who had never studied neuroscience except on my own," he says. He therefore had not been inculcated with the conventional wisdoms of the field, one of the most important of which dated back to classical experiments by the British neurophysiologist Sir Charles Sherrington.

In 1895 Sherrington, working with F. W. Mott, reported the results of a now-classic experiment in which he deafferented a single upper forelimb or lower limb of rhesus monkeys. Sherrington pioneered deafferentation experiments, painstakingly severing the sensory nerves but leaving motor nerves intact, much as Taub would some sixty years later. Sherrington and Mott sought to investigate whether the animals would continue to use the deafferented limb. They found, instead, that after they cut the sensory nerves, the monkeys stopped using the affected limb. The animals were completely unable voluntarily to grasp, to support their weight, or to ambulate with the deafferented limb. The "volitional power," as Sherrington and Mott called it, "for grasping with the hand etc. had been absolutely abolished." Even when Sherrington restrained a monkey's remaining good arm, deferred feeding time, and then put a morsel within reach, the monkey did not use its deafferented arm to reach for the food. The only movements it seemed willfully capable of were crude rapid jerks, which were induced in the monkeys by causing them to "struggle," as when they tried to free themselves while being held awkwardly. Sherrington attributed these motor actions to reflex effects triggered by movements in intact parts of the body. Since the motor nerves were still intact, why should somatosensory deafferentation abolish the ability of the monkey to move that arm? This was even more perplexing given that stimulating the motor area of the cerebral cortex elicited totally normal movement of the affected limb. Reflecting on the 1895 results in 1931, Sherrington said this served as "a caveat against accepting the movement excited electrically at the motor cortex as any close homologue of a willed one."

What the disconnection of sensory nerves did, Sherrington argued, was abolish a critical "influence of sensation upon voluntary movement," thus interfering with a basic mechanism necessary for the expression of "volitional power." As we will see, this observation somewhat overstated the case. In any event, working with Derek Denny-Brown, Sherrington replicated the deafferentation study, publishing virtually identical results in 1931, a year before he won the Nobel Prize.

Researchers as late as the mid-1950s continued to report that sensory deafferentation led to loss of motor ability. These results and others led Sherrington to conclude that the modulation of reflex pathways is the basis for purposive behavior: the brain's motor cortex taps into preexisting reflex circuitry to carry out its commands. In other words, all voluntary behavior is built on integrated, hierarchical reflexes. An animal moves and the movement produces sensory feedback; feedback plus learning guide the next movement; this produces its own feedback, which, again in conjunction with learning, eventually produces, after countless iterations, purposive, sequential movement. This theory came to be called "Sherringtonian reflexology."*

"Reflexology was the dominant view in neuroscience, even more dominant than the idea that there is no plasticity in the adult brain," recalls Taub.

> *The idea was that if you were interested in voluntary behavior, which was thought to be just an elaboration of a simpler phenomenon, it made more sense to study the simpler phenomenon. At this point it is hard to grasp how influential Sherrington's views were in psychology and certainly in neuroscience. Since we were psychologists, God help us, we decided [in the mid-1950s] that we would use the relatively new techniques of conditioned response to reevaluate the Sherringtonian canon, not because we had any reason to think it was wrong but because we could apply these new techniques to his ideas, which were so dominant.*

But something else was lurking in the back of Taub's mind. While doing the usual literature search for previous experiments in preparation for his own, Taub happened on a 1909 book by one H. Munk called *Ueber die Functionen von Hirn und Rückenmark* (On the functions of the brain and spinal cord). In it, the German scientist recounted how he, too, had performed unilateral deafferentation experiments on monkeys. Yet his results differed dramatically from Sherrington's of fourteen years earlier. Munk claimed that it was possible to induce a hungry monkey to lift food to its mouth with the unfeeling arm under two conditions: if the intact arm were

*By 1947, however, Sherrington was arguing that the power of the mind could markedly influence these reflexes. "The psychical may influence the physical act," he said. "Further, it is claimed that the psychical can increase the reactivity of the body's physical system. . . . [I]t is clear that the body/mind liaison provides in a largely physical world the means of giving expression to the psychical."

restrained, and if the initial halting attempts by the deafferented arm were immediately rewarded. Until Taub stumbled on them, Munk's observations had been essentially ignored and, in effect, lost to science.

In 1957 Taub and two colleagues at the Brooklyn hospital began a series of experiments designed to test Sherrington's theory that sensation is necessary for intentional, goal-directed movement. They confirmed that monkeys whose arm had been deafferented failed to use it if given any say in the matter. But Taub suspected that the deafferented animals retained what he called "a latent capacity for purposive movement." The experiments he designed were meant to "force the expression" of that capacity; in other words, the monkeys would be induced to use their "useless" limb. What was required, Taub suspected, were three things: motivating the monkeys, keeping the motor tasks simple, and repeating the motor trials.

Taub provided the motivation in the very first set of experiments, conducted two weeks after deafferentation surgery. The monkey sat, immobilized, in a "restraining chair." He heard a tone. If he flexed his deafferented right arm so that it broke a light beam five inches above a waist-high board within 3.5 seconds of hearing the tone, nothing else happened. But if he did not move his senseless arm, usually by flexing the elbow and shoulder, to interrupt the beam of light, he experienced an intense electric shock that lasted up to 3.5 seconds. This type of behavioral conditioning is called *avoidance conditioning;* it uses what B. F. Skinner (the psychologist who first systematically described it) termed "primary negative reinforcers," such as loud noise or electric shocks, to teach an organism new patterns of behavioral response. As Skinner put it, "A negative reinforcer strengthens any behavior that reduces or terminates it." Conditioning of this sort was widely used in animal (and even human) research in the 1960s and 1970s, the period of Skinner's greatest influence.

In the training paradigm Taub used, the conditioned stimulus was the buzzer. Because it was paired with a primary negative reinforcer (the shock), it became, in Skinnerian lingo, "a conditioned negative reinforcer." The conditioned response or "operant behavior" to be performed to avoid the shock was flexing the deafferented arm. Each training session lasted twenty trials. The entire training series, Taub said, "was to be of long duration, if necessary." In fact, it would typically require more than nine weeks of testing, five days a week, spent conditioning and conditioning the animal through electric shocks. Within a few weeks Taub had collected results that threatened a key finding of the Sherringtonian canon. "What do you know: the monkeys with the deafferented arms could learn new conditioned responses," he recalls more than thirty years later. To avoid the shock, the monkeys could move their deafferented arms.

Taub had another way to motivate the monkeys. In the second set of experiments, he and colleagues put six monkeys in straitjackets to restrain their good arm. This made the monkeys very upset—they either struggled obsessively to wriggle out of the constraint or refused to move at all. Nevertheless, five of the six animals got the point: they figured out that if they wanted to reach the food placed outside their cage, they would have to extend the deafferented arm. Two of the six also used their bad arm to support themselves, to move around "in the limited confines of their cages," as Taub put it, and even to feed themselves a peanut. "If you use simple restraint of the unaffected limb, within a couple of hours the monkeys will begin using the affected limb," Taub says, looking back on his results. "The use of that limb is clumsy, but it is permanent. This demonstrated unequivocally that the Sherringtonian reflexology position was incorrect." Clearly, volitional movement did not require sensory feedback.

"I couldn't imagine the incredible response when we began publishing in the early 1960s," Taub recalls. His discovery that sensory feedback is not necessary for movement contradicted the position of a professor of his, one of the Columbia psychology department's leading researchers. Taub was about to experience the first, but far from the last, consequence of bucking the conventional scientific wisdom. When it came time for him to defend his proposal for his doctoral thesis, the professor whose views his work undercut showed up at an event usually attended only by the student's thesis committee. "He was very angry," Taub recalls. After debating Sherringtonian reflexology with Taub, the professor stormed out. Soon after, Taub learned that he had failed a course (taught by the same professor) required for his Ph.D. because he had not taken the final. He had expected to receive an incomplete, which could usually be converted into a grade as soon as a student took the exam. Instead, the professor told Taub he had failed—because of his "insolence."

In 1962 Taub transferred to New York University. He continued his deafferentation work, under an NIH grant to the senior scientist in the group, A. J. Berman. Within a few years they found that monkeys could flex the fingers of a deafferented arm if properly motivated. To provide the motivation, Taub strapped a monkey into a restraining chair and taped a fluid-filled plastic cylinder into its bad hand. If the monkey squeezed the cylinder within a given amount of time, nothing happened to him; if he failed to do so, he received an electric shock. Just as the monkeys conditioned to break a light beam after hearing a tone did, these monkeys, too, learned (eventually) to grasp the container in order to avoid an electric shock, Taub reported in 1966.

By 1968 he had concluded that a monkey will not use a deafferented arm if it can get along reasonably well with the other three limbs, especially since trying to use an arm without feeling can lead to uncoordinated movement, falling, and dropping food. The monkey succumbs to what Taub dubbed *learned nonuse*. But conditioning (getting an electric shock if it doesn't use the bad arm) and restraint (leaving the monkey no choice if it wants to eat or walk but to use the bad arm) force it to use the arm or to be subjected to electric shock or go hungry. The motivation to use the limb is increased, so the monkey uses it. "With specific training," Taub concluded, "deafferented monkeys can learn to perform almost any sequence of movements of which a normal monkey is capable, except the most precise."

Something else was happening. Animals with two deafferented arms behaved quite differently from those deafferented of a single arm: the bilaterally deafferented monkeys were able, soon after surgery, to use both arms to grasp, walk, and climb, Taub and Berman reported in 1968. This counterintuitive result—the more crippling the surgery, the better the monkey did—was, Taub decided, "one of the central enigmas" in the field. Somehow, a smaller lesion (single limb deafferentation) was producing worse crippling than a lesion twice the size. When a single limb is deafferented, the monkeys don't use the arm; when both limbs are deafferented, the monkeys exhibit almost normal movement. This, Taub decided, was "a paradoxical inversion of results."

Twenty years later, Taub still professes shock at the reaction his results elicited.

I am a psychologist, and was just calling the data as I saw them. I had never gone to school in neuroscience so I had no idea what the violence of the reaction from the traditionalists would be. But the people in control of neuroscience then were all students of Sherrington [who had died in 1952]—people like Sir John Eccles and Ragnar Granit [who won Nobel Prizes in 1963 and 1967, respectively]. They were very upset. I was invited to give a colloquium at NIH and Granit was there. He spoke English [in addition to his native Finnish] reasonably well. But at the end of my talk he got up and began to question me; he got so angry that at first his face got incredibly red, and then he lost his English. He denied that Sherrington had ever said what I said he said, or believed it that strongly. I said okay, but really, it was so obvious, there was a reason the position was called Sherringtonian reflexology. . . . These people really despised me. I couldn't imagine where all this emotion was coming from. Not that I was an easy person. I grant that I was a very difficult young man. I was very insecure, and that translated into unconscious arrogance, and I was certainly convinced of my interpretation. People didn't like me; I was going up against all these distinguished people. I didn't have any mentors of my own, since I was in behavioral psychology rather than neuroscience. And that's probably why I was able to do the research that I did, because I had no preconceptions.

In 1970 Taub finished his Ph.D. thesis in experimental psychology, "Prism Adaptation and Intermanual Transfer: An Application of a Learning Theory of Compensation for Sensory Rearrangement." (The thesis described how monkeys whose view of the world was skewed as a result of looking through a prism could still direct a deafferented arm to a target, even though the prism made objects appear displaced.) In the fall of 1969 he had been offered a research job at the Institute for Behavioral Research in Silver Spring. There Taub decided to continue his deafferentation experiments, focusing on which movements would or would not be impaired after sensory deprivation, and under what conditions.

At IBR, Taub carried out numerous variations on the deafferentation experiments that would change neuroscience. From the first, he observed that immediately after surgery a deafferented monkey gets by with only its good arm. That made sense, for recovering function in a deafferented limb requires time. Unless the experimenter intervenes to induce use, "the monkeys never learn that, several months after the operation, the limb has become potentially useful," Taub explained. Putting the good arm of unilaterally deafferented monkeys in a straitjacket for nine full weeks after the surgery, he found, overcame this learned nonuse. The monkeys continue to use the deafferented limb after being freed from the restraint. All the animal needed was some motivation to prevent him from coddling his deafferented arm. Depriving monkeys of the use of their one good arm, or subjecting them to electric shocks until they learned to use the bad arm, did the trick. "Everything we did," Taub said more than twenty years later, "was to answer an important question." Eventually, monkeys could use their bad arm to climb to the top of an eight-foot bank of cages, as well as climb laterally and pick up raisins.

One problem was starting to crop up, however. "Deafferented monkeys have a tendency to sustain severe damage to their affected extremities, frequently as the result of self-mutilation," Taub reported in 1977. "This tendency toward injury, self-inflicted or otherwise, constitutes one of the major difficulties in carrying out deafferentation experiments with monkeys." "Difficulties," he would find out, was putting it mildly.

So far, all of Taub's experiments had been carried out with adolescent monkeys, who of course had been using their limbs for years before Taub severed their sensory nerves. That raised a key question: was somatic sensation early in life necessary for developing normal coordination? How early could an animal lose sensation to a limb and still learn, or relearn, to use it? Were certain movements hard-wired into the brain, or did they require the sensory feedback that Sherrington posited? If the latter, how much sensory feedback was necessary—days, or weeks, of movement in utero? To answer this, in the early 1970s Taub began to deafferent monkeys on the day of their birth. Amazingly, their ability to walk, climb, and reach was as good at three months as that of monkeys that had not been operated on.

The next logical step was to deafferent fetal monkeys. In 1975, Taub and colleagues at IBR performed the delicate surgery outside the womb, with the fetuses in a warm saline solution bath, and then returned the tiny things to the uterus. Unfortunately, the experiment had a high mortality rate: in one batch, six out of eleven fetuses died. Not even the survivors fared well. All were quadriparetic; the best-off could stand for a few moments, but do little else. Their impairment worsened; one died; and the rest were "sacrificed" at five to twelve months of age. But the reason for their immobility, Taub learned when he autopsied the animals, was that the surgery had made the fetuses' vertebrae flare in such a way as to damage the spinal cord. After using a different surgical procedure to deafferent fetal monkeys, Taub found that (in the two monkeys that survived) the deafferented arm was impaired but not useless: the monkeys supported themselves, walked, and reached with it. Volitional movement, Taub concluded, was not dependent on sensory feedback; rather, it was preloaded into an animal's brain like Windows XP on a laptop.

In the late 1970s, Taub began an experiment designed to test his hypothesis of learned nonuse directly: he restrained the deafferented arm of his monkeys for three months after surgery. This way, a monkey would never learn that the limb was useless during this period; it would simply assume that the restraint was holding it hostage. Taub also restrained the animal's good arm, so the monkey would not become adept at living one-handed, carrying out its daily activities with a single forelimb. So for three straight months, monkeys were strapped into straitjackets, with their arms crossed over their chest, pinned to the side of their body, or tied behind the back. The limb position was changed every other day.

Almost as soon as the restraints came off, the monkeys managed to use the deafferented limb.

Taub had demonstrated the phenomenon, and the power, of learned nonuse. In the immediate aftermath of an injury, the animal learns to avoid using the affected extremity, because doing so gives the monkey only "negative feedback" when it attempts to walk, climb, or grasp with the arm. That is, those attempted movements are clumsy or ineffectual. At the same time, the monkey learns compensating moves that are, in stark contrast, successful and rewarding: it works around the injury, much as stroke patients with a useless arm learn to do. The combination of negative feedback if he tries to use the affected extremity and reward if he develops compensatory moves suppresses use of the affected extremity. Although the condition is normally permanent, Taub had seen hints in the Silver Spring monkeys that learned nonuse of the affected limb could be reversed by restraining the intact limb so the monkey was compelled to use the deafferented arm. "The simplest method for evoking purposive use of a single deafferented limb is prolonged impersonal restraint of the intact limb," Taub concluded. Failure to use a deafferented limb reflected learned helplessness, not a motor incapacity. It was 1980.

In a chapter he wrote for a book that year, Taub argued forcefully that his work with deafferentation pointed the way toward testing whether learned nonuse accounted for a stroke patient's inability to use an arm, and he laid out a training procedure to overcome it. He emphasized that motivation was crucial but allowed that the electric shocks he used with the monkeys would probably not be necessary in stroke patients. (Praise, and perhaps tokens good for privileges or favorite foods, might be adequate, he suggested.) "I had had these data for 10 years, yet I never before thought of applying learned non-use to stroke rehabilitation," Taub recalls. "In neuroscience then, you just didn't think of applications to human beings. It didn't occur to anybody."

Actually, it did occur to one person. In 1967 Larry Anderson visited Taub's laboratory at the Institute for Behavioral Research. There, he observed some of the conditioned-response experiments with the monkeys and asked Taub whether he thought something similar might work in stroke patients. "I said sure, try it," Taub recalls.

Anderson did, with three stroke patients. He tied down the unaffected arm of each patient, leaving only the "paralyzed" arm free. Anderson then sounded a tone. Patients who failed to move the seemingly immobilized arm at this signal received a mild electric shock. Amazingly, the stroke patients learned to move an arm that they had thought they would never use again. For one of the patients, this demonstration led to a significant improvement in the life situation; for two of them, without follow-up therapy, the demonstration did not translate into real life.

Anderson's boss decided to scale up the experiment and recruited twenty-four stroke patients. Immobilizing the good arm, and motivating patients to use their "useless" arm, produced substantial improvement in all twenty-four. "There are those two papers in the rehab literature," Taub says,

> *but they were so far out in left field that I don't believe there has been a single reference to either except in my own articles. It was as if they had never been written. They were just too different from the traditional view of what is feasible and appropriate for treating stroke patients. For myself, I was fully engaged in working with monkeys, and my plate was full. I was a pure scientist, and you didn't do that then—rush to apply a finding in basic science to medicine. The message in the field was that if you were not doing pure research, you were tarnished. It took me so long to think about applying the monkey results to stroke patients because one just didn't think along those lines. It was only 10 years later that it occurred to me to try this in stroke patients.*

That was the radical proposal in his 1980 paper. He was a year into this work when Alex Pacheco asked whether Taub could use some help around the lab.

The month after the September raid, the National Institutes of Health suspended the rest of Taub's grant. The decision reflected a simple political calculus: although the agency knew that to withdraw funding from a researcher embroiled in a controversy—indeed, a court case—over his use of lab animals would ignite the wrath of many in the biomedical community, NIH also recognized that it would otherwise have been impossible to maintain its credibility with the public and Congress. Throughout the saga of the Silver Spring monkeys, NIH would be caught in a crossfire between its core constituency—biomedical researchers—and the public.

At Taub's November 1980 trial, Pacheco testified that cages were cleaned only infrequently, that cockroaches had the run of the lab, and that if caretakers failed to show up, the monkeys could go two or three days without food. He testified to the animals' self-mutilation, describing how Billy chewed off eight of his ten fingers and Paul tore off all five fingers of one hand. Courtroom exhibits featured gruesome photographs of the macaques with chewed-off fingers and bandaged arms. Five of the nine deafferented monkeys had mutilated themselves; open sores drained the length of their arms. Several had bone fractures; one suffered from osteomyelitis. Defense witnesses, and Taub himself, testified to the scientific merit and promise of his work and argued that deafferented monkeys are notoriously difficult to care for: with no feeling in the limb, they treat it as a foreign object, mutilating it and chewing off digits. The main point, and the only fair indicator of whether conditions in the lab were acceptable, argued the defense, was that the monkeys were healthy. As for the feces and other filth, they argued, monkeys are well known for fouling their cages.

"Nobody ever saw the conditions that Pacheco photographed," Taub maintains; he had long been convinced that Pacheco staged at least two of the photos introduced into evidence.

In the long history of researchers' administering electric shocks to animals, operating on them without anesthesia, and, of course, "sacrificing" them by the truckload, Taub had the distinction of being the only scientist ever hauled up on criminal charges for what he had done. People were aghast to learn that, as part of the experiment, day-old monkeys had their eyelids sewed shut. Yet when Harvard's David Hubel and Torsten Wiesel did the same to newborn kittens, as described in the previous chapter, their research won them a Nobel Prize. Many scientists therefore viewed Taub as a victim. They aided his defense in the belief that an unfavorable verdict would be the leading edge of a drive by antivivisectionists (the term most scientists preferred to "animal rights activists") to outlaw all animal experiments. Supporters such as Edward Coons, Jr., of NYU, Neal Miller of Rockefeller University, and Vernon Mountcastle of Johns Hopkins University raised more than $2,500 for Taub's defense; they believed that he had been set up and that NIH's suspension of his grant reflected nothing but a cold political calculus. Yet the biomedical community was clearly fractured by the case of the Silver Spring monkeys. NIH officials, writing in *Neuroscience Newsletter,* noted that deafferented monkeys kept at NIH "have not developed lesions comparable to those in five of the nine deafferented monkeys from IBR. . . . [F]ractures, dislocations, lacerations, punctures, contusions, and abrasions with accompanying infection, acute and chronic inflammation, and necrosis are not the inevitable consequences of deafferentation." Their animals, implied the NIH officials, had been given proper, humane care—unlike the Silver Spring monkeys.

In late November 1981, a district court judge found Taub guilty of six counts of failing to provide veterinary care for six monkeys (Paul, Billy, Domitian, Nero, Big Boy, and Titus) who had, among other injuries, massive scar tissue on their open wounds. The judge dismissed the other 113 counts. Taub was defiant throughout. He insisted that the animals had suffered no pain and after the verdict declared, "What has happened to my work harks back to the Middle Ages, and to the period of religious inquisition, when scientists were burned at the stake." The $3,000 fine the court imposed belies the true price that Taub paid. His NIH grant would never be reinstated; he lost his job at IBR; his research came to a standstill. He would spend several years trying to write up his work on deafferentation, under a $20,000 grant he received from the Guggenheim Foundation in 1985.

Taub appealed the verdict to the circuit court in Rockville,

Maryland. A jury there cleared him of all but one misdemeanor count of animal cruelty on July 2, 1982, sustaining the conviction only for Nero, whose arm had required amputation after the raid because it developed a massive infection. Two months later, Judge Calvin Sanders ordered Taub to pay the maximum fine, $500, but added, "I hope and trust this will not deter you from your efforts to assist mankind with your research." On August 10, 1983, however, the Maryland Court of Appeals unanimously overturned that one-count conviction, ruling that a federally funded researcher was not subject to state laws on animal cruelty. Taub described himself as "delighted to be exonerated . . . delighted on behalf of science." Two weeks later, as he addressed the American Psychological Association's annual meeting in Anaheim on "Tactics of Laboratory Attacks by Anti-Vivisectionists," nearly 200 animal rights demonstrators burned him in effigy.

As for the monkeys, their saga was far from over. Immediately after their seizure they were housed in the basement of the home of a PETA member in Rockville. There, vets cleaned and bandaged the animals' wounds, volunteers groomed them with toothbrushes, and a television was installed so they could watch soap operas, which they seemed to love. Just days after the seizure, however, a judge ordered the monkeys returned to Taub's lab, where a court-appointed vet would supervise their care. After mysteriously vanishing for several days (no one ever owned up to kidnapping them, but after prosecutors explained to Pacheco that they could not make a case against Taub without their star evidence, the monkeys reappeared), the seventeen monkeys were trucked back to IBR. Their stay was brief, however. Six days after, Charlie was found dead in his cage, apparently of cardiac arrest suffered after surgery to repair damage he had sustained in a fight with Nero. The judge was singularly unamused at this turn of events. That day he reversed his order and ordered the sixteen surviving monkeys sent to NIH's primate facility in nearby Poolesville. NIH took custody even though the animals remained the property of the Institute for Behavioral Research, an arrangement that would cause problems.

PETA sued in U.S. District Court to have the monkeys transferred from Poolesville to a primate sanctuary called Primarily Primates, in San Antonio, Texas, but the court ruled that PETA lacked legal standing. In 1986 PETA persuaded Representative Robert C. Smith of New Hampshire and Representative Charlie Rose of North Carolina to draft a petition calling on NIH to send the monkeys to the sanctuary; 252 members of the House signed it. (Smith even offered to buy the monkeys himself.) Pundits weighed in; James J. Kilpatrick wrote that the monkeys "deserve a break that the law won't give them. . . . Why can't a just and humane court . . . let the monkeys go?" In a letter, James Wyngaarden, the director of NIH, promised that he would indeed allow the monkeys to be moved from Poolesville; he also promised, "These animals will not undergo invasive procedures for research purposes." Any experiments, he continued, would occur only after their "natural death." On June 13 of that year Wyngaarden repeated his promise in testimony before Congress: the animals would never again undergo invasive procedures as part of research. By this time, investigating panels from the Society for Neuroscience, the American Psychological Association, and the American Physiology Society had all cleared Taub of animal cruelty charges. The Society for Neuroscience even contributed $5,000 toward Taub's legal bills.

Whatever promise animal rights activists thought they had wrested from Wyngaarden, the Silver Spring monkeys were not moved to a sanctuary. NIH had begun to feel the wrath of the biomedical community—its constituency—over the perception that it was caving in to "antivivisectionists." In 1984, researchers at the University of Pennsylvania had been caught, on videotapes that

PETA stole, dangling baboons by crippled hands, even propping up one trembling, brain-damaged baboon, turning the camera on him, and asking in voice-over, "Look, he wants to shake hands. Come on . . . he says, 'You're gonna rescue me from this, aren't you? Aren't you?' " That was too much even for NIH. The following summer, the secretary of health and human services (HHS) ordered the Penn lab shut down, an action that triggered a flood of furious calls and letters from scientists to NIH. In retrospect, it seems very likely that the firestorm over the Penn closing and other perceived cave-ins to animal rights activists "sealed the fate," as the *New Yorker* put it, of Billy, Sarah, and the other Silver Spring monkeys. NIH was accused of pandering to "animal lunatics" and was feeling the bitter backlash of a scientific community convinced the agency had sold out one of their own. NIH was going to take a stand in favor of using animals in biomedical research, and that stand would be on the Silver Spring monkeys.

So over a June weekend in 1986, the NIH assistant director, William Raub, contacted two of the country's leading primate facilities, Yerkes Regional Primate Center in Atlanta and the Delta Regional Primate Center, located across Lake Ponchartrain from Tulane University's main campus in New Orleans. Yerkes wanted no part of the symbol-laden animals. But Delta's director, Peter Gerone, was game. On June 23, the fifteen surviving monkeys (Hard Times had been euthanized at Poolesville in 1982) were moved to Delta, deep in the tranquil woods of Covington, Louisiana, surrounded by magnolias, sweet gums, and pine trees. Within a week of the animals' arrival, protesters were blocking the entrance road. Alex Pacheco felt betrayed; he had made arrangements for the monkeys' transfer to Primarily Primates, going so far as to outfit a mobile home as an animal clinic where the animals could be cared for. Instead, the monkeys were housed at Delta, in double-decker stainless steel cages that lined the walls of a nine- by twelve-foot concrete-block room. Brooks died a few months after he arrived; five of the control monkeys—Chester, Sisyphus, Adidas, Hayden, and Montaigne—were sent to the San Diego Zoo in the summer of 1987. That left Sarah, plus the eight male macaques that had undergone deafferentation—Augustus, Domitian, Billy, Big Boy, Titus, Nero, Allen, and Paul. Gerone refused to allow anyone from an animal rights group, or even newspaper reporters and photographers, to see the animals.

Although PETA continued pleading for the monkeys to be moved to a sanctuary, in April 1987 the Supreme Court upheld lower court rulings that PETA lacked legal standing to sue for custody. The following month, less than a year after taking custody of the animals, Gerone recommended that eight be put to death as "the humane thing to do." PETA and its allies were outraged. Newkirk charged that the monkeys "had been through hell and back" and deserved to be with "people who care about them." NIH rebuffed the requests throughout 1988—for it had been presented with an intriguing proposal.

In connection with a paper sent to the *Proceedings of the National Academy of Sciences* on February 22, 1988, the neuroscientists Mortimer Mishkin and Tim Pons of the National Institute of Mental Health suggested that the Silver Spring monkeys perform one last service for science. When humane considerations require that one of the animals be euthanized, they said, let scientists first examine its brain in search of evidence that the cortex had reorganized after twelve years of being deprived of sensory input from one limb or more. The Silver Spring monkeys, which had been deafferented when they were three or four years old, were a unique resource, the scientists argued. As NIH's William Raub told a reporter, "The Silver Spring monkeys were the first animals ever . . . in which so large an area of the brain—namely, the region corresponding to the map of an entire forelimb—had been devoid of its normal sensory input for as long as a decade." Throughout the 1980s and even earlier, as we'll see in the next chapter, scientists had been documenting cortical remapping in the brains of adult primates. *Cortical remapping* is what happens when an area of the brain that once processed sensation from, say, a thumb now processes input from a finger. In earlier studies, Pons and Mishkin found that the brains of seven macaques had been remapped in that the cortical representation of the hand had been taken over by the foot. But the remapping generally being reported by other studies was minuscule: the distance between the old representation and the new was usually on the order of only a couple of millimeters. By examining the brains of the Silver Spring monkeys, Pons and Mishkin hoped to determine whether cortical remapping occurred to a greater extent than anyone had previously reported.

NIH, of course, was not stupid enough to call for the monkeys' deaths in order to let scientists saw open their skulls and examine their brains. Instead, the institutes decided that when the animals became so ill that they had to be put down, then scientists could—with the animals under deep anesthesia—examine their brains, just before the animals were sacrificed. On July 1, 1988, William Raub wrote that the monkeys "are likely to require euthanasia eventually and that some almost surely would reach that stage this year." NIH had therefore prepared a plan, he wrote: the deafferented animals would undergo a procedure, while still alive but before being anesthetized, in which scientists would remove part of their skull and probe their brain for signs of cortical reorganization. Only after this would a monkey be euthanized.

Alex Pacheco and PETA were livid. Even Representative Rose wrote, in a scathing letter to NIH, that experimenting on the animals would be "a very serious violation of a commitment to me, to the Congress and to the public." The (first) Bush administration received thousands of letters protesting the decision; the first lady alone got 46,000, apparently from people who thought this kindly looking white-haired grandmother would intervene on behalf of the crippled monkeys. In 1988, animal rights groups successfully sought a restraining order prohibiting euthanasia if the brain surgery were to accompany it. By this point, Pacheco so mistrusted NIH that he suspected they would find any excuse to kill the animals in order to carry out the brain experiments.

Just when it seemed that matters could not get any worse, in the winter of 1989 Paul began to die. He started chewing apart the arm that had been deafferented; there was, of course, no feeling to signal him to stop. (Experience had shown that animals quickly maul any protective covering.) He actually cracked the bones in his hand. After the Tulane vets amputated half the arm and put him back in his cage, Paul stopped eating. Although the caretakers tried to soothe him, rubbing his back and offering him treats like peanut butter and sliced bananas, he refused all food. He began ripping apart his stump; gangrene streaked what remained of the limb. On the Fourth of July, vets amputated the rest of the arm at the shoulder. Even with force-feeding, Paul wasted away, finally dying on August 26, 1989, on the floor of his cage, with his head tucked beside the only arm that scientists had left him. He weighed seven pounds, compared to his original twenty. Throughout the ordeal, PETA had refused to acquiesce to euthanasia, convinced that Tulane's description of Paul's condition was an exaggeration, if not an outright lie.

Then it was Billy's turn. Although he had two deafferented arms, he managed to scoot around his cage with grim determination. But his odd locomotion caused pressure wounds on the backs of his hands and made his spine curl. After developing a bone infection that failed to respond to antibiotics, he was reduced to huddling in a corner. Tulane asked PETA for permission to put Billy down, to spare him the tortured and drawn-out death that Paul had suffered. Although PETA's own vet agreed that Billy should be euthanized, Pacheco rejected the advice. He didn't believe Billy was suffering, or about to die, especially since Tulane refused to allow him or anyone else from PETA to see for himself. By this time Billy's spine was fused in a curve and he was immobilized. "We had a crisis at Christmastime," said Peter Gerone, who directed the primate lab at the time. "He stopped eating." Although the International Primate Protection League, citing a state animal-protection statute, asked for and received a temporary restraining order from the U.S. District Court for the Eastern District of Louisiana that held up experiments on the seven surviving monkeys, on January 10, 1990, Tulane won an order from the U.S. District Court of Appeals allowing scientists to carry out the brain experiment before euthanizing Billy.

On January 14, 1990, Billy became the first of the Silver Spring monkeys to undergo neurosurgery before being put to death. After anesthetizing him with ketamine hydrochloride, neuroscientists led by Pons and Mishkin administered a mixture of isoflurane gas and oxygen, a deep anesthetic. Placing his head in a frame to hold it steady, the scientists drilled through the skull covering the cortex opposite the deafferented limb. Then, using tungsten microelectrodes, they recorded from brain areas approximately 0.75 millimeter apart across the region of the somatosensory cortex, to measure the activity that occurred in Billy's brain when they gently stroked different parts of his body with a camel's-hair brush or cotton swab. The goal was to determine where, in the somatosensory cortex, the brain processed each sensory input. In particular, the researchers hoped to determine whether the region of the somatosensory cortex that had originally received sensory input from Billy's arms, but that had been deprived of this normal input for more than twelve years as a result of the deafferentation, had changed. In macaques, earlier studies had established, the arm representation in the somatosensory cortex lies between the representation of the trunk and the representation of the face. The representation of the chin and lower jaw abuts the representation of the hand. In Billy, the zone representing the fingers, palm, lower and upper arm of the deafferented limb, remember, was not receiving any sensory input. It would not be far off to call this "deafferentation zone" the zone of silence: it was a radio dish tuned to a station that was no longer broadcasting.

Or so everyone thought. But when Pons took electrical recordings from the deafferentation zone, he found that the entire region "responded to stimulation of the face." Touching or brushing Billy's face, or even gently moving his facial hair, produced vigorous neuronal responses in the supposedly silent zone. Apparently, having waited so long for signals to arrive from the arm and hand, this region of cortex had done the neural equivalent of moving its antenna slightly to pick up signals from a different transmitter altogether. The cluster of neurons in the somatosensory cortex that responded to stimulation of the face had pushed so far into the once-silent zone—which originally received inputs from the deafferented arm—that it abutted the somatosensory representation of the monkey's trunk. Indeed, all 124 recording sites in the "silent zone" now responded to light stimulation of the face. After the experiment, Billy was given an overdose of pentobarbital and put to sleep.

That month, an editorial in the journal *Stroke* argued against the relevance of such animal work to humans: "Each time one of these potential treatments is observed to be effective based upon animal research," it said, "it propagates numerous further animal and human studies consuming enormous amounts of time and effort to prove that the observation has little or no relevance to human disease." But Louis Sullivan, then secretary of HHS, glimpsed in the experiment a ray of hope for brain-injured people: "The investigators entered uncharted territory when they studied the brain of the first of the [Silver Spring] primates to be euthanized for humane reasons," he declared. On July 6, 1990, Augustus, Domitian, and Big Boy were also experimented on and then euthanized. An appeal by PETA to the Supreme Court, asking the justices to block the euthanasia of Titus and Allen, was denied on April 12, 1991. Titus was put to sleep at 2:00 P.M. that day. Allen was put under deep surgical anesthesia as part of a four-hour experiment; he never awoke.

The researchers reported their findings from four monkeys in the journal *Science* in June 1991. (Taub's name was also on the paper, but only because he had overseen the original deafferentation experiments more than twelve years before.) They found, they said, that the deafferented region, which included primary somatosensory maps of the fingers, palm, arm, and neck, was not the nonfunctional desert they expected; rather, the entire zone had responded when the researchers brushed the animal's face. "Deafferentation zone," then, was a misnomer: although part of the monkeys' somatosensory cortex had been deprived of its original afferent input, from the arm, over the course of the previous dozen years it had been innervated by neurons from the face—specifically, the part of the face from the chin to the lower jaw. The part of the cortex that usually received sensory input from the monkey's arm did not simply go out of business. Instead, neuronal axons from adjoining cortical regions had grown into it. The result was a rezoning of the monkey's somatosensory cortex. Virtually the entire hand region, measuring 10 to 14 millimeters across, had been invaded by neurons of the face area. Like an abandoned industrial neighborhood that has been rezoned for residential use, the monkeys' somatosensory cortex had been rezoned so that the arm region now received input from the face. The scientists had discovered, they wrote, "massive cortical reorganization" that was "an order of magnitude greater than those previously described." Pons made it clear why they were able to discover what they did. "It was, in part, because of the long litigation brought about by animal-rights activists that [made] the circumstances extremely advantageous to study the Silver Spring monkeys," he told the *Washington Post*.

All these years later Taub, who had been hired by the psychology department of the University of Alabama at Birmingham (UAB) in 1986, makes an admission. "Nothing was lost to science" as a result of the raid on his lab, he concedes. "Just a few years later Mike Merzenich [at the University of California, San Francisco] made the discoveries that we were headed toward." He pauses. "Though I must say, I wouldn't have minded making those discoveries myself." Instead, Taub was unable to conduct research for six years. Journals that once published his work wanted no part of him; agencies that once funded him turned down his grant proposals.

Taub's experiments on deafferentation in monkeys generated two complementary lines of research. One was called *constraint-induced movement therapy*. It grew out of Taub's discovery that animals with bilateral forelimb deafferentation eventually use their limbs extensively, whereas those with unilateral deafferentation—a lesion only half as extensive— have a virtually useless arm. The lack of purposeful movement, Taub concluded, reflected learned nonuse. For more than twenty years, until the 1981 raid on his lab, Taub had sought ways to overcome learned nonuse, motivating his monkeys through hunger or the desperate desire to avoid electric shock to use an arm they were otherwise content to leave hanging uselessly at their side. At UAB he would finally take up the idea he had broached so long ago, in that chapter he had written in 1980, on whether learned nonuse might explain a stroke patient's inability to use a limb, and whether behavioral therapy might overcome it. Taub would not starve, let alone shock, his patients at UAB. He would simply put their good arm in a sling, and their good hand in an oven mitt, so that if they wanted to hold something, or feed themselves, or get dressed, or do the laborious rehabilitation exercises he put them through, they would have to use their "useless" arm. He called it constraint-induced movement therapy, or CI therapy for short. It was the work with the deafferented monkeys that had demonstrated to him that behavioral intervention might help patients overcome the learned nonuse of a limb affected, for instance, by stroke. In November 1992, a year after the experiments on the Silver Spring monkeys demonstrated massive cortical remapping, UAB granted Taub $25,000 to study whether stroke patients could be taught to overcome learned nonuse of a limb. This is the subject of Chapter 5.

The other avenue of research was more purely scientific. Pons and Mishkin had shown that deafferentation results in cortical reorganization or remapping. The plasticity of the adult brain overturned an entrenched paradigm, opening the door to a greater understanding of the brain's capacities. Cortical remapping became the first example of neural plasticity in the adult brain. As these things tend to do, the two lines of research would meet up again, eventually, when Taub discovered that the brains of rehab patients had changed as a result of constraint-induced movement therapy.

For almost twenty years the Silver Spring monkeys were famous not for what they had done, but for what had been done to them. But, looking back, it is clear that they left a double legacy. Their case prompted revision of the Animal Welfare Act in 1985, requiring that researchers reduce unnecessary suffering among lab animals. It made PETA a force in animal rights: the group went from "five people in a basement," as Ingrid Newkirk puts it, to a national movement. It put biomedical researchers on notice that the rules had changed, that complying with the lenient animal use standards would not be enough to insulate them from the fury of animal rights advocates. "Until the Silver Spring monkeys," says Newkirk, "people thought, yes, animals are used in labs, but there is nothing I can do about that. But then they saw the animals' faces, and their suffering, and realized that there are things ordinary people can do. The animals came out of the lab for the first time, and people saw their suffering. After the Silver Spring monkeys, nothing was ever the same." At Poolesville, which houses more than 1,000 monkeys, animals are now kept in large social groups rather than solitary cages, "since they are social creatures by nature," says J. Dee Higley, who joined the facility in 1989 and studies violence associated with alcoholism. "When I first got here, hundreds of animals were kept in single cages. Now we know that if you keep a primate in a single cage, you are very likely to have an abnormal animal."

But the Silver Spring monkeys also changed forever the dogma that the adult primate brain has lost the plasticity of childhood. Instead, a new paradigm was beginning to emerge.

{ FIVE }

THE MAPMAKERS

Although the content of consciousness depends in large measure on neuronal activity, awareness itself does not. . . .
To me, it seems more and more reasonable to suggest that the mind may be a distinct and different essence.

—*Wilder Penfield, 1975*

Looking back on it, there had been hints for decades. At the end of the nineteenth century, long before Allen and Domitian and Big Boy had their cortices mapped, long before the brains of OCD patients changed in response to therapy, scholars generally agreed that the adult brain is not immutable. To the contrary: most believed that learning physically alters the brain. As neuronal pathways are repeatedly engaged, the psychologist William James argued in the nineteenth century, those pathways become deeper, wider, stronger, like ruts in a well-traveled country road. In the chapter on habit in his magisterial 1890 work *Principles of Psychology*, James had this to say:

> *Plasticity, then, in the wide sense of the word, means the possession of a structure weak enough to yield to an influence, but strong enough not to yield all at once. Each relatively stable phase of equilibrium in such a structure is marked by what we may call a new set of habits. Organic matter, espe-*

> *cially nervous tissue, seems endowed with a very extraordinary degree of plasticity of this sort; so that we may without hesitation lay down as our first proposition the following, that the phenomena of habit in living beings are due to the plasticity of the organic materials of which their bodies are composed.*

It was an idea that reflected the spirit of its age. With the scientific revolution of the eighteenth and nineteenth centuries, notions that had once existed solely as abstract hypotheses—electrons, atoms, species—were being shown to have a physical reality, a reality that could be quantified, measured, and probed. Now it was the mind's turn. Farewell to the airy notion that our habits, to take James's example, were patterns whose basis floated above the physical realm. Now theorists proposed that the experiences of our lives leave footprints in the sands of our brain like Friday's on Robinson Crusoe's island: physically real but impermanent, subject to vanishing with the next tide or to being overwritten by the next walk along the shore. Our habits, skills, and knowledge are expressions of something physical, James and others argued. And because that physical foundation can change, so, too, we can acquire new habits, new skills, new knowledge.

Experimentalists soon vindicated that theory. In the early twentieth century neuroanatomists began discovering something odd. They were investigating so-called movement maps of the brain, which show which spot in the motor cortex corresponds to moving which part of the body. The maps, more often than not, turned out to vary among individual animals: electrical stimulation of a particular spot in the motor cortex of one monkey moved the creature's index finger, but stimulation of the same spot in another monkey moved the hand. You couldn't even think of drawing a tidy movement map for, say, the "typical" squirrel monkey. Sure, you could draw a map for this monkey. But it would be different from the map for that monkey.

In 1912 T. Graham Brown and Charles Sherrington, the British neurophysiologist we met in the last chapter, decided to see whether this variability in movement maps reflected mere experimental sloppiness or something real. In landmark but long-forgotten experiments, the duo methodically applied surface electrical stimulation to lab animals' motor cortices and observed which muscles responded. It was true: movement maps were as individual as fingerprints. Stimulating one animal's motor cortex here produced a twitch of a cheek muscle; stimulating another animal in the exact same spot twitched a different muscle. What was the basis for this variability? Unlike fingerprints, the scientists concluded, the cortical representations of movements are not inborn. Instead, they reflect the history of use of the motor system—the footprints in the sand. Enduring changes in the complex neural circuits of our cerebral cortex, they proposed, must be induced by our behaviors. To take a fictitious example, a monkey in the habit of holding its fruit with its thumb and pinky would have a movement map in which the spots of the cortex moving those two fingers lie close together. If the monkey switched to habitually using its thumb and forefinger, then the brain would eventually shift too, rezoning the motor cortex so that neurons moving the thumb lay beside those moving the forefinger, with the pinky representation shunted aside. Sherrington's and Brown's work provided the earliest empirical evidence that, as James had guessed, habits are behavioral expressions of plastic changes in the physical substrate of our minds.

And it launched what would be a blossoming of research into neuroplasticity. Three years after the work on monkeys' movement maps, a neurologist named S. Ivory Franz compared movement maps in the primary motor cortices of macaques. He, too, found high variability and concluded that the differences probably reflect the motor experiences and skills of the different monkeys. In 1917, Sherrington himself described "the excitable cortex of the chimpanzee, orang-utan and gorilla," documenting great variation in the movement areas of the cortex. The brain, he concluded, is "an enchanted loom, where millions of flashing shuttles weave a dissolving pattern, always a meaningful pattern, though never an abiding one."

In 1923 Karl Lashley, a former colleague of Franz, added his voice. His work was a departure from that of his predecessors, who compared one animal to another. Logically, the differences they discovered between movement maps need not have been the result of the animals' different life experiences; the idiosyncrasies might have been inborn. To rule out that explanation, Lashley derived four movement maps over the course of a month from the same adult rhesus monkey. If differences in the maps reflect only inborn differences, then the map of that monkey's cortex today should be the same as its map last week. But it was not. Each time Lashley worked out the monkey's movement map, he found that it differed in detail from the previous one, and even more from maps derived earlier. There must be, he surmised, a general "plasticity of neural function" that allows the movement map in the motor cortex to change throughout life, remodeling itself continually to reflect its owner's motor experiences. Crucially, Lashley concluded that muscles that move more receive a greater cortical representation than muscles that move less. That bears repeating: the more a creature makes a movement, the larger the cortical area given over to that movement. Each time Friday walks his favorite route in the wet sands at the water's edge, he leaves new imprints, fresh and sharp. If he walks the same route, his footprints become ever deeper, while those on the route less traveled fade away, until they barely dimple the sands.

By the middle of the twentieth century, there was a compelling body of evidence that the cerebral cortex is dynamic, remodeled continually by experience. Thus when Donald Hebb postulated coincident-based synaptic plasticity in 1949 ("Neurons that fire together, wire together," as discussed in Chapter 3), he didn't regard his proposal as particularly revolutionary: the notion that coincident inputs strengthen synapses was, he thought, generally acknowledged. But there had always been voices of dissent over the notion of a plastic brain. In 1913 the great Spanish neuroanatomist Ramón y Cajal had argued that the pathways of the adult brain are "fixed, ended, immutable." Although he also posited that "absolutely new relations between previously nonconnected neurons are elicited by learning," by the 1950s the "immutable" paradigm had become the conventional wisdom in neuroscience. The theories and experimental findings of Sherrington, Franz, and Lashley were swept aside and largely forgotten. According to the prevailing camp at midcentury, the brain establishes virtually all of its connections in such primary systems as the visual cortex, auditory cortex, and somatosensory cortex in the first weeks of life. The groundbreaking work on the visual system by Hubel and Wiesel in the 1960s, as discussed in Chapter 3, seemed to establish once and for all the principle that, after a critical period early in life, experience can no longer change the brain much. The mature cortex is fixed and immutable. This became a tenet of neuroscience.

The few experiments that continued to mine the vein that Sherrington and his successors had opened therefore made all the impact of a whisper at a rock concert. Take the rats, for instance. Researchers reported in 1976 that the amount of auditory cortex given over to neurons that process a tone used in Pavlovian conditioning increases: the more the rat uses those neurons, the more space they occupy in the auditory cortex. Lashley would have been pleased. Or take the cats. In 1979, the neuroscientists John Kalaska and Bruce Pomeranz reported that denervation of the paws of kittens and adult cats causes the "paw cortex" in the brain to respond to stimulation of the felines' forearm instead, suggesting that the forearm representation creeps into the paw representation once paw neurons no longer send signals to the cortex. (As you'll recall from Chapter 4, *representation* is the space in the cortex devoted to processing particular sensory inputs or movement outputs.) This was precisely what Tim Pons and his team had found in the Silver Spring monkeys: if an animal stops receiving sensory input from one part of its body, the area of somatosensory cortex that used to process that input remaps itself. Instead of wasting valuable processing space on the sounds of silence, the area starts listening to a part of the body that is still transmitting signals to headquarters. And don't forget the raccoons (though neuroscientists did). In 1982, after amputating a raccoon's fifth digit (pinky), Douglas Rasmusson found that its somatosensory cortex reorganized, reassigning the cortical region that used to handle incoming signals from the pinky to a part of the body (the fourth digit) that was still transmitting. Andrew Kelahan and Gernot Doetsch also found somatosensory reorganization in the cortices of raccoons after amputation of a digit.

But it is a rare neuroscientist who pays much attention to raccoon experiments. No one exactly rewrote the textbooks on the basis of these rats, cats, or raccoons. Their brains were assumed to be too simple to serve as models for the human brain. As a result, neuroscientists largely ignored experiments that, in the late 1970s and early 1980s, began raising questions about the permanence of the brain's zoning maps, suggesting instead that the cortex is highly plastic and driven by experience. A loud silence greeted Patrick Wall's prescient suggestion of the physical basis for such rearrangements and expansions. In a 1977 paper in *Philosophical Transactions of the Royal Society of London (Biological Sciences)*, Wall wrote, "There are substantial numbers of nerve terminals which are normally ineffective. . . . If the normally functioning afferent nerve fibres are blocked or cut . . . large numbers of cells begin to respond to new inputs. The presence of ineffective synapses in the adult offers . . . a possible mechanism to explain plasticity of connections in adult brains." Little wonder scientists failed to pick up on Wall's suggestion of a mechanism for neural plasticity. After all, the phenomenon wasn't even supposed to exist.

What everyone "knew" to be true can still be seen in any lavishly illustrated brain book. There, in full-color diagrams, the structures of the brain are clearly mapped and labeled: areas that control language and areas that receive visual input, areas that process auditory input and areas that sense tactile stimulation of the left big toe or the right elbow. The thing resembles nothing so much as a zoning map produced by the most rigid of land-use boards. Every bit of real estate is assigned a function; and territory given the job of, say, processing sensations from the lower leg seem no more able to start recording feelings from the cheek than a plot of land zoned residential could suddenly become the site of a tractor factory. This view of the brain dates back to 1857, when the French neurosurgeon Paul Broca discovered that particular regions are specialized for particular functions. Throughout the nineteenth century neuroscientists had a field day demonstrating that different clusters of neurons located in well-defined places assumed specific functions. The neuroanatomist who determined the function of a region first was often awarded (or claimed) pride of nomenclature: thus we now have Broca's region (speech), for instance, and Wernicke's region (language comprehension).

The discovery of links between structure and function gave rise to a view that became axiomatic: namely, that different parts of the brain are hard-wired for certain functions. Nowhere was this clearer than in every medical illustrator's favorite brain structure, the somatosensory cortex. A band that runs from about halfway along the top of the brain to just above each ear, the somatosensory cortex processes feelings picked up by peripheral nerves. Every surface of the body has a corresponding spot on this strip of cortical tissue, called a representation zone, as the Canadian neurosurgeon Wilder Penfield found in his experiments in the 1940s and 1950s, reviewed in Chapter 1. While patients were under local anesthesia for brain surgery, Penfield, who studied under Sherrington, stimulated spots on the surface of the exposed brain with a tiny electrode. Then he asked his conscious subjects what they felt. They didn't hesitate: depending on which spot Penfield's electrode tickled on the somatosensory strip, the patient would report feeling a sensation in the fingers, lips, feet, or other part of the body.

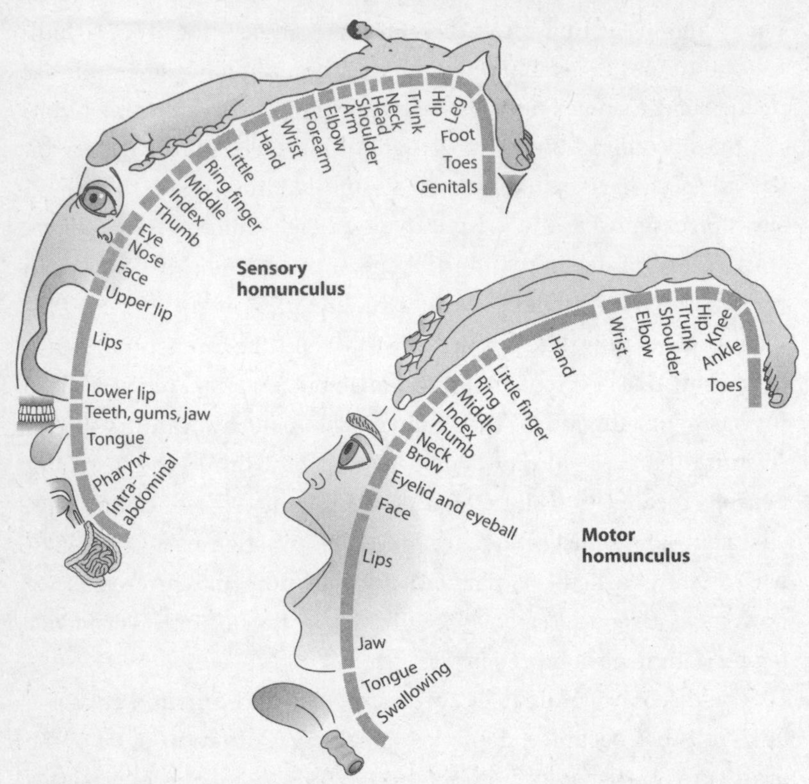

Figure 6: A. The sensory homunculus depicts the location and amount of cortical space devoted to processing tactile signals from different places on the body. Sensitive regions such as the lips and genitals command a great deal of cortical space. **B.** The motor homunculus shows the amount of cortical space devoted to controlling the movement of different regions of the body. Muscles involved in speech and hand movements receive a great deal of cortex, while less dextrous regions such as the shoulder receive very little.

But it was an odd map. True, the part of the somatosensory cortex that registers sensation from the lips lies between the regions that register sensation from the forehead and the chin. So far, so good. The cortical representation of one finger is positioned relative to those of the other fingers, reflecting the arrangement of the fingers on the hand. Also good. But beyond these basics, the cortical representations of different regions of the body are arranged in a way that makes you suspect nature has a bizarre sense of humor. The somatosensory representation of the fingers, for instance, sits beside the face. The representation of the genitals lies below the feet. The reason for this arrangement remains lost in the mists of evolution. One intriguing hypothesis, however, is that it reflects the experience of the curled-up fetus: in utero, our arms are often bent so that our hands touch our cheeks, our legs curled up so that our feet touch our genitals. Perhaps months of simultaneous activation of these body parts, with the corresponding synchronous firing of cortical neurons, results in those cortical neurons' "being fooled" into thinking that these body parts are contiguous. It would be another example of coincident input's producing coherent structures during prenatal development, as discussed in Chapter 3.

The other oddity of the somatosensory cortex is easier to explain. The amount of cortical territory assigned to a given part of the body reflects not the size of that body part but its sensitivity. As a consequence, the somatosensory representation of the lips dwarfs the representation of the trunk or calves. The result is a homunculus with dinner-plate lips. Our little man also has monstrous hands and fingers: the touch-sensitive neurons on the tip of your index finger are fifteen times as dense as those on, for instance, your shin, so the homunculus's index finger receives more cortical real estate than a whole leg. The density of touch receptors on the tongue is also more than fifteen times as great as that of those on the back of your hand. Place the tip of your tongue under your front teeth and you'll feel the little ridges; but place the back of your hand against the teeth and all you're likely to feel is a dull edge.

The motor cortex, which controls the voluntary actions of muscles moving every part of the body, is also laid out like a homunculus. Here, the amount of neural territory assigned to moving such dexterous parts as the hands dwarfs the amount given to moving, say, the ears. The lips get more motor cortex than the leg; we are, after all, the ape that speaks. The torso is dwarfed by the representations of the face, tongue, and hands. The amount of motor cortex devoted to moving the thumb is as great as the amount zoned for moving the entire forearm: the former is capable of much finer movements than the latter. But the motor homunculus is as jumbled as his somatosensory brother. Penfield, again using mild electrical stimulation of the exposed brains of surgical patients, discovered that the motor cortex maps out a body plan as cartoonish as the somatosensory cortex does. The representation of the leg sits near the center of the motor cortex, at the crown of the head; working outward, the arm (including hand and fingers), head, and face follow.

Despite a contradictory experiment here and an iconoclast there, for decades it had been axiomatic that there was no plasticity in the somatosensory or motor cortex of the adult brain. The only form of plasticity allowed into the textbooks was that based on Hebbian remodeling, in which neurons that fire together wire together. Since Hebb's 1949 paper, many studies had demonstrated this limited kind of cortical plasticity, but plasticity in the sense of extensively rezoning the cortex, so that a region that originally performed one function switches to another, was unheard of.

This dogma had profound real-world consequences. It held that if the brain sustained injury through stroke or trauma to, say, a region responsible for moving the left arm, then other regions could not step up to the plate and pinch-hit. The function of the injured region would be lost forever. Observations that challenged this paradigm were conveniently explained away. Faced with the fact that stroke-related brain injury, for instance, is not always permanent—someone who suffers an infarct in the region of the right motor cortex responsible for moving the left leg might nevertheless regain some control of the left leg—the antiplasticity camp didn't budge. No, it isn't possible that another region of the motor cortex assumes control of the left leg in such cases, they argued. At best, lower and more primitive regions such as the basal ganglia, which encode grosser patterns of movement, might take over some of the functions of the injured region. But recovery from brain injury, held this camp, in no way undermined the paradigm that neural circuitry in the adult is fixed (except for memory and learning through Hebbian processes). The possibility that the adult brain might have the power to adapt or change as the result of experiences was dismissed. Sherrington's "enchanted loom" weaving a "dissolving pattern" seemed to be a whimsical illusion of a more naïve age.

As an undergraduate at Oregon's University of Portland in the early 1960s, Michael Merzenich was pretty sure he wanted to become a physician. But he stumbled onto a different vocation. A Portland alumnus had founded a scientific equipment company called Tektronix; over the years, the alum had contributed entire rooms full of gadgets and gizmos to support his alma mater. Because almost no one knew how to use it all, though, the stuff sat largely untouched. Almost on a lark, Merzenich and a friend decided to see what they could make of it. Even though they were "almost entirely ignorant about what we were doing," as Merzenich recalls, after a lot of fiddling around they actually managed to accomplish something: recording the electrical activity in the neurons of insects. A professor suggested that Mike call the med school; with luck, he might find someone who would take pity on him and his coconspirator and supervise their Tektronix exploits. Making a cold call, Merzenich suddenly had John Burkhardt on the line. President of the Physiological Society, Burkhardt was a lion of neuroscience. Surprised and impressed at what Merzenich had been able to accomplish, he decided to take the young man under his wing. Eventually, Burkhardt made a few calls for Merzenich; without even applying, Merzenich found that both Harvard and Johns Hopkins University would be delighted to have him enroll in their graduate school. Merzenich headed for Hopkins, whose department had a strong reputation for research into awareness and perception. Although barely into his twenties, Merzenich already knew that his interest in neuroscience stemmed from more than a passionate desire to work out, say, the neuron-by-neuron circuit that enables a fly to move its right front leg. "I had been interested in philosophy," Merzenich says, "and I looked at neuroscience as a way to address questions of philosophy from a scientific perspective."

After finishing graduate school in 1968, Merzenich began a postdoctoral fellowship at the University of Wisconsin. There, he focused on how information from peripheral nerves is represented in the brain, and how that representation might change. In his experiment, he cut ("transected") the sensory nerves in one hand of each of six macaque monkeys. Later, after the tiny, peripheral nerve branches had atrophied, he surgically reconnected each severed nerve where it had been cut. The peripheral branches were left to grow back on their own. The result: skin "addresses" in the brain were shuffled like a deck in Vegas. What happened was that the branches of the sensory nerves grew back into the skin almost randomly, and not necessarily to their original sites, Merzenich reported in 1972. "They sort of meandered," he explains. The poor brain was hoodwinked. A nerve that used to carry information from, say, the tip of the forefinger had instead grown back to the middle segment of that finger. When a signal arrived in the brain via that nerve, the brain naturally figured it was hearing from the fingertip, when it fact the transmission came from a few millimeters away. Something similar happened at the other end, too: nerves from some skin surfaces took over the cortical representation zones originally occupied by others. As a result, a single skin surface (such as the tip of a finger) came to be represented across several small, separate patches in the cortex, rather than the usual continuous swatch, as its nerves grew back to different regions of the cortex. Normally, adjacent regions within a parcel of somatosensory cortex represent adjacent skin surfaces. But now the skin inputs to these adjacent cortical regions were all messed up.

But not necessarily forever. With enough use of their rewired hands, Merzenich's monkeys could achieve near-total correction of the scrambled brain addresses. The brain sorted out the new pattern of connections—okay, I keep receiving input from these two nerves at the same time; I'm going to guess that they come from adjacent areas of skin—and remade the somatosensory cortex accordingly. In other words, the brain registers which skin sites fire simultaneously. Through such coincident sensory ("afferent") input, the cortex creates functionally coherent receptive fields, a dramatic example of what has come to be called activity-dependent cortical reorganization.

"I knew it was astounding reorganization, but [back in the 1970s] I couldn't explain it," says Merzenich. "It was difficult to account for the emergence of such orderly receptive fields when we shuffled the sensory input so drastically. Looking back on it, I realized that I had seen evidence of neuroplasticity. But I didn't know it at the time. I simply didn't know what I was seeing." Merzenich pauses. "And besides, in mainstream neuroscience, nobody would believe that plasticity was occurring on this scale." Although scientists in James's and Sherrington's day had debated and speculated about brain remodeling, by the time Merzenich got interested, the idea had pretty much been run out of town on a rail. Those tidy diagrams assigning one function to this patch of brain and another to that one—here some language comprehension, there some lip sensation—proved too compelling: neurons of the brain, held the dogma, figure out early what they're going to be and stick to it for life.

Merzenich wasn't persuaded. He determined to see just how extensively the cortex could reorganize after new patterns of input. What he needed were brains in which the somatosensory cortex is spread over a flat surface, rather than being plagued by fissures and sulci, simply so he could see the thing better. While at Wisconsin, he had struck up a friendship with Jon Kaas, also a postdoctoral fellow there. When Merzenich went off to the University of California, San Francisco (UCSF), in 1971, Kaas joined Vanderbilt University, where he was doing experiments with little New World primates called owl monkeys (*Aotus trivirgatus*); their somatosensory cortex was perfect for what Merzenich had in mind. The squirrel monkey (*Saimiri sciureus*), too, had an easy-to-map somatosensory cortex and would also prove popular in neuroplasticity investigations. In both species, the map of the hand takes up roughly eight to fourteen square millimeters of cortical real estate. Merzenich and Kaas began to investigate how surgically disconnecting sensory input alters animals' brains.

"We decided to re-do the experiment we had started together at Wisconsin," recalls Kaas, "in which we cut one of the monkey's peripheral nerves, let it grow back, and then examined the somatosensory cortex to see if there had been any changes." They started with what they figured would be the control experiment: severing the median nerve of an adult monkey's hand and not reconnecting it (left alone, severed nerves do not mend themselves). Once the monkeys had lived with their severed nerve for several months, Merzenich took a sabbatical from UCSF and joined Kaas at Vanderbilt. The next step was to probe how the surgery altered the animals' brains. To do that, they recorded the activity in hundreds of locations in the monkeys' somatosensory cortices. Thanks to new anesthetics, which did not render the cortex unresponsive as old barbiturates did, the team was able to put the animals under anesthesia but still get readings. "We were in the lab all the time," recalls Kaas.

> *Mapping takes hours and hours, so we would start in the morning and not leave until two days later, once we had finished all the recordings—or else got too goofy to work. No one wanted to miss out on what we were finding. There was a feeling that you didn't know what would be seen next, and if you weren't right there you wouldn't believe it. I remember taking a candy break one midnight after a pretty successful run, and eating a Payday. When I finally finished at 6 A.M. one day, I broke out a beer. It was called "Quitting Time."*

Their findings were worth more than a cheap beer, for what the researchers assumed would be the control experiment—preventing the cut nerve from reconnecting—turned out to be a neuroscience landmark. "Quite unexpectedly, the cortex that had received input from the severed nerve, and which should now have been silent, responded to stimulation of other parts of the hand," Kaas recalls. Within three weeks after they had severed the median nerve, which carries signals from the thumbward half of the monkey's palm and fingers, new inputs from the radial and ulnar nerves—which serve the pinky side and the back of the hand, respectively—had completely annexed the median nerve's cortical territory. After four and a half months, the new maps were as refined as the original: "A beautiful, complete topographic representation of the dorsal hairy fingers [and ulnar palm] emerges," Merzenich later wrote with his UCSF collaborator, William Jenkins, "almost equal in detail to the representation . . . that it supplanted." As the investigators put it in 1983, "These results are completely contrary to a view of sensory systems as consisting of a series of hardwired machines."

The result was greeted with outright hostility. Most of the neuroscience community regarded the finding as somewhere between unlikely and impossible. "Whenever I talked about the extended implications of this, people were very antagonistic," Merzenich recalls. "Hubel and Wiesel's work had shown just the opposite: that after a critical period early in life the brain does not change as a result of changes in sensory input." At scientific meetings, critics heaped scorn on the idea. The peer reviewers of the 1983 paper seemed astonished and doubted its validity. The prevailing view, that the adult brain is fixed and immutable, was so strong that Kaas and Merzenich's work didn't come close to toppling it. "No one had really thought about activity-dependent reorganization in adult animals until Merzenich and Kaas's work," says Terry Allard, who would later spend four years in Merzenich's lab. "Even after this work, it seemed like no one really wanted to."

Kaas found that out the hard way. In another study, he cut some of the retinal nerves in lab animals. After a while, the surviving nerves filled in the area in the visual cortex that the damaged nerves had once delivered inputs to ("so that there were no holes in the vision field," as Kaas puts it). He submitted a paper describing that result to the journal *Science*. An anonymous reviewer dismissed it out of hand, because "everyone knew" that the visual system was not plastic in the adult. Hubel and Wiesel had shown that. Kaas was incredulous. How can you say that, he asked, when the experiment had never been done until now?

Slicing up monkeys' nerves was a pretty drastic way of inducing neuroplasticity, of course. Might the brain manage the feat more, well, naturally? In 1987 Merzenich and Kaas found out. They conducted, in adult owl and squirrel monkeys, experiments resembling Graham Brown and Sherrington's of three-quarters of a century before: comparing cortical maps of the hand in monkeys of about the same size and age. The representation of the hand in the primary somatosensory cortex, they found, varied in size by more than a factor of 2. Representations of individual fingers or segments of digits varied upward of threefold; representation of the back of the hand sometimes occupied half the hand-zone area and sometimes just a small percentage of it. Differences between individuals often swamped differences between species averages—not that averages were looking very meaningful at this point. The different maps, Merzenich suspected, likely reflected the unique life history of each animal. The way the monkey ordinarily used its hands and fingers left an imprint on its brain. As they said, "We propose that the differences in the details of cortical map structure are the consequence of individual differences in lifelong use of the hands."

In another tip of the hat to classic experiments, Merzenich and Kaas mapped the hand representations in the somatosensory cortices of monkeys two to four times. Between mappings, the monkeys lived their normal laboratory life. "Each time we did it the map was unequivocally different," says Merzenich.

I realized that we had inadvertently repeated that 1923 experiment of Karl Lashley, from which he argued that if you make a map of the motor cortex it would be different every time. He believed that the motor cortex is dynamic, reflecting the movements of the body part each spot represents. We were

> *mapping somatosensory cortex, of course, and he was mapping motor cortex. But the conclusion was the same: the cortex is not static, but dynamic. Each time we mapped it, it was different. So what, we asked, was driving this dynamism? It could only have been behavior.*

The brain's response to messages from its environment is shaped by its experiences—experiences not only during gestation and infancy, as most neuroscientists were prepared to accept, but by our experiences throughout life. The life we live, in other words, shapes the brain we develop. To Merzenich, the real significance of the findings was what they said about the origins of behavior and mental impairments. "This machine we call the brain is being modified throughout life," he mused almost twenty years later. "The potential for using this for good had been there for years. But it required a different mindset, one that did not view the brain as a machine with fixed parts and defined capacities, but instead as an organ with the capacity to change throughout life. I tried so hard to explain how this would relate to both normal and abnormal behavior. But there were very few takers. Few people grasped the implications." For a while, it appeared that the monkeys' brains were a lot more adaptable than the research community's.

In an effort to break through, Merzenich decided to pose what he calls "a greater challenge to the brain." Until now, he had typically altered sensory input by transecting a nerve; cutting the nerve to the palm, for example, resulted in an expansion of cortical areas dedicated to the hand's hairy surfaces. But critics suggested that the hairy surfaces might have been connected to the palm area of the cortex all along. According to this line of argument, there was no true cortical remapping, in which neurons carrying signals from the back of the hand invaded the palm's representation zone after its own input was cut off. Instead, maybe back-of-hand neurons had always been present, though silent, in the palm-side representation and were merely being "unmasked" once input from the palm vanished. To (he hoped) overcome such objections, Merzenich and his UCSF team decided to go beyond nerve transection. They amputated a single finger in owl monkeys, removing all possibility of sensory input from the digit, by any route.

Two to eight months after the surgeries, the researchers anesthetized each animal and carefully recorded electrical activity in the somatosensory cortex. They found that the cortical representation of the hand had reorganized. Skin of the palm and of the still-intact fingers adjacent to the amputated finger had taken over the cortical representation of the missing finger, invading the "amputation zone." Put another way, in the monkey version of the somatosensory homunculus, the little guy had lost his middle finger but grown a larger second finger. When the researchers stimulated the monkeys' second digit, the region of the somatosensory cortex that registered sensation in that digit fired, as expected. But so did the representation of what had been the area for the amputated digit, they reported in 1984. When his second finger was touched, the monkey responded as if the scientists were touching his missing finger.

"The amputation work was regarded as the breakthrough experiment," says Ed Taub, now more than a decade past his Silver Spring monkey trials. "Until the mid-1980s, it was an axiom of science that there was little or no plasticity in the adult nervous system. For that reason Merzenich's data aroused a great deal of interest."

Interest, however, is one thing; acceptance is another. The existing paradigm, denying the possibility of such cortical reorganization, would not die easily. The cortical reorganization that Merzenich and his colleagues reported was taking place over only two millimeters of cortical space—the distance, in the owl monkey's brain, that neurons from the second digit had spread in the cortex after amputation of the third digit. Even when Merzenich performed two-digit amputations, to see whether the cortex could remodel over even greater distances, reorganization was confined to a region no larger than a few millimeters. To those reluctant to accept the implications, this degree of rewiring seemed insignificant, perhaps even an error of measurement.

In 1984 Terry Allard, with a fresh Ph.D. from the Massachusetts Institute of Technology, arrived as a postdoc in Merzenich's lab, where he teamed up with Sharon Clark, a talented microsurgeon. Their assignment was an experiment in artificial syndactyly. (*Syndactyly* is a birth defect in which the fingers are joined together, as if in a fist; in artificial syndactyly, two adjacent fingers are sewn together.) What inspired the experiment was a simple enough question: what creates separate representations, in the somatosensory cortex, of the five digits? Merzenich's team hypothesized that the distinct representations reflect differences in the timing of their sensory input: because fingers receive noncoincident sensory stimulation, they develop discontinuous representations. If so, then surgically fusing the digits should eliminate separate representations. "I had basically no background in this," says Allard, "but Mike was very convincing. If the somatosensory map is truly activity-dependent, he convinced me, then artificial syndactyly should be reflected in a new cortical map."

To test their guess, the scientists first had to determine the lay of the land in the brains of adult owl monkeys before their fingers were fused. After anesthetizing each monkey, Bill Jenkins exposed its cortex and then carefully moved the animal to a large camera stand so he could take a four- by five-inch Polaroid of the surface of its brain. He marked several hundred spots on the photo—the places where he would check for activity by positioning electrodes there. Then he gently brushed a spot on the animal's hand or fingers. Through the electrodes inserted into the marked spots, he determined which spot responded to the stimulus. "It was hugely time-consuming," Jenkins recalled. "Constructing a hand map would take, typically, eight hours. It would usually be me and a couple other people, me looking through the microscope and positioning the electrodes, and someone else defining the receptive fields based on the electrodes' response."

Once they had their baseline map, Sharon Clark split the skin of the ring finger and the middle finger of the owl monkeys and then sewed together the dorsal and ventral surfaces. Recalls Allard, "After that, the monkeys just lived their life in the cage. We didn't do anything additional to drive stimulation. But after two or three months, we found that the cortex had been remapped. The very first monkey we did, there was no question the brain had reorganized." Whereas before the surgery the monkeys' fingers transmitted nonsimultaneous signals to the cortex, with the result that the cortex devoted separate little islands to receive input from each separate finger, once the fingers began sending only joint messages (since whenever one finger touched an object, so did the other, as if they were a single digit), the brain reassessed the situation. It seemed to figure that it needed only a single receiver rather than two. What had been separate representations of the fingers became a single, continuous, overlapping representation, they reported in 1988. "We felt we had found the language of the somatosensory cortex, the input that determines how it is organized," says Allard. "We had a sense that we were part of something important, discovering an aspect of the brain that hadn't been recognized before—this whole dynamic aspect of the brain." Years later, researchers in New York would find that the same principle applied to people. Surgeons operated on two patients to separate their congenitally fused fingers. Before the surgery, the cortical map of their digits was shrunken and disorganized. But when the fused digits were separated, the brain quickly created separate receptive fields for the two digits.

Back at Vanderbilt, Kaas knew that no matter how many such breakthroughs were reported, mainstream neuroscience was not about to abandon Hubel and Wiesel's antiplasticity paradigm—at least not until someone challenged their findings head-on. So Kaas and his team turned their attention to the visual cortex of cats, the very animals and the very system that the earlier scientists' Nobel-winning work had characterized as plastic only in infancy. "The organization of the visual cortex has been considered to be highly stable in adult mammals," Kaas's group declared, with some understatement. But when the researchers created small lesions in the cats' retinas, the representation of the retina in the visual cortex shifted. Cortical neurons that formerly received input from the now-lesioned regions did the equivalent of changing pen pals after the original correspondent stops writing. With no input arriving from the lesioned areas of the retina, the cortex began processing inputs from parts of the retina surrounding the lesions. The adult visual cortex seemed just as capable of reorganizing itself as other areas of the brain were.

There was entrenched opposition even to considering whether the cortical reorganization that Merzenich, Kaas, and their colleagues had found in owl monkeys might be applicable to cortical injuries in people—in particular, injuries from stroke. "The reason people were interested but not excited was that the results did not seem to have the potential for recovery of function, because the region involved was too small," recalls Taub. "Even if you extrapolated this to human beings, you were still talking about only 3 to 4 millimeters." Although Merzenich and Kaas were by now convinced that the brain is dynamic and adaptive, creating its maps of the body on the basis of the inputs it receives, and changing those maps as the input changes, critics still dismissed the extent of reorganization they were finding as simply too small to have any significance.

But then Pons and Mishkin got permission to experiment on four of the Silver Spring monkeys. Their 1991 discovery that the deafferentation zone—the part of the somatosensory cortex that originally processed signals from the entire upper limb—was not silent at all, but was instead receiving input from the macaques' faces, changed everything. Merzenich's amputation experiments had documented reorganization of the somatosensory cortex in adult owl monkeys of a millimeter or so; in the Silver Spring monkeys, cortical reorganization spanned a distance an order of magnitude greater, between one and two centimeters. And the reorganization was very complete: every single neuron from 124 recording sites tested in the deafferentation zone had a new connection. "This generated a great deal of excitement," says Taub. "It had the odor of being involved in recovery of function. With this result, it began to look like you could get cortical reorganization on a massive scale, and that might mean something."

At this point, however, there had never been a demonstration of cortical reorganization in people. That was about to change. As soon as the neurologist V. S. Ramachandran read the Silver Spring monkeys study, it "propelled me into a whole new direction of research," he recalled. "My God! Might this be an explanation for phantom limbs?" If touching the faces of the Silver Spring monkeys could excite the somatosensory cortex representation of what was once their arm, Ramachandran wondered, might his amputees' homunculi have been rearranged, too, in a way that would explain the phenomenon of phantom limbs? After all, in the human homunculus, the hand and arm are also near the face.

Although the term *phantom limb* had been around since just after the Civil War, when it was coined by Dr. Silas Weir Mitchell, it had remained a medical conundrum. In 1866 Mitchell had first published his description of it—under a pseudonym. Even when he went public with his finding in 1871, he eschewed the medical journals in favor of the pop magazine *Lippincott's Journal*, the better to insulate himself from the expected derision of colleagues. The phenomenon has struggled to earn respect, or even recognition of its physical reality. As recently as the 1980s researchers (in the *Canadian Journal of Psychiatry*) ascribed phantom limb to wish fulfillment. Just as one might imagine hearing the voice of a recently deceased loved one, went their reasoning, so might an amputee feel a recently lost limb.

Ramachandran immediately phoned colleagues in orthopedic surgery and asked whether they had any recent amputees. They did: Victor Quintero, seventeen, who a month before had lost his left arm just above the elbow in a car crash. Victor swore up and down that he could still feel the missing appendage. Ramachandran enlisted him for an experiment. With Victor sitting still with his eyes closed tight, Ramachandran lightly brushed the boy's left cheek with a cotton swab just as Pons's team had the Silver Spring monkeys. "Where do you feel that?" Ramachandran asked. On my left cheek, Victor answered—and the back of my missing hand. Stroking one spot on the cheek produced the sensation of his absent thumb's being touched. Touching the skin between his nose and mouth created the sensation that his phantom index finger was being brushed. The somatosensory remapping was so fine that when Ramachandran stroked a spot just below Victor's left nostril, the boy felt a tingling on his left pinky. And in perhaps the most peculiar result of somatosensory remapping, when Victor felt an itch in his spectral hand, scratching his lower face produced relief. (Victor was delighted at this, since now, whenever his missing fingers itched, he knew where to scratch.) In a final test, Ramachandran dribbled warm water down Victor's left cheek—and the young man, incredulous, felt a warm feeling in the ghost of his amputated hand. The feeling was so powerful that he actually double-checked that his arm was still gone.

There are some 4 million amputees in the United States. For nearly 70 percent of them their missing arms, hands, legs, or feet continue to experience all-too-real feelings of pressure, pain, warmth, cold, tingling, or other sensations—including Victor's itching. Human amputees, Ramachandran told a 1993 scientific meeting in Santa Fe, experienced cortical reorganization similar to that found in the Silver Spring monkeys: stimulation of the face produced an electrical response in both the somatosensory representation of the face and the amputation zone representing the now-missing arm, as if facial nerves had invaded that region. Brain neurons that originally received input from a limb, it seems, react much as the Silver Spring monkeys did to the decrease in sensory input: rewiring themselves to receive input from other sources. Phantom sensation arises from neuroplastic changes in the brain. Neurons in regions that originally fired in response to stimulation of a now-missing body part look for new work, as it were, and instead respond to peripheral neurons that are still in the game. Just as people in Times Square on New Year's Eve push into any suddenly vacant spot, so surrounding neurons push into the otherwise-silent region of cortex. And also like the New Year revelers, neurons immediately adjacent to a cortical area are most likely to get first dibs at any vacancies.

Which part of the upper quadrant of the body invades the amputation zone therefore turns out to be somewhat random. After a hand is amputated, either the face or the trunk can invade its somatosensory representation. And because the representations of the feet and genitals abut, some people who have suffered the loss of a leg report feeling phantom sensations in the missing limb or limbs during sex: the somatosensory map of the leg, starved of sensation as a result of losing its original input, can be invaded by nerves from the genitals. Similarly, a man whose cancerous penis is amputated may, if his foot is stimulated, have sensations of a phantom penis. (This proximity may help explain why some people find feet erogenous: not merely because the foot unconsciously reminds some people of the penis, as Freud suggested, but also because the somatosensory representation of the foot lies beside the representation of the genitalia.)

The amputation zone, it appeared, was akin to the deafferentation zone in the brains of the Silver Spring monkeys. Monkeys, being somewhat less verbal than your typical amputee, had not been able to tell Pons that cortical remapping produced perceptual effects. Thus Ramachandran's was the first report of a living being's describing the effect of his own brain rewiring.

One of those attending the 1993 Santa Fe meeting at which Ramachandran presented his data was Edward Taub. Taub's rehabilitation into the world of science began in 1986, when Carl McFarland, chairman of the psychology department, recruited him to the University of Alabama, Birmingham (UAB). Taub started work in 1987. The city was trying to shake its history as a citadel of racism and turn itself into a research powerhouse. Taub had an office and a research home. He even had a salary. But he had no "real" money—no research grants. "When I came here I had zero, and not only did I have zero but I couldn't get anything," Taub recalls. "It wasn't the Silver Spring situation," as he calls it, but the sheer unacceptability of his views on neuroplasticity. Soon after he arrived in Birmingham he gave a presentation on the deafferentation data. After methodically describing how the monkeys would resume using their supposedly useless, deafferented arm if their good arm were constrained, he boldly suggested that a similar approach—constraining the movement of the unaffected arm of stroke patients—might restore the use of the affected arm. After all, there was little to lose. No physical or occupational therapy had really been effective in chronic stroke patients, those whose stroke was years in the past and who were thus past the point of spontaneous recovery.

That amounted to millions of people. Every year, at least 600,000 Americans suffer a stroke, which works out to one victim every fifty-two seconds. Of the 440,000 who do not die immediately, 300,000 are left seriously disabled. Thanks to the graying of America, the personal and social toll from stroke is on the increase, with the prevalence of cerebrovascular accident survivors—the technical term—projected to double by 2050. "I just laid it out, not being antagonistic, and of course I didn't know anything about the rehabilitation community," Taub recalls of that first presentation. "I was stepping on everyone's toes with this. The head of the rehabilitation center literally began to stammer, and his face became purple, and he said, 'Are you trying to tell me that a behavioral intervention has an ameliorative effect on a neurological injury of the central nervous system?!' I said, 'But, after all, what is physical therapy if not a behavioral intervention?' He went ballistic. You still have this orientation in the medical community that behavior isn't real."

Taub wasn't the only one whose work connecting plasticity to rehab fell on deaf ears. In 1981 Steve Wolf took up a suggestion Taub had made the year before (Taub himself was still unable to conduct research at this point). Wolf had twenty-five patients with brain damage, most due to stroke, wear a sling on their unaffected arm all their waking hours, except for a half-hour exercise period, for two weeks. He did nothing else. Consistent with Taub's findings on the deafferented monkeys, however, the patients' speed and strength of movement in the disabled arm showed significant improvement on lab motor function tests. Although the effect was small (mostly because Wolf did not use intensive training of the patients' disabled arms), it seemed worth following up. Yet for years no one did. At UCSF, Merzenich and Jenkins had had a similar inspiration. In 1987, they independently proposed that the plasticity of the cortex in response to sensory input, experience, and learning was relevant for stroke rehab. But no one beat down their doors to follow up on the suggestion. After all, "the rehab community was united in opposition to the idea that therapy after a stroke could reverse the neurological effects of the infarct," Taub recalls. "The official position of the American Stroke Association was that rehab for patients with chronic stroke only increases a patient's muscular strength and confidence."

Others were more open-minded. One was the behavioral neuroscientist Niels Birbaumer of Germany's University of Tübingen. At a 1991 presentation in Munich, he heard Taub propose adapting the therapy he had used on the Silver Spring monkeys—constraining their good arm, forcing them to use their "useless" one—to stroke patients. Birbaumer invited him to set up a stroke program in Germany. Taub arrived in Tübingen soon after the 1993 Santa Fe meeting and had lunch with the German psychologist Herta Flor. He told her about Ramachandran's study, describing Ramachandran's claim that touching the face of someone whose arm has been amputated can evoke the feeling that the missing arm is being touched, and suggested that it needed to be verified. Flor responded, "No problem—why don't we do it? I'll just call up my friend Thomas Elbert who has an MEG [*magnetoencephalograph*, which records magnetic changes in neurons that correspond to neuronal activation] and we'll run some patients." "I said, 'fine,'" recalls a still-startled Taub. And thus was born a collaboration that would influence the entire landscape of neuroplasticity. As we learned in Chapter 4, by this time the deafferented Silver Spring monkey research had given rise to two parallel research tracks. One was large-scale cortical reorganization, which Pons and colleagues had put on the map with their experiment on the monkeys. The other was constraint-induced movement (CI) therapy, which as long ago as 1980 had been a glimmer in Taub's eye, but a glimmer extinguished by the debacle of Silver Spring.

As early as 1987 at least some of Taub's colleagues at Birmingham had come around to the notion that behavior can leave footprints on the brain, including the injured brain—well, they'd come around enough to collaborate with him. That year Taub and some UAB colleagues began a pilot experiment. They started working with four patients who were in the top quartile of stroke survivors in terms of ability to move their affected arm: they were able to extend their wrist a minimum of twenty degrees and to flex each finger a minimum of ten degrees. Restraining the intact arm of the Silver Spring monkeys or training the deafferented arm had induced the creatures to use that deafferented arm. Taub suspected that the same two procedures applied to a stroke patient would coax movement out of the affected one—especially training the affected arm. The same general techniques that accomplished that in the deafferented monkeys, Taub maintained, "should be equally applicable following other types of neurological injury, including stroke."

In constraint-induced movement therapy, stroke patients wear a sling on their good arm for approximately 90 percent of waking hours for fourteen straight days. On ten of those days, they receive six hours of therapy, using their seemingly useless arm: they eat lunch, throw a ball, play dominoes or cards or Chinese checkers, write, push a broom, and use standard rehab equipment called dexterity boards. "It is fairly contrary to what is typically done with stroke patients," says Taub, "which is to do some rehabilitation with the affected arm and then, after three or four months, train the unaffected arm to do the work of both arms." Instead, for an intense six hours daily, the patient works closely with therapists to master basic but crucial movements with the affected arm. Sitting across a pegboard from the rehab specialist, for instance, the patient grasps a peg and labors to put it into a hole. It is excruciating to watch, the patient struggling with an arm that seems deaf to the brain's commands to extend far enough to pick up the peg; to hold it tightly enough to keep it from falling back; to retract toward the target hole; and to aim precisely enough to get the peg in. The therapist offers encouragement at every step, tailoring the task to make it more attainable if a patient is failing, then more challenging once the patient makes progress. The reward for inserting a peg is, of course, doing it again—and again and again. If the patient cannot perform a movement at first, the therapist literally takes him by the hand, guiding the arm to the peg, to the hole—and always offering verbal kudos and encouragement for the slightest achievement. Taub explicitly told the patients, all of whose strokes were a year or more in the past, that they had the capacity for much greater use of their arm than they thought. He moved it for them and told them over and over that they would soon do the same.

In just two weeks of constraint-induced movement therapy with training of the affected arm, Taub reported in 1993, patients regained significant use of a limb they thought would forever hang uselessly at their side. The patients outperformed control patients on such motor tasks as donning a sweater, unscrewing a jar cap, and picking up a bean on a spoon and lifting it to the mouth. The number of daily-living activities they could carry out one month after the start of therapy soared 97 percent. That was encouraging enough. Even more tantalizing was that these were patients who had long passed the period when the conventional rehab wisdom held that maximal recovery takes place. That, in fact, was why Taub chose to work with chronic stroke patients in the first place. According to the textbooks, whatever function a patient has regained one year after stroke is all he ever will: his range of motion will not improve for the rest of his life.

"It's true, spontaneous recovery of function usually stops between three and twelve months," Taub says. But his constraint-induced movement therapy picked up where spontaneous recovery stopped. "We got a large effect in the lab and a huge effect in the life situation," Taub says. Two years after treatment ended, the constraint patients were still outperforming controls, brushing their teeth, combing their hair, eating with a fork and spoon, picking up and drinking from a glass.

That fell short of winning over the establishment, however. Throughout 1992 and 1993, Taub recalls, he was rejected for funding by NIH "right and left" because his proposed stroke therapy was so beyond the pale. But as he and his colleagues ran more and more patients, and as other labs replicated their work, it became clear that his hunch, and his hope, were correct.

The Department of Veterans Affairs (Veterans Administration), which has a large population of elderly stroke survivors, finally awarded Taub a grant to extend his research beyond the top-functioning stroke patients to lower-functioning ones. In 1997 he found that patients in the top three quartiles exhibited significant improvement on standard tests of motor ability. The constraint-induced movement therapy worked for them, too, though not as well: the more-affected patients improved by a score of 1.7 on a scale of motor ability, compared to a change of 2.2 for higher-functioning patients. Patients who were functioning best before therapy retained most of their gains even two years afterward; second- and third-quartile patients lost a small fraction of their gains after two years, suggesting the need for what Taub calls "brush-up" training. But the point had been made. The therapy has restored function to patients who had their stroke as much as forty-five years before. "CI therapy appears to be applicable to at least 75 percent of the stroke population," concluded Taub.

The VA also supported an extension of Taub's work to stroke patients who had lost the use of a leg. In this case, constraining the unaffected limb isn't part of the therapy. Patients walk on a treadmill, wearing a body harness for support if necessary, to give them the confidence that they will not collapse as they try to use a leg that they had dismissed as hopelessly impaired. They walk up and down the hall of the Birmingham VA hospital. They rise from a sitting position, climb steps, and do balance exercises. They work for seven hours a day for three weeks. In Taub's first group of sixteen stroke patients with lower-limb impairment, four had not been able to walk at all without support. Two of them learned to walk independently, if awkwardly. Two learned to walk again with only minimal assistance. Of the twelve less-impaired patients, all improved substantially.

What might be the basis for the improvement? In 1998 and 1999 two important studies on patients who underwent the arduous regimen of constraint-induced movement therapy began to provide the answers. In the first, Joachim Liepert and Cornelius Weiller of Friedrich-Schiller University in Jena, Germany, led an investigation of brain changes in six chronic stroke patients. They evaluated the patients before and after they were treated with fourteen days of CI therapy. All six showed significant improvement of motor function. Moreover, all six also showed "an increase of excitability of the neuronal networks in the damaged hemisphere," they found.

"Following CI therapy, the formerly shrunken cortical representation of the affected limb was reversed. . . . [O]nly two weeks of CI therapy induced motor cortex changes up to seventeen years after the stroke." Taub's method of stroke rehabilitation had resulted in a clinically meaningful "recruitment of motor areas adjacent to the original location" involved in control of the limb.

In 1999, Taub and his German collaborators reported on four patients whose strokes had left the right arm extremely weak. The patients again underwent two weeks of CI therapy. All improved significantly. Then, three months later, the scientists recorded changes in the brain's electrical activity. In the most striking finding, when the patients moved their affected arm, the motor cortex on the same side crackled with activity. Ordinarily, the left motor cortex controls the right side of the body, and vice versa. But in these patients, the motor cortex on the same side as the affected arm "had been recruited to generate movements of [that] arm," Taub says. This suggests that the healthy side of the brain had been drafted into service by the patient's continued use of the affected arm. Normally, activity in one hemisphere suppresses the activity of the mirror-image region on the other side, apparently through the bundle of connecting nerves called the corpus callosum. But when activity in the original region is silenced, as by a stroke, that suppression is lifted. Something more than the absence of suppression was needed, however. The increase in the use of the affected arm had, through sustained and repeated movements, "induced expansion of the contralateral cortical area controlling movement of the . . . arm and recruitment of new ipsilateral area." Taub, adopting Mike Merzenich's term, called it *use-dependent cortical reorganization*. He suspected that it served as the neural basis for the permanent improvement in function of what had been thought a useless limb.

One of the patients Taub is proudest of is James Faust, who lives in Calera, Alabama. After a stroke damaged the left side of his cortex, Faust's right arm was so completely paralyzed that he even thought about asking a surgeon to cut off the useless appendage. But hearing about Taub's CI movement therapy, Faust enrolled. After only a few weeks the change was astounding. One evening, when Faust and his wife were having dinner at a restaurant, she looked across the table at him. Her jaw dropped. James was holding a steak knife in his right hand and slicing away as if the stroke had never happened. That was all the encouragement he needed. From that evening on, he began using his right hand as much as he did before the stroke, even more so than he did with the at-home exercises Taub had prescribed: Faust had overcome the "learned nonuse" that Taub had first seen in his monkeys. Success bred success. The more Faust used his right arm and hand, the greater the cortical area the brain presumably devoted to their movement; the greater the cortical area devoted to their movement, the better they moved. Faust is now able to tie his shoes, shave, brush his teeth, and drive.

These two studies were the first to demonstrate a systematic change in brain function in stroke patients as a result of CI therapy. They documented that treatment produces a marked enhancement in the cortical areas that become active during movement of a muscle of an affected limb. Through CI therapy, the brain had recruited healthy motor cortex tissue in the cause of restoring movement to the stroke-affected hand. "Repetitive use of the affected limb induces an extremely large use-dependent cortical reorganization," says Taub. "The area that is responsible for producing movements of the affected arm almost doubles in size, and parts of the brain that are not normally involved, areas adjacent to the infarct, are recruited. You also get recruitment of parts of the brain that are not usually involved in generating movement in the affected arm—that is, areas on the other side of the brain."

The results Taub was obtaining with his stroke patients, corroborated in labs adopting his constraint-induced movement approach, made people more willing to accept such explanations of how and why that therapy worked at a neurological level. In 1999 his UAB team and Emory University received funding from the National Institutes of Health for a national clinical trial of constraint-induced movement therapy at six sites. It would be the first national clinical trial for stroke ever funded by NIH. Sadly, no previous therapy had achieved results sufficient to warrant one. The record of smaller clinical trials for ischemic stroke, as the UCLA neurologist Chelsea Kidwell put it in 2001, was "remarkably dismal."

In the spring of 2000, Taub and his colleagues reported on thirteen more stroke patients in what would be the definitive paper on the power of CI therapy. The thirteen had been living with their disabilities for between six months and seventeen years. They underwent twelve days of CI therapy. When it was over, the amount of motor cortex firing to move the disabled hand had almost doubled. Rehab, it seemed, had recruited new enlistees as effectively as anything the army has ever tried: huge numbers of previously uninvolved neurons were now devoted to moving the stroke-affected hand. Constraint-induced movement therapy had produced cortical remapping. And the improvements in function that accompanied these brain changes remained when the scientists tested the patients after four weeks, and again after six months. "This is the first time we have seen, in effect, the re-wiring of the brain as a result of physical therapy after a stroke," said Dr. David Goode of Wake Forest University.

It was the result that Taub had been working toward from his days with the Silver Spring monkeys and thus, for him, a personal vindication. It was, more than any other, the breakthrough that brought him in from the cold, and almost made up for his period in the wilderness, for the trial, for the fact that his name would forever be associated with the most notorious animal cruelty trial in the history of American research. Few people outside the animal rights community even remembered the Silver Spring monkeys. Those who did hardly cared. In November 2000, at the annual meeting of the Society for Neuroscience, Taub could mention before a roomful of reporters "some monkeys that lived for more than twelve years after deafferentation" without eliciting a single curious inquiry.

Cortical regions supporting sensory and motor functions are better understood, with their little homunculi, than are areas underlying memory and language, two functions whose loss after a stroke can be most devastating. It might seem almost natural, if the region of the motor cortex that once controlled the hand were damaged, for hand control to be taken up by the region that once controlled the shoulder. It's all motor cortex, after all, and therefore not so different from, say, one clothing boutique's blowing through a wall to annex the adjoining haberdashery. But can the same approach apply to higher-level functions? Taub was sure it could, probably through cortical reorganization like that in motor cortex. "If a stroke knocks out your Broca's region, I am suggesting, you can in effect grow a new Broca's region," he says. "That's the whole point. Functions are assigned in the brain in a very general way based on genetics, but they can be co-opted by new patterns of use. If you increase the use you create a competition for available cortical space, which is won by the function that is being most used. That's what we demonstrated in the motor cortex in stroke. So why shouldn't it be applicable in speech? It's just brain." Taub made good on this prediction in 2001, when a similar therapy was used successfully to treat patients who had been left aphasic—unable to speak—by a stroke.

Neurologists had debated for more than a century what lay behind spontaneous (that is, not in response to therapy) language recovery after stroke. One school held that unaffected language regions in the (otherwise damaged) left hemisphere begin playing a greater role. Another, more proplasticity school suspected that regions in the right hemisphere, which in most people are not specialized for language, suddenly undergo a midlife career change. In 1995 researchers led by Cornelius Weiller addressed this question. They studied six men whose devastating left-hemisphere stroke had largely destroyed their Wernicke's area. This region, lying near the junction of the left temporal and parietal lobes, is critical to understanding speech. The men had serious impairments in their ability to use and comprehend spoken words. Over time and with intensive therapy, however, all six largely regained their ability to speak and communicate. What happened? To find out, the researchers scanned the patients' brains with positron emission tomography (PET) while they carried out two word exercises. The PET scans showed that regions in the right hemisphere, corresponding in position to the left cortex's Wernicke's area and other language centers, became active. Recovery, it seemed, had been accompanied by cortical reorganization. Right brain areas analogous to the left brain's damaged language zones had taken over their function.

The next year, Randy Buckner and colleagues in Saint Louis reported a similar finding. They studied a patient who had suffered a lesion to a small area in the left frontal lobe that plays a role in tasks like completing words from three-letter fragments. In normal subjects, turning letter strings such as *cou-* into words like *courage* activates this region. Although the patient was initially unable to master many language functions, within six months of his stroke and with no specific therapy he was performing at almost normal levels on this test. Brain scan results showed that, although the left frontal lobe region normally used to carry out this verbal task was quiet and dark (having been knocked out by the stroke), the mirror-image spot in the right frontal lobe was working away. As the investigators described it, "a pathway similar to that of normal subjects was activated except that, instead of left prefrontal cortex, [our patient] activated right prefrontal cortex." How could this be? Just as in the Weiller study, damage to the original language region in the left hemisphere apparently lifted the suppression of the corresponding region on the right, allowing it to step in and assume the functions of its impaired counterpart.

More support for the "It's all just brain" school of thinking emerged in 1996 from Mark Hallett's lab at NIH. They studied people who had been blind from an early age. In such patients, the primary visual cortex does not receive input from the expected sources, namely, the retina via the optic nerve. But it doesn't take this silence as a license to retire. Instead, Hallett found, reading Braille and performing other fine tactile discrimination tasks activate the visual cortex. But "reading" Braille, of course, means running fingers over raised dots, a task usually handled by the somatosensory cortex. From an early age, it seems, the visual cortex recognizes that it is not receiving signals from the eye. So it switches jobs, taking up tactile processing. The result is that a brain area usually dedicated to vision starts working on the sense of touch, a process that may explain the superior tactile sense of the congenitally blind. This is called *cross-modal functional plasticity*: brain areas that were thought to be genetically "hard-wired" for one function take on totally different functions.

Can such functionally significant brain reorganization be directly influenced by therapy? As we have seen, for people with OCD the answer is yes. The latest evidence, as mentioned, shows that therapy can help stroke patients regain not only the use of a limb, as Taub showed, but the use of language, too. In 1999, researchers in Germany led by Mariacristina Musso and Cornelius Weiller of Friedrich-Schiller University reported brain changes in four stroke patients suffering from aphasia as a result of lesions in Wernicke's area. They designed their study to see whether brief intensive therapy can reorganize the brain and improve language comprehension. The patients had eleven short training sessions in which they had to point to a particular picture on the basis of an oral description, for instance, or indicate on the basis of a picture which of three oral sentences was accurate. The training was intended to stimulate the conscious processes that access linguistic knowledge. In twenty pilot patients, performance on a series of increasingly more complex commands, from "Pick up the yellow triangle" or "Pick up the white circle" (the patients had an array of tokens in front of them) to "Put the red circle on the green triangle" or "Put the white rectangle behind the yellow circle," improved significantly. PET scans on four of the patients provided systematic evidence of brain reorganization. A region of the right hemisphere, in a spot that was the mirror-image of the damaged Wernicke's region of the left, showed significantly increased activation, echoing the 1995 findings. But there was a critical difference. This study showed, for the first time, that the clinical recovery of language performance caused by the hard work of training is functionally related to brain reorganization. The increased activation in the right cortex, compensating for the functional loss of the left-brain homologues, reflected the training each stroke patient had. As with my OCD patients, it was yet more evidence that functional recovery reflects brain reorganization.

These stroke studies finally toppled the old dogma that when a stroke damages a neural network, function (whether speech- or movement-related) is lost because no other system in the brain knows how to perform the same job. The spontaneous recovery of such lost function, sometimes in only a few weeks, had always mystified neurologists. With the dawn of the new century it became clear that neuroplasticity, especially when nudged by effective therapy and mental effort, allows the brain to reassign tasks. The power of plasticity distinguishes the nervous system from every other system in the body. Although plasticity still seems to be greatest from infancy through early adolescence, it was now evident that the brain retains some plasticity throughout life, offering possibilities undreamed of just a few short years ago.

Now it is up to the rehabilitation community to use the findings for the good of millions of stroke patients. Some recover spontaneously, and some suffer damage so extensive that even intense therapy cannot reweave the torn threads in their neural tapestry. For the former, little or no therapy is necessary; for the latter, therapy that teaches how to compensate and cope is about all that one can hope for. But for a large middle group, therapy to induce directed neuroplasticity offers promise of independence and recovery.

Inducing neuroplasticity through a decrease in sensory input such as that after deafferentation, amputation, or stroke was the first challenge to the tenet that the adult human brain is incapable of reorganization. In fact, something close to the opposite is true. "I always think of the Balkans—all those countries that have come and gone and changed their boundaries over the 20th century," says Jordan Grafman of NIH. Through rehabilitation that exploits the vast potential of directed neuroplasticity, stroke patients can now learn to perform actions and carry out tasks once lost to their brain injury. In many of these recoveries, functional reorganization is induced not only through the effort of repeated movement but also by sensory-input increase—something that, in many cases, is under the willful, conscious control of normal, healthy individuals. If the brain is like a map of lived experience, then the mind can, with directed effort, function as its own internally directed mapmaker. This is the subject to which we now turn.

{ SIX }

SURVIVAL OF THE BUSIEST

What is it but a map of busy life,
Its fluctuations, and its vast concerns?
—*William Cowper, "The Task"*

Laura Silverman started playing the piano at age six, and for thirteen years she practiced with a dedication that astonished her teachers. Every day after school, on weekends, and even during the muggy days of a Saint Louis summer Laura would settle in at the upright in the living room for at least a couple of hours. When an important recital loomed, soon after Laura turned eighteen, she stayed at the keyboard for eight hours a day, fanatically rehearsing Mozart's Twentieth Piano Concerto in D Minor, a piece renowned for its difficult arpeggios and a demanding andante movement that requires the fingers to fly over the keys in a blur. But suddenly, one afternoon, Laura's fingers not only failed to find the correct notes; they could barely reach the keyboard. If before they seemed possessed by the musical muse, now they seemed in the thrall of a darker demon. The thumb on her left hand drooped and refused to rise to the keyboard. Other fingers, on both hands, would rise unbidden when a neighboring digit reached for a note. Terrified, Laura canceled the concert and stayed away from the piano from May until August.

When she dared to try some scales after her three-month hiatus, the problem returned in force: her fingers still would not obey her mind's commands. She told her instructor what had happened, but he had never heard of such a bizarre condition and had no idea what to suggest, or even where Laura might seek help. Laura quit taking lessons. She tried to play on her own, but her fingers defied her mind's attempts to control them. At the typewriter, her fingers rebelled just as they did at the piano.

The following spring, Laura's mother saw an article about a famous hand doctor in Boston who specialized in working with musicians. Laura went to see him, only to be told, "I don't know what's wrong with your hand, but I suspect it's all in your head." He sent her to a psychiatrist, one who helped musicians overcome problems such as performance anxiety. Laura played the piano for him, but with only her right hand; by now, her left was musically useless. The psychiatrist dismissed her, saying she must have always had the problem. Laura visited several more doctors in and around Boston. Those who didn't say it was all in her head ventured that she might be suffering the beginnings of multiple sclerosis.

Desperate now, Laura tried alternative medicine . . . acupuncture . . . Alexander technique . . . yoga . . . breathing exercises. She was finishing her sophomore year in college; she would spend her junior year in Japan. After finishing her degree she entered Harvard graduate school to study comparative religion. By this time she could hardly take notes, and her condition only worsened as she pushed herself to function despite it. Every once in a while she would try another doctor, with the same results: *It's all in your head*.

One, who asked her to go to his house to demonstrate how she played, or didn't play, the piano, finally offered a diagnosis no one else had. He told Laura, "You have focal hand dystonia." This usually painless condition, marked by a loss of control of individual finger movements, most often affects the third to fifth fingers (the middle finger to the pinky), and usually involves two or more adjacent fingers. It can strike pianists, flutists, guitarists, and other string players and is believed to reflect the brain's response to the many hours of daily practice that serious musicians engage in, often from a young age. "You can spend the rest of your life doing therapy," the doctor told Laura, "but there is no cure for what you have. Doctors are trying psychotherapy, biofeedback, prolonged rest, steroids, and even botulinum toxin for focal hand dystonia, but nothing works. I suggest you give up piano and start singing." Laura left his house devastated. That week, she faced final exams. As she filled out her blue book in religion class, she stopped halfway through, marched up to the professor, and said she had decided to drop out. She walked out of the exam room and was at Logan Airport the next morning awaiting a flight home—and throwing up in a garbage can. Although she soon reconsidered and returned to Harvard, when she began telling her professors about her disability many were incredulous. "I see you can still hold a book up," one shot back. Many assumed she was malingering. One blasted her for sloppy penmanship.

Laura got a job in New York with *Newsweek Japan* in May 1997 and started seeing a physical therapist every third week. The sessions consisted of typing without thumbs, or with only the forefinger, or very, very slowly. Although she saw no results for more than a year, in the summer of 1999 she began to regain control of her fingers and to type almost normally. She even bought a piano for her apartment. By the end of 1999, Laura had progressed: she could once again hold up her wrists and execute the proper keyboard fingering. "Bach is great because there are a lot of distinct finger movements on both hands, and a lot of mirroring, in which you play with the right and then the left hand," Laura says. "Speed is your enemy." Although the therapist didn't explain the underlying science to Laura, she was drawing on the latest findings in neuroplasticity: that coincident sensory input, such as touching the fingertips to piano keys, can alter the brain circuits responsible for moving those fingers.

Laura and other patients with focal hand dystonia illustrate, painfully, a route to cortical reorganization that is the polar opposite of the sensory input decrease caused by stroke or amputation: sensory input increase. In monkeys with deafferented arms and humans with amputations or strokes, cortical remapping follows a reduction of sensory input, as discussed in the previous chapter. As a result of the reduction of sensory input after such injuries, regions of the somatosensory cortex representing other parts of the body invade areas that represented the now-absent or sensory-deprived part of the body. But less than ten years after his work on amputee owl monkeys jump-started the field of neuroplasticity, Michael Merzenich demonstrated another avenue to cortical reorganization: a concentrated dose of stimulation, analogous to what Laura got through constant piano practice. It is called either sensory input increase cortical reorganization or, preferably, *use-dependent cortical reorganization.*

For more than a decade Merzenich had investigated cortical remapping through the scientific equivalent of brute force, transecting animals' nerves or amputating their fingers. The UCSF team was about to embark on something subtler. William Jenkins had joined Merzenich in January 1980 as a postdoctoral fellow. After spending his first years studying the auditory cortex, mapping which spots respond to which sound frequencies, in 1983 he read the landmark study in which Merzenich and Kaas showed that, after peripheral nerve cuts, the cortex reorganizes. "In that paper, Mike had used the term 'differential use,'" Jenkins recalls. "As a behaviorist, I was really struck by that, because until then only surgical interventions of one kind or another were producing the cortical remapping that Mike was reporting. I went to him and said this was a fascinating idea, with huge implications: if true, we should be able to drive competition for cortical space behaviorally, just as he and Kaas had done surgically." Merzenich agreed. "Cortical remodeling after injury, whether amputation, nerve transection, or lesion, is fascinating," he remembers thinking at the time, "but it's a sidelight to what I regarded as the real issue: how the brain remodels itself in response to behavioral demands." The experiments in which Sharon Clark created artificial syndactyly were an early shot in the battle to convince the neuroscience establishment that the brain creates representations of the body based on the input it receives. But that reflected a decrease in input, as two separate digits were reduced to one larger one. Would the brain also rezone itself to reflect additional input?

In 1985, a new postdoc arrived in Merzenich's lab. Randolph Nudo had been lured away from Florida State University, where he had completed his Ph.D., with the promise that Merzenich would extend his investigation of cortical reorganization from the somatosensory cortex to the motor cortex, Nudo's specialty. "I told Randy that this brain plasticity work Mike was doing was really exciting," recalls Bill Jenkins. "We were studying plasticity in the somatosensory regions, so I told him he should come out and do it for the motor cortex. 'You can own this,' I told him." But when Nudo arrived, it was not without qualms. "People who knew the motor cortex told me that was the last place they'd look for plasticity," Nudo recalls. "Everything in M1 [the primary motor cortex] was supposed to be hard-wired. It was the wrong place to look, they said. There's no question that when I joined Mike I was keeping an open mind, but I was definitely a skeptic."

Nudo came by his interest in wiring, plus some skills in electronics, honestly: he is the son of a TV repairman. He immediately put his talents to good use. In his initial experiment at UCSF, he compared the motor maps on the two sides of the brains of squirrel monkeys. The brain has a left and a right primary motor cortex, each controlling voluntary movements on the opposite side of the body. To map the motor cortex, Nudo inserted tiny stimulating electrodes into scores of locations in that part of the brain and noted which muscles moved when a particular area fired. The motor cortex's representations of movements varied greatly between one monkey and the next, he found, as did the representations of the two hands: specific movements of the hand a monkey preferred to use for retrieving small objects took up more cortical area than maps of the same movements in the nonpreferred hand. Says Nudo, "The motor cortex controlling the preferred hand was bigger and more spatially complex. This suggested that the brain map was a function of the animal's experience, though at that point we had no way of telling whether it reflected recent experience or life experience." In a cart-before-the-horse move, it was only after Nudo had his results that he scoured the old literature to see whether anyone else had noticed such an asymmetry. That's when he discovered S. I. Franz's old paper, mentioned in the last chapter, which found an asymmetry in the movement maps incised in the right and left hemispheres of macaque monkeys. After another experiment, Nudo again discovered that someone else had beaten him to the punch. "It became a game," Nudo recalls. "Could I do anything original at all, or was I doomed to repeat what people had done 80 years ago?"

Nudo then launched an experiment that he would complete only after he had left UCSF: a study of the effects of motor skill learning on the motor cortex. It was the obvious next step, investigating whether everyday experiences, with no surgery to complicate the picture, would trigger reorganization of the motor cortex. To figure out how the monkeys' brains were zoned before training, Nudo needed to determine the so-called movement map of the hand in a monkey's motor cortex. He anesthetized the first of what would eventually be four squirrel monkeys. While monitoring the monkey's vital signs, Nudo spent some two hours preparing for the actual mapping: surgically photographing the brain's surface and setting up the electronics. He then inserted, one by one, tiny stimulating electrodes into the region of the brain that controls the voluntary movement of the forearm, wrist, and digits. The mapping itself, noting which spots in the motor cortex moved which fingers in which ways, took ten to fifteen hours. Often only one scientist— "Me," Nudo says—performed the mapping, once staying up forty-eight hours straight. (Because an animal's skull can't be repeatedly opened and closed, mapping sessions often take on this marathon nature.) "Apart from the exhaustion, the experiment was quite straightforward," he remembers. "We were surprised we could show anything with such a small number of monkeys, but we got clear results with only four."

Whenever a small pulse of current moved, for instance, a digit, Nudo noted the location of the cortical neurons causing the movement. In this way, he created a brain map of the neurons that controlled how a monkey moved its hands. Now he wanted to see whether that map could be rezoned. He placed, outside each monkey's cage, a series of four food wells, ranging in diameter from 25 to 9.5 millimeters. In the wells, he placed banana-flavored food pellets one at a time. Each subsequent well was smaller than its predecessor, and so progressively more difficult to retrieve a food pellet from. To pick up the morsel, the monkey had to extend his arm fully, drop a finger or two into the shallow well, palpate the pellet at the bottom, remove it, grasp it, and get it to his mouth. For the three larger wells, the monkeys didn't have much of a problem getting the pellet. But the 9.5-millimeter well was a different story. At first, the creatures fumbled a lot and almost invariably couldn't quite grasp the pellet. But after a few hundred tries over the course of several days or (in the case of some slow learners) weeks, their performance became nearly flawless, and they were able to pick up their daily allotment of 600 or so tiny pellets fluidly, decisively, and confidently, as if they'd never known any other way to dine.

No wonder: the animals had a new brain to go with their new skill, as Nudo discovered when he went back in to their motor cortices and repeated the arduous brain mapping. What he found was that retrieving 600 pellets during two thirty-minute sessions every day for eleven to fifty days produced dramatic remodeling of the monkeys' motor cortices. The area that became active when a monkey moved his digits, wrist, and forearm had doubled, compared to the motor cortex representations in animals not trained to retrieve food pellets from annoyingly narrow food wells. The neurons controlling the busy fingers had undergone the neural equivalent of suburban sprawl, colonizing more space in the motor cortex, crowding out neurons that controlled some other part of the body (though with no obvious effect on other body parts). As a monkey mastered the smallest well, the area of activation expanded, and the representation in the motor cortex of some movements increased several-fold. In a related experiment, done to see whether the forearm representation could be selectively increased, one of the four monkeys was trained to turn a key-shaped bolt with a twisting movement of its arm. Sure enough, there was a marked expansion in the forearm area of the motor cortex. The motor cortex, concluded the researchers, "is alterable by use throughout the life of an animal." Much as during brain development in childhood, experience changes the connectivity of neurons in response to the circuits that are the most active. Learning a new skill strengthens billions of synaptic connections. "This dramatic brain remodeling almost certainly is the cortical part of the skill acquisition," Merzenich and colleagues concluded later that year. Sherrington and Lashley were undoubtedly smiling down from scientific heaven.

Even more remarkable was what happened after the researchers made small lesions in the monkeys' brains—this time, in the area of the somatosensory cortex that represented the tip of the finger used to palpate the pellet. At first the monkeys became real klutzes at retrieving the pellets from the cups, but after several days they got the hang of it again. You can guess the punchline: when the scientists mapped the somatosensory representation of the fingertip, they found that it had changed again, as alternative response zones emerged to do the job. Although the area of the brain that originally received sensory input from the fingertip had been knocked out of service, other regions took over that function because signals kept arriving from that fingertip. Similarly, when the researchers destroyed the part of the motor cortex that controlled these deft finger movements, the monkeys were once again all thumbs. But after they practiced and practiced the move, the representation of that movement reemerged in areas of motor cortex that formerly represented movement of the hand or forelimb. "Functional recovery can be accounted for by the reemergence of the representation of functions critical for the behavior, in cortical zones that were concerned primarily with other . . . activities prior to the lesion," the scientists concluded.

Several years before Nudo focused on rezoning in the motor cortex, Merzenich had turned on other postdocs to the somatosensory cortex. The UCSF team trained lab animals in a behavior that, by driving sensory input, might change the representation of skin surfaces in the brain. To determine whether any change occurred, of course, the scientists first had to know how the monkeys' brains were wired at the start. Hands are so important to owl monkeys (as to humans) that a significant amount of the brain's real estate is devoted to them. To construct what they called a premap of the animals' cortex, Jenkins and Terry Allard labored for three days straight, in a tag-team approach, starting at 9 A.M. and working through the night. They took photographs of the monkey's exposed cortex, rushed off to the darkroom to develop them, and figured out what was what from vascular landmarks. One scientist sited the electrodes; one recorded activity in response to light brushes on the fingers. "We'd put three chairs together so one of us could sleep while the other worked," recalls Allard.

Then it was time to teach the old (or at least adult) owl monkeys new tricks. In what's fondly called the spinning disk experiment, Jenkins trained them to reach through the bars of their cage and keep a couple of digits in contact with wedge-shaped grooves in a four-inch disk that was spinning like an old LP. The monkeys had to modulate carefully the force they applied to the disk: too little, and their fingers would lose contact with the disk; too much and their fingers would ride along as if on a carousel. But if the animals did it just right, maintaining contact without getting taken for a ride, they were rewarded with a banana-flavored pellet. "I'd sit there for hours, hand-training a hungry monkey until he got it," says Jenkins. Then, some 500 times a day, the monkeys practiced the move; if successful, they got a pellet. "We made sure the monkeys were hungry, and put the disk near them," recalls Allard. "Once they had mastered the task and were performing it hundreds of times a day for several weeks, we went in to their brains. We found a fourfold increase in the area of the somatosensory cortex responding to signals from these fingers." This wasn't a response to something as traumatic as an amputation, a lesion, or a nerve transection, as the earlier work had been. The researchers didn't have to cut the animal to get a change in its brain: the rezoning was purely a response to purposeful behavior.

Greg Recanzone had arrived as a graduate student in Merzenich's lab in 1984. Allard taught him how to train monkeys as well as how to carry out cortical cartography—to determine which minuscule spots in an animal's somatosensory cortex process sensory input from which spots on the animal's body. The first time Recanzone went into the lab, Allard and Jenkins were running the final monkey in the spinning disk experiment. "Mike had a really small lab," recalls Recanzone. "To get from A to B, you had to walk past other people. There was always someone working at a computer, making cortical recordings from an animal. Getting one little piece of equipment took an hour as you stopped and chatted, or looked at what people were doing. Bill had just finished the rotating disk experiment. The obvious question to ask next was, how do cortical changes relate to performance?"

In a landmark follow-up to the spinning-disk experiments, the UCSF researchers embarked on what they called flutter-vibration studies. "This was the first time we used psychophysical behavioral techniques to see whether a frequency discrimination ability gets better with practice," explains Jenkins. They taught seven adult owl monkeys to discriminate among vibrational frequencies applied by a mechanical device to a single spot on one finger. The flutter vibration felt like the flapping wing of a bird. The frequency, but not the location, of the stimulus varied. To train the monkeys, Jenkins and Recanzone arrived early every morning for 200-plus days to run through the same drill: set up the equipment, retrieve a monkey from the basement room where the colony lived, put the animal in a booth for a couple of hours so it could practice the task, eat lunch, then repeat with a second monkey. "We put them in the apparatus every morning and adjusted the stimulation based on their performance," recalls Jenkins. "We ran the animals seven days a week. Because we used food for a reward, we couldn't skip any days, or they'd get out of practice." Six of the seven trained monkeys got better at recognizing when the frequency changed. At first they could detect a change only when frequencies differed from a 20-hertz (twenty flutters per second) standard by at least 6 to 8 hertz. But over the course of training the monkeys learned to discriminate differences as small as 2 or 3 hertz.

Then the researchers compared the sensory representations of the trained hand of the monkey to those of the untrained hand, each in the opposite hemisphere of the cerebral cortex. On a photograph or computer image of each monkey's cortex, they would carefully draw the receptive fields for the various skin surfaces. Bending over a light table and using colored pens, they meticulously mapped the cortical representation of fingers and hands in the somatosensory cortex of both hemispheres. The differences between the two sides of the brain were dramatic. "The cortical representations of the trained hands were substantially more complex in topographic detail than the representations of unstimulated hands," concluded the team. The size of the cortical maps of the skin patches that had felt the flutter were up to three times larger than the maps of the analogous, unstimulated skin surfaces.

But there was a wild card. The use-dependent neuroplasticity that Recanzone and Jenkins found occurred only when the monkeys were attentive to frequency changes, as they would see in their next experiment. This time, they applied the flutter vibration to monkeys' fingers but used sound to distract the animals. To make the distraction effective, the scientists rewarded the monkeys with a food pellet whenever they correctly responded to a tone. The monkeys attending to the tones thus served as "passive controls" for the monkeys attending to skin vibrations: they felt the same flutters but were paying attention to sounds rather than tactile stimuli. The distracted monkeys had no meaningful brain changes. As it turns out, it's not the good vibrations: it's the attention that counts. If the monkeys' attention was focused elsewhere while they received the same tactile stimulation that had otherwise produced massive cortical remapping, no such reorganization occurred.

In a curtain raiser for experiments that would make Merzenich's name, Recanzone then switched to the primary auditory cortex, investigating whether that, too, exhibited plasticity. He trained adult owl monkeys for several weeks to discriminate small differences in the frequency of tones. The monkeys all became progressively better at the job. At the end of training, Recanzone recorded which clusters of neurons fired in response to different tones. Compared to that in control monkeys who were distracted by paying attention to tactile stimulation while hearing the same tones, in the trained monkeys the cortical area representing the behaviorally relevant frequencies had enlarged several-fold, he reported in 1993. This increase in cortical area was the only change that correlated with improved performance.

The emerging picture was dramatic: the brain's representations of the body, of movements, and of sounds are all shaped by experience. Our brain is marked by the life we lead and retains the footprints of the experiences we have had and the behaviors we have engaged in. "These idiosyncratic features of cortical representation," Merzenich said in a model of understatement, "have been largely ignored by cortical electrophysiologists." As early as 1990 Merzenich was floating a trial balloon: maybe, just maybe, the behaviorally based cortical reorganization he was documenting supported functional recovery after brain injury such as that caused by a stroke, which until then (and to some extent, even now) was attributed to entirely different causes. And maybe focal dystonias like the one that thwarted Laura Silverman reflect use-driven changes in cortical representations, degrading fine-grained cortical maps like the colors of a Mondrian bleeding into each other after a rain.

Merzenich's wasn't the only lab probing the power of neuroplasticity. In 1993, Alvaro Pascual-Leone, then at the National Institute of Neurological Disorders and Stroke, reported one of the earliest studies in human neuroplasticity. Are there people, he asked, who habitually experience powerful stimuli to a given portion of their bodies? The answer came almost immediately: the blind, who read Braille with their fingertips. Pascual-Leone therefore studied the somatosensory cortex of fifteen proficient Braille readers. He gave weak electrical shocks to the tip of the right forefinger (the "reading" finger) and then recorded which somatosensory regions fired in response. He compared the brain response to that when the left index finger—a nonreading finger—was stimulated. The area of the brain devoted to the reading finger of expert Braille readers was much larger than that of the nonreading finger, or of either index finger in nonreaders, Pascual-Leone found. It was a clear case of sensory input increase, with the person paying close attention, leading to an expansion of the brain region devoted to processing that input.

At about the same time Edward Taub, although deep in his studies of how constraint-induced therapy might enable stroke patients to regain use of a limb, was also pursuing another goal: to determine how increases in sensory input affect the brain's organization. In the spring of 1995 Taub and his German collaborators had reported that arm amputation produces extensive reorganization in the somatosensory cortex. Soon after, Thomas Elbert of the University of Konstanz, who was about to embark on a major collaboration with Taub, joined Taub and his wife for dinner. Is there any normal human activity in which there is much more use of one hand than of the other hand? Taub asked. Elbert thought a bit and said, "Well, yes, pianists." But that wasn't right; pianists use both hands. But Taub's wife, Mildred Allen, a lyric soprano who had been a principal artist at the Metropolitan Opera in New York and a leading singer at the Santa Fe Opera, chimed in, "Oh, that's easy; use the left hand of string players." When a right-handed musician plays the violin, four digits of the left hand continuously finger the strings. (The left thumb grasps the neck of the violin, undergoing only small shifts of position and pressure.) The right, or bowing, hand, undertakes far fewer individual finger movements. Might this pattern leave a trace on the cerebral cortex?

The scientists recruited six violinists, two cellists, and one guitarist, all of whom had played the instrument for seven to seventeen years, as well as six nonmusicians who had no experience with stringed instruments. For the study, the volunteers sat quietly while a pneumatic stimulator applied light pressure to their fingers. A magnetoencephalograph positioned over their skulls recorded neuronal activity in the somatosensory cortex.

There was no difference between the string players and the nonmusicians in the representation of the digits of the right hand, the researchers reported in 1995. But there was a substantial cortical reorganization in the somatosensory map of the fingers of the left hand. The researchers concluded, "[T]he cortical territory occupied by the representation of the digits [of the left hand] increased in string players as compared with that in controls." The brain recordings showed that the increase in cortical representation of the fingering digits was greater in those who began to play before the age of twelve than in people who took up an instrument at a later age. When the results were published, the attendant publicity trumpeted that as the revelation—missing the point entirely, to Taub's frustration. The much greater discovery was the unmistakable evidence of cortical reorganization in all the string players. The surprise was not that the immature nervous system is plastic, as "everyone knew," says Taub, but that plasticity persists into adulthood. "Even if you take up the violin at 40, you still get use-dependent cortical reorganization," says Taub.

That same year Merzenich's lab reported, using their old reliable adult owl monkeys, that applying tactile stimuli to the fingers changes the maps of the hand surface in the somatosensory cortex—a lab version of the real-world changes Taub and his team found in the string players. Despite dramatic results like these, even into the 1990s Merzenich still had not won over a lot of neuroscientists. When Xiaoqin Wang, then a graduate student at Johns Hopkins University, was weighing a postdoctoral fellowship in Merzenich's lab, he kept the idea from his thesis advisor. "His colleagues were very skeptical of the work Mike started," Wang recalls. He joined Merzenich's lab anyway and immediately set to work testing whether another nonsurgical, purely behavioral intervention would alter an animal's brain. Sure, Merzenich and Kaas had shown that lesions bring about a reorganization of the cortex. But "Mike's most important contribution was to argue that reorganization in the cortex is not solely a product of lesions," says Wang. "Brilliantly, he saw that the cortex adapts to its environment, and responds to a changing environment—including the behavioral environment."

Wang trained monkeys in a task that would finally shed light on what had gone wrong in the brains of people like Laura Silverman and other victims of focal hand dystonia. The monkeys placed one hand on a form-fitting handgrip containing two little metal bars. One bar, perpendicular to the fingertips, stimulated the tips of the second, third, and fourth fingers simultaneously. A second bar stimulated the same three fingers just above the knuckles. To make sure the monkeys were paying attention to the alternating stimuli (as we have seen, attention is a prerequisite for use-dependent brain changes), the scientists rewarded the animals for responding whenever two consecutive stimuli were applied by either bar. The monkeys underwent the behavioral training for some 500 trials day in and day out, for six and sometimes seven days a week. "I was supposed to go to my Hopkins graduation that May [of 1991], but I skipped it because I was training the monkeys," Wang recalls. "I didn't have the heart to leave in the middle."

After four to six weeks of training the monkeys, Wang mapped their brains. To produce a map with the necessary high resolution, he recorded with microelectrodes from 300 locations, each just a few micrometers apart in the somatosensory cortex. The goal was to see which clutches of neurons fired in response to a light touch on a finger. Since both monkey and human fingers usually feel stimuli nonsimultaneously (we tend not to move our fingers as one, unless we are waving bye-bye to a toddler), the intense synchronous input across the monkeys' digits was expected to produce changes in the brain. And it did. "Individual fingers were no longer differentiated," says Wang. The normal segregation of fingers in primary somatosensory cortex "was completely broken down." In control animals whose fingers had been stimulated asynchronously, the brain represented each finger discretely. But when digits were stimulated synchronously, their representation in the brain fused, much as had happened to Laura Silverman: a single region of somatosensory cortex responded to the touch of two or even three fingers. Simultaneous stimulation, whether by bars in a lab or by too many andante movements, fools the brain's primary somatosensory cortex into thinking that different fingertips are part of a single unit. This discovery strongly reinforced Merzenich's findings nearly a decade earlier that fusing fingers to create syndactyly breaks down the discrete representations of fingers, and that separating long-fused digits reseparates the fused representation in the somatosensory cortex. It all flows from the basic Hebbian principle: Cells that fire together wire together. In this way our brain, it seems, contains the spatial-memory traces of the timing of the signals it receives and uses temporal coincidence to create and maintain its representations of the body.

The motor cortex, you'll recall, is arranged like a little homunculus. But it is hardly a static layout. From day to day and even moment to moment, the motor cortex map changes, reflecting the kinds of movements it controls. Complex movements result in outputs from the motor cortex that strengthen some synapses and weaken others, producing enduring changes in synaptic strength that result in those things we call motor skills. Learning to ride a bicycle is possible, in all likelihood, not merely because of something called muscle memory but also because of motor-cortex memory.

In 1995, Alvaro Pascual-Leone, following up on his Braille study, conducted an experiment that, to me, has not received nearly the attention it deserves. This one modest study serves as the bridge between the experiments on humans and monkeys showing that changes in sensory input change the brain, on the one hand, and my discovery that OCD patients can, by changing the way they think about their thoughts, also change their brain. What Pascual-Leone did was have one group of volunteers practice a five-finger piano exercise, and a comparable group merely think about practicing it. They focused on each finger movement in turn, essentially playing the simple piece in their heads, one note at a time. Actual physical practice produced changes in each volunteer's motor cortex, as expected. But so did mere mental rehearsal, and to the same degree as that brought about by physical practice. Motor circuits become active during pure mental imagery. Like actual, physical movements, imagined movements trigger synaptic change at the cortical level. Merely *thinking about* moving produced brain changes comparable to those triggered by actually moving.

These were the opening shots in what would be a revolution in our understanding of the origins of human disabilities as diverse as focal hand dystonia, dyslexia, and cerebral palsy. Merzenich firmly believed that the focus of the previous two decades—attributing neurological illness (especially developmental abnormalities) primarily to molecular, genetic, or physical defects—had missed the boat. Instead, he suspected, it is the brain's essential capacity for change—neuroplasticity—that leaves the brain vulnerable to such disabilities. But if that is true, Merzenich persisted, surely the reverse would hold as well: if neuroplasticity opens the door to disabilities, then maybe it can be harnessed to reverse them, too—just as it reversed the "errors" caused by the ministroke in the food-pellet-retrieving monkeys. Just as a few thousand practice trials at retrieving pellets resulted in a new brain that supported a new skill in the monkeys, so, too, might several thousand "trials" consisting of hearing spoken language imperfectly, or playing the same piano notes over and over, result in a new brain—and possibly a new impairment—in people. The brain changes causing these impairments could become so severe that Merzenich coined a term to capture their magnitude: *learning-based representational catastrophe,* as he characterized it to a scientific meeting in the late fall of 2000.

If an increased cortical representation of the fingering hand of string players, and a strengthening of motor circuits in the brains of people (and lab animals) learning a motor skill, is the positive side of use-dependent cortical reorganization, then focal hand dystonia is the dark side. In fact, Laura's doctors were not far off when they said that her condition was all in her head. "The musician loses control over one or more of the digits of one hand, and that usually terminates his or her professional career," says Taub. A pianist, or a typist, loses the ability to make rapidly successive movements with two (usually adjacent) fingers: when the index finger rises, for instance, the middle finger follows uncontrollably. "There is a fusion of the representation of the fingers in the dystonic hand," says Taub. "We think it has something to do with simultaneous excitation of the fingers, typically from playing rapid passages forcefully."

In 1990 Merzenich's group was already suggesting, on the basis of their monkey findings, that focal hand dystonia reflects brain plasticity. In the early 1990s Merzenich hooked up with Nancy Byl, director of the graduate program in physical therapy at UCSF, for a study in which they simulated writer's cramp in two adult owl monkeys by training them to grasp a handgrip that repeatedly opened and closed, moving their fingers about a quarter-inch each time, up to 3,000 times during a one- or two-hour daily training session. To keep the monkeys focused on the task, Byl rewarded them with food pellets for holding onto the hand grip. After three months of this for one monkey and six months for the other, the animals could no longer move their fingers individually. In the brain, the receptive field of fingers' sensory neurons had grown ten- or twentyfold, often extending over multiple fingers. "Rapid, repetitive, highly stereotypic movements applied in a learning context can actively degrade cortical representations of sensory information guiding fine motor hand movements," Byl told the 1999 meeting of the Society for Neuroscience. "Near-simultaneous, coincident, repetitive inputs to the skin, muscles, joints and tendons of the hand may cause the primary sensory cortex in the brain to lose its ability to differentiate between stimuli received from various parts of the hand." A patient with focal hand dystonia may feel a touch of her fingertip as a touch of another finger. She may have trouble identifying objects by feel. Fishing keys out from the bottom of a bag becomes hopeless.

If focal hand dystonia arises from highly attended, repetitive, simultaneous sensory input to several fingers, then the logical treatment is obvious. Correcting the problem, Merzenich believed, "requires a re-differentiation . . . of these cortical representations," through highly attended, repetitive, nonsimultaneous movements. In early 1999 Byl and colleagues therefore launched small-scale studies based on this premise, with the goal that patients with focal hand dystonia would remap their own somatosensory cortex. They had the patients carry out tasks that demand acute sensory discrimination, such as reading Braille or playing dominoes blindfolded, all the while focusing their attention like a laser beam on the task. Byl encouraged them to use mental imagery and mentally to practice using the disabled hand and fingers; just as Pascual-Leone found that mentally practicing a piano exercise produces brain changes comparable to those produced by actually hitting the ivories, so patients with focal hand dystonia, she suspected, might break apart the merged representation of their fingers by imagining moving each finger individually. It would be wrong to minimize the challenge of this therapy, however. Merzenich's findings suggest that lab animals need something like 10,000 to 100,000 repetitions to degrade the initial representation of a body part; Byl therefore suspects that people require a comparable number of repetitions of a therapeutic exercise to restore normal representation. Her early findings look encouraging. In 2000, she reported an 85 to 98 percent improvement in fine motor skills in three musicians with focal hand dystonia after they took part in her "sensorimotor retraining program." Two of three returned to performing. The implication? In at least some patients with focal hand dystonia, the degraded cortical representation can be repaired.

In 1998, after confirming that in focal hand dystonia the somatosensory representations of the affected digits are fused, Taub and Elbert's team in Germany also developed a therapy based on the finding. To come up with an appropriate therapy, a grad student, Victor Candia, applied Taub's constraint-induced approach to restrain the movement of one or more less-dystonic fingers. The researchers recruited professional musicians with focal hand dystonia: five pianists (all soloists except one chamber music player) and two guitarists. Despite their disability, five of the musicians were still concertizing, masking their dystonia in some cases through atypical fingerings that avoided the dystonic finger. Taub and his colleagues thought they could do better. The scientists therefore restrained one or more of the healthy, less-dystonic fingers. The subject then used his dystonic finger to perform instrumental exercises, under a therapist's supervision, for one and a half to two and a half hours each day for eight straight days, followed by home exercises of one hour or more a day. The exercises consisted of sequential movements of two or three digits, including the dystonic one, followed by a brief rest and then another sequence. If the subject's ring finger was dystonic, for instance, and the pinky had been compensating for its neighbor's impairment, then the researchers restrained the pinky and had the patient run through the exercise index-middle-ring-middle-index. In simple terms, this separate stimulation teaches the brain that the ring finger is a separate entity, distinct from its digital neighbors. All five pianists were successfully treated, though one who did not keep up his exercises regressed. Two resumed concertizing without resorting to the fingering tricks they had used before. Four of the original seven played as well as they had before the dystonia struck. "Our suspicion was that we were breaking apart the fusion of the brain's representation of three and sometimes four fingers," says Taub.

The plasticity of the motor cortex might even underlie something so common, unremarkable, and seemingly inevitable as the tentative gait that many elderly people adopt. With age, walking becomes more fraught with the risk of a spill, so many people begin to walk in an ever-more constrained way. Old people become erect and stiff, or stooped, using shorter steps and a slower pace. As a result, they get less "practice" at confident striding—bad idea. Because they no longer walk normally and instead "overpractice" a rigid and shuffling gait, the motor-cortex representation of fluid movement degrades, just as in monkeys that stop practicing retrieving little pellets from wells. The result: we burn a trace of the old-folks' walk into our brain, eventually losing the ability to walk as we once did. It is the sadder facet of the neural traces burned into our brain at the beginning of life. There is, though, a bright side: there is every reason to believe that practicing normal movements with careful guided exercise may help prevent, or even reverse, the maladaptive changes.

Tinnitus, or ringing in the ears, is characterized by the perception of auditory signals in the absence of any internal or external source of sound. It strikes an estimated 35 percent of the population at some point in life. In about 1 percent, the condition is severe enough, and maddening enough, to interfere with daily life. The source of the problem had remained a mystery for centuries: half of the investigators interested in tinnitus thought the central nervous system was involved, and half didn't. Taub and Thomas Elbert were squarely in the first camp, suspecting that tinnitus reflects cortical reorganization that is the a result of sensory input increase. Taub and Elbert again teamed up, this time with Werner Mühlnickel, a grad student. They compared ten subjects with tinnitus (in the range of 2,000 to 8,000 hertz) to fifteen without it. To the healthy subjects, they played four sets of pure tones, of 1,000, 2,000, 4,000, and 8,000 hertz. The tinnitus subjects heard the tone that matched their tinnitus frequency (determined by having subjects move a cursor on a computer screen that varied the tone of the sound output, until they reached the one that they always heard), and then the three standard tones (usually 1,000, 2,000, and 8,000 hertz). Usually, sound frequencies are represented in the auditory cortex according to a logarithmic scale: the lowest frequencies are near the surface of the brain, and higher frequencies are toward the interior. But in tinnitus sufferers, the scientists reported in 1998, the tinnitus tone had invaded neighboring regions. "The tonotopic map was normal except at this frequency, where there was a huge distortion, with more area given over to the tinnitus tone," says Taub. "Not only do you get cortical reorganization, but the strength of tinnitus is related to the amount of cortical reorganization." Increased sensory input to the auditory cortex at a particular frequency had apparently produced use-dependent cortical reorganization. And that suggests a therapy for what had been an untreatable syndrome: if patients attend to and discriminate acoustic stimuli that are near the frequency of the tinnitus tone, that might drive cortical reorganization of the nontinnitus frequencies into the cortical representation of the tinnitus tone. That should reduce the tinnitus representation, diminishing the sense that this tone is always sounding.

It is worth pausing here to address what neuroplasticity is not: just a fancy name for learning and the formation of memories. This not-infrequent criticism of the excitement surrounding the neuroplasticity of the adult brain is reminiscent of the old joke about how new ideas are first dismissed as wrong, and then, when finally accepted, as unimportant. In the case of neuroplasticity, the criticism goes something like this: the idea that the adult brain can rewire itself in some way, and that this rewiring changes how we process information, is no more than a truism. If by neuroplasticity you mean the ability of the brain to form new synapses, then this point is valid: the discovery of the molecular basis of memory shows that the brain undergoes continuous physical change. But the neuroplasticity I'm talking about extends beyond the formation of a synapse here, the withering away of a synapse there. It refers to the wholesale remapping of neural real estate. It refers to regions of the brain's motor cortex that used to control the movement of the elbow and shoulder, after training, being rewired to control the movement of the right hand. It refers to what happens when a region of the somatosensory cortex that used to register when the left arm was touched, for example, is invaded by the part of the somatosensory cortex that registers when the chin is gently brushed. It refers to visual cortex that has been reprogrammed to receive and process tactile inputs. It is the neural version of suburban sprawl: real estate that used to serve one purpose being developed for another. Use-induced cortical reorganization, says Taub, "involves alterations different from mere learning and memory. Rather than producing just increased synaptic strength at certain junctions, which is believed to underlie learning, some unknown mechanism is instead producing wholesale topographic reorganization." And more: we are seeing evidence of the brain's ability to remake itself throughout adult life, not only in response to outside stimuli but even in response to directed mental effort. We are seeing, in short, the brain's potential to correct its own flaws and enhance its own capacities.

The existence, and importance, of brain plasticity are no longer in doubt. "Some of the most remarkable observations made in recent neuroscience history have been on the capacity of . . . the cerebral cortex to reorganize [itself] in the face of reduced or enhanced afferent input," declared Edward Jones of the University of California, Davis, Center for Neuroscience, in 2000. What had been learned from the many experiments in which afferent input to the brain increased? Cortical representations are not immutable; they are, to the contrary, dynamic, continuously modified by the lives we lead. Our brains allocate space to body parts that are used in activities that we perform most often—the thumb of a video-game addict, the index finger of a Braille reader. But although experience molds the brain, it molds only an attending brain. "Passive, unattended, or little-attended exercises are of limited value for driving" neuroplasticity, Merzenich and Jenkins concluded. "Plastic changes in brain representations are generated only when behaviors are specifically attended." And therein lies the key. Physical changes in the brain depend for their creation on a mental state in the mind—the state called attention. Paying attention matters. It matters not only for the size of the brain's representation of this or that part of the body's surface, of this or that muscle. It matters for the dynamic structure of the very circuits of the brain and for the brain's ability to remake itself.

This would be the next frontier for neuroplasticity, harnessing the transforming power of mind to reshape the brain.

{SEVEN}

NETWORK REMODELING

The mind is its own place, and in itself
Can make a heaven of hell.
—*John Milton,* Paradise Lost

In the previous two chapters, we examined the brain's talent for rewriting its zoning laws—or, to be more formal about it, the expression of neuroplasticity that neuroscientists call cortical remapping. We've seen how a region of somatosensory cortex that once processed feelings from an arm can be rezoned to handle input from the face; how the visual cortex can stop "seeing" and begin to "feel"; how the motor cortex can reassign its neuronal real estate so that regions controlling much-used digits expand, much as a town might expand a playground when it enjoys a baby boom. In all these cases, brain plasticity follows an increase or decrease in sensory input: an increase, as in the case of violin players' giving their fingering digits a workout, leads to an expansion of the cortical space devoted to finger movement, whereas a decrease in sensory input, as in the case of amputation, leads to a shrinkage. But there is another aspect of neuroplasticity. Rather than a brute force expansion or shrinkage of brain regions zoned for particular functions, this form of neuroplasticity alters circuitry within a given region. And it results not from a change in the amount of sensory input, but from a change in its quality.

By the mid-1990s, Michael Merzenich and his UCSF team had two decades of animal research behind them. In addition to all the studies they had made of how changing levels of sensory stimulation altered the somatosensory cortex, they had shown that auditory inputs have the power to change the brain, too: altering sound input, they found, can physically change the auditory cortex of a monkey's brain and thus change the rate at which the brain processes sounds. The researchers began to suspect that the flip side of this held, too: a brain unable to process rapid-fire sounds, and thus to recognize the differences between sounds like *gee* and *key*, or *zip* and *sip*, may be different—physically different—from a brain that can. Across the country, at Rutgers University in New Jersey, Paula Tallal and Steve Miller had been studying children who had specific language impairment (SLI). In this condition, the kids have normal intelligence but great difficulty in reading and writing, and even in comprehending spoken language. Perhaps the best-known form of specific language impairment is dyslexia, which affects an estimated 5 to 17 percent of the U.S. population. When Tallal began studying dyslexia in the early 1970s, most educators ascribed it to deficits of visual processing. As the old (and now disproved) stereotype had it, a dyslexic confuses *p* with *q*, and *b* with *d*. Tallal didn't buy it. She suspected that dyslexia might reflect a problem not with recognizing the appearance of letters and words but, instead, with processing certain speech sounds—fast ones.

Her hunch was counterintuitive—most dyslexics, after all, have no detectable speech impediments—but it turned out to be right. Dyslexia often does arise from deficits in phonological processing. Dyslexics therefore struggle to decompose words into their constituent sounds and have the greatest trouble with *phonemes* (the smallest units of oral speech) like the sounds of *b*, *p*, *d*, and *g*, all of which burst from the lips and vanish in just a few thousandths of a second. In these dyslexics the auditory cortex, it seems, can no more resolve closely spaced sounds than a thirty-five-millimeter camera on Earth can resolve the craters and highlands of the Moon. They literally cannot hear these staccato sounds. How might this happen? Pat Kuhl's work, discussed in Chapter 3, shows how infants normally become attuned to the sounds of their native language: particular clumps of neurons in the auditory cortex come to represent the phonemes they hear every day. But consider what would happen if this input were somehow messed up, if the brain never correctly detected the phoneme. One likely result would be a failure to assign neurons to particular phonemes. As a result, dyslexics would be no more able to distinguish some phonemes than most native Japanese speakers are to distinguish *l* from *r*. Since learning to read involves matching written words to the heard language—learning that *C A T* has a one-to-one correspondence with the sound *cat*, for instance—a failure to form clear cortical representations of spoken language leads to impaired reading ability.

Merzenich knew about Tallal's hypothesis. So at a science meeting in Santa Fe, they discussed her suspicion that some children have problems hearing fast sounds, and her hunch that this deficit underlies their language impairment and reading problems. You could almost see the light bulb go off over Merzenich's head: his plasticity experiments on monkeys, he told Tallal, had implications for her ideas about dyslexia. Might reading be improved in dyslexics, he wondered, if their ability to process rapid phonemes were improved? And could that be done by harnessing the power of neuroplasticity? Just as his monkeys' digits became more sensitive through repeated manipulation of little tokens, Merzenich thought, so dyslexics might become more sensitive to phonemes through repeated exposure to auditory stimuli. But they would have to be acoustically modified stimuli: if the basis of dyslexia is that the auditory cortex failed to form dedicated circuits for explosive, staccato phonemes, then the missing circuits would have to be created. They would have to be coaxed into being by exposing a child over and over to phonemes that had been artificially drawn out, so that instead of being so staccato they remained in the hearing system a fraction of a second longer—just enough to induce a cortical response.

Tallal, in the meantime, had received a visit from officials of the Charles A. Dana Foundation, which the industrialist David Mahoney was leading away from its original mission of education and into neuroscience. But not just any neuroscience. Incremental science was all well and good, Mahoney told Tallal, but what he was interested in was discovery science, risk-taking science—research that broke paradigms and made us see the world, and ourselves, in a new light. "Put your hand in the fire!" he encouraged her. The upshot was the launch of a research program on the neurological mechanisms that underlie reading and on how glitches in those mechanisms might explain reading difficulties. Rutgers and UCSF would collaborate in a study aimed at determining whether carefully manipulated sounds could drive changes in the human auditory cortex.

In January 1994, Merzenich, Bill Jenkins, Christoph Schreiner (a postdoc in Merzenich's lab), and Xiaoqin Wang trekked east, and over two days Tallal and her collaborators told the Californians "everything they knew about kids with Specific Language Impairment," recalls Jenkins. "We sat there and listened, and about halfway through I blurted out, 'It sounds like these kids have a backwards masking problem'—a brain deficit in auditory processing. That gave us the insight into how we might develop a way to train the brain to process sounds correctly." Two months later, the Dana Foundation awarded them a three-year grant of $2.3 million.

The UCSF and Rutgers teams set to work, trying to nail down whether a phonological processing deficit truly underlies dyslexia and whether auditory plasticity might provide the basis for fixing it. They started with the hypothesis that children with specific language impairment construct their auditory cortex from faulty inputs. The kids take in speech sounds in chunks of one-third to one-fifth of a second—a period so long that it's the length of syllables, not phonemes—with the result that they do not make sharp distinctions between syllables. It's much like trying to see the weapons carried by troops when your spy camera can't resolve anything smaller than a tank. So it is with this abnormal "signal chunking": the brains of these children literally do not hear short phonemes. *Ba,* for instance, starts with a *b* and segues explosively into *aaaah* in a mere 40 milliseconds. For brains unable to process transitions shorter than 200 milliseconds, that's a problem. The transition from *mmm* to *all* in *mall*, in contrast, takes about 300 milliseconds. Children with specific language impairment can hear *mall* perfectly well, but *ba* is often confused with *da* because all they actually hear is the vowel sound. There are undoubtedly multiple causes of this processing abnormality, including developmental delays, but middle ear infections that muffle sounds are a prime suspect. These deficits in acoustic signal reception seem to emerge in the first year of life and have profound consequences. By age two or three, children with these deficits lag behind their peers in language production and understanding. Later, they often fail to connect the letters of written speech with the sounds that go with those letters. When *ba* sounds like *da*, it's tough to learn to read phonetically.

If language deficits are the result of abnormal learning by the auditory cortex, then the next question was obvious: can the deficits be remedied by learning, too? To find out, Rutgers recruited almost a dozen kids with SLI and set up experimental protocols; UCSF developed the acoustic input, in the form of stretched-out speech, that they hoped would rewire the children's auditory cortex. But from the beginning Mike Merzenich was concerned. The auditory map forms early in life, so that by the time children are two they have heard spoken something like 10 million to 20 million words—words that, if the hypothesis about phonemic processing deficits was correct, sounded wrong. He knew that cortical representations are maintained through experience, and experience was what these kids had every time they misheard speech. "How are we going to undo that?" he worried. And worse, although the kids would hear modified speech in the lab, they would be hearing, and mishearing, the regular speech of their family and friends the rest of the time. That, Merzenich fretted, would reinforce all of the faulty phonemic mapping that was causing these kids' problems. Short of isolating the children, there was no way around it: the researchers would simply have to take their best shot at rebuilding a correct phonemic representation in the children's brains, competing input be damned.

As luck would have it, Xiaoqin Wang had joined Merzenich's lab in the early 1990s after finishing his Ph.D. at Johns Hopkins, where he had studied the auditory system. Although reading a book on the brain had lured him into neuroscience, Wang's first love had been information processing: he had earned a master's degree in computer science and electrical engineering. That experience had given him just the signal-processing knowledge that Paula Tallal and Merzenich needed to produce modified speech tapes that would, they hoped, repair the faulty phonemic representations in the brains of SLI children. Wang was reluctant to enlist in the project, because he was so busy with the experiments on cortical remapping of monkeys' hand representations. "But Mike is someone you just can't say no to," he recalls. "So we took this idea of Tallal's that if you slow down rapid phonemes the kids will hear them. What I managed to do was slow down speech without changing its pitch or other characteristics. It still sounded like spoken English, but the rapid phonemes were drawn out." The software stretched out the time between *b* and *aaah*, for example, and also changed which syllables were emphasized. To people with normal auditory processing, the sound was like an underwater shout. But to children with SLI, the scientists hoped, it would sound like *baa*—a sound they had never before heard clearly. When Tallal listened to what Wang had come up with, she was so concerned that the kids would be bored out of their minds, listening to endless repetitions of words and phonemes, that she dashed out to pick up a supply of Cheetos. She figured her team would really have to bribe—er, motivate—the kids to stick with the program.

And so began Camp Rutgers, in the summer of 1994. It was a small study with a grand goal: to see whether chronic exposure to acoustically modified phonemes would alter the cortical representation of language of an SLI child and help him overcome his impairment. The scientists' audacious hope was that they could retrain neurons in the auditory cortex to recognize lightning-fast phonemes. The school-age kids would show up every weekday morning at eight and stay until eleven. While their parents watched behind a one-way mirror, the children donned headphones. Using tapes of speech processed with Wang's software, they were coached in listening, grammar, and following directions (a novelty for some, since they'd never understood many instructions in the first place). For example, "Point to the boy who's chasing the girl who's wearing red," intoned the program over and over, the better to create the cortical representations of phonemes. To break up the monotony, the scientists offered the kids snacks and puppets, frequent breaks—and in one case, even handstand demonstrations. Steve Miller recalls, "All we did for three hours every day was listen. We couldn't even talk to the kids: they got enough normal [misheard] speech outside the lab. It was so boring that Paula had to give us pep talks and tell us to stop whining. She would give us a thumbs-up for a good job—and we'd give her a different finger back." In addition to the three hours listening to modified speech in the lab, every day at home the children played computer games that used processed speech.

As the children progressed, the program moved them from ultra-drawn-out phonemes through progressively less drawn-out ones, until the modified speech was almost identical to normal speech. The results startled even the scientists. After only a month, all the children had advanced two years in language comprehension. For the first time in their life, they understood speech as well as other kids their age.

"So we had these great results from a small group of kids," Steve Miller says. "But when Paula went to a conference in Hawaii, people jumped all over her, screaming that we couldn't make these results public. They pointed out that we had no controls: how did we know that the language improvement didn't reflect simply the one-on-one attention the kids got, rather than something specific to the modified speech?" Merzenich was hugely offended. He was itching to get the results to people who would benefit from them. But he agreed to keep quiet. "Sure, we had these great results with seven kids," says Bill Jenkins. "But we knew no one would believe it. We knew we had to go back," to get better data on more children.

So they did. The following summer, they held Camp Rutgers II. For twenty days, twenty-two SLI kids aged five to nine played CD-ROM games structured to coax the brain into building those absent phonological representations. One game, for instance, asked the child to "point to rake" when pictures of a lake as well as a rake were presented, or to click a mouse when a series of spoken *g*'s was interrupted by a *k*. At first, the computer voice stretched out the target sounds: *rrrrrake*. The usual 0.03-second (30-millisecond) difference between *day* and *bay*, for instance, lasted several times that long. The modified speech seemed to be recruiting neurons to make progressively faster and more accurate distinctions between sounds. When a child mastered the difference between *ba* and *pa* when the initial phoneme was stretched to 300 milliseconds, the software shortened the transition to, say, 280 milliseconds. The goal was to push the auditory cortex to process faster and faster phonemes. The kids also took home books like *The Cat in the Hat* on tape, recorded in processed speech. Again the results were striking: a few months after receiving twenty to forty hours of training, all the children tested at normal or above in their ability to distinguish phonemes. Their language ability rose by two years. Although the research did not include brain scans, it seemed for all the world that Fast ForWord (as the program was now called) was doing something a bit more revolutionary than your run-of-the-mill educational CD: it was rewiring brains.

In January 1996, the Rutgers and UCSF teams reported their results in the journal *Science.* Modified speech, they concluded, had altered the children's brains in such a way that they could now distinguish phonemes and map them correctly onto written words. Just as Greg Recanzone showed that when monkeys pay attention to a frequency-discrimination task their auditory cortex changes and their ability to hear tiny differences in tone improves, so SLI children who had received intensive training in discriminating acoustically modified phonemes seemed to have undergone cortical reorganization in the region of the brain that carries out auditory processing. "You create your brain from the input you get," says Paula Tallal.

"We realized we had a tiger by the tail," says Jenkins. Merzenich fretted that once the results were out, everyone with a language-impaired child would want it. He was right. The work was covered in newspapers and magazines, and in just ten days some 17,000 people had jammed the Rutgers switchboard, blowing out the e-mail and phone systems, in an effort to get hold of the miraculous CD-ROM that seemed to conquer dyslexia. Desperate parents awakened Merzenich at 2 A.M. (his phone number was listed), imploring him to help their children. Representatives from the venture capital firm E. M. Warburg, Pincus & Co. descended on Tallal's lab to figure out whether she had the basis for a profit-making business. Fellow scientists were appalled. Some even suspected that the Rutgers/UCSF team had manufactured all the media interest, as if in a replay of the cold fusion claims of the 1980s.

A Rutgers regent offered advice on how they might license the CD-ROM, but Merzenich, who had been on the team that developed the cochlear implant for hearing loss, was convinced that if you license a scientific discovery, "you lose all control over it." He and Bill Jenkins discussed the dilemma endlessly. "We were afraid that if we just licensed the software to Broderbund or the Learning Company or something they wouldn't understand the scientific complexity of it, and wouldn't implement it right," says Jenkins. "And if that happened, the opportunity would be lost. We wanted to make sure the science got properly translated." So the month after the *Science* publication, Merzenich, Paula Tallal, Bill Jenkins, Steve Miller, and recruits from the business world raised enough private financing to form Scientific Learning Corp., the first company dedicated to making money from neuroplasticity.

Merzenich told colleagues that forming a business was the only way to get the benefits of neuroplasticity out of the lab and into the hands—or brains, actually—of the people it could help. When Ed Taub once expressed frustration about how slow the rehabilitation community was to embrace constraint-induced movement therapy for stroke, Merzenich responded that only the profit motive was strong enough to overcome entrenched professional interests and the prejudice that the brain has lost plasticity after infancy. By October 1996 Merzenich and his partners had secured venture capital funding from E. M. Warburg, and the next month Scientific Learning conducted its first public demonstration of Fast ForWord, at the annual meeting of the American Speech-Language-Hearing Association. "No one would be using Fast ForWord if there were not a commercial force driving it into the world," Merzenich said four years later. "The nonprofit motive is simply too slow."

By unleashing the force of commercialism, Merzenich is convinced, Fast ForWord reached more children than it would have if he and Tallal had simply sung its praises from the offices of their ivy-walled universities. In October 1997 19 schools in nine districts across the country participated in a pilot program using Fast ForWord, enrolling more than 450 students with specific language impairment. Within four years some 500 school systems had learning specialists trained to use Fast ForWord (soon renamed Fast ForWord Language), and by the year 2000 25,000 SLI children had practiced on it for at least 100 minutes a day, five days a week. Once the children master recognition of the stretched-out phonemes, the program speeds them up until eventually the children are hearing ordinary speech. After about four weeks, the kids can process phonemes pronounced at normal speed. After six to eight weeks, "90 percent of the kids who complete the program have made 1.5 to two years of progress in reading skills," says Tallal. Scientific Learning itself graduated, too, financially speaking: in July 1999 it announced its initial public offering. Anyone who believed in neuroplasticity, and its power to turn a profit, could now ante up.

Scientific Learning has hardly won universal acceptance. Critics say it is too expensive for most schools. Some also say the system is being rushed to market before its stunning claims have been proved in independent tests. The claim that Fast ForWord reshapes the brain has been the target of the most vituperation. In one representative comment, Dr. Michael Studdert-Kennedy, past president of the Haskins Laboratories, a center for the study of speech and language at Yale University, told the *New York Times* in 1999 that inducing neuroplasticity was "an absurd stunt" that would not help anyone learn to read.

Yet only a year later, researchers reported compelling evidence that Fast ForWord changes the brain no less than Taub's constraint-induced movement therapy or Merzenich's monkey training does. Merzenich, Tallal, and colleagues had teamed up with John Gabrieli of Stanford University to perform brain imaging on dyslexic and normal adults. For the first time, brain scans would be used to look for changes that accompanied the use of their learning program. Using functional magnetic resonance imaging (fMRI), the researchers first ascertained that their eight adult dyslexics and ten matched controls differed in their processing of rapid acoustic stimuli. When they heard rapid, computer-generated nonsense syllables (designed to mimic the consonant-vowel-consonant pattern of English, but without being real words), the brains of nine of ten normal readers showed greater activation (compared to that triggered by slower sounds) in the left prefrontal region of what are called Brodmann's areas 46/10/9. Only two of eight dyslexics showed such left prefrontal activity when they heard rapid acoustic signals. The highly significant difference between the groups strongly suggests that the response in that region had been disrupted in the dyslexics. Since this area is thought to be responsible for the processing of staccato sounds, its lack of activity could explain dyslexics' inability to hear those sounds.

The researchers next performed fMRIs on the three dyslexics who had undergone Fast ForWord training for 100 minutes a day, five days a week, for about thirty-three sessions. Two of the three showed significantly greater activation in this left prefrontal region. These were also the subjects who showed the greatest improvements, after Fast ForWord, in their processing of rapid auditory signals and language comprehension. (The dyslexic whose brain showed no such change also did not improve on auditory processing.) Training in rapid acoustic discrimination can apparently induce the left prefrontal cortex, which is normally attuned to fast-changing acoustic stimuli but is disrupted in dyslexics, to do its job. The region, even in adults, remains "plastic enough . . . to develop such differential sensitivity after intensive training," the scientists concluded.

The discovery that modified speech can drive neuroplasticity in the mature brain is just the most dramatic example (so far) of how sensory stimuli can rewire neuronal circuits. In fact, soon after Merzenich and Tallal published their results, other scientists began collecting data showing that, as in my own studies of OCD patients, brain changes do not require changes in either the quantity or the quality of sensory input. To the contrary: the brain could change even if all patients did was use mindfulness to respond to their thoughts differently. Applied mindfulness could change neuronal circuitry.

It seemed to me that if the mindfulness-based Four Steps had any chance of finding applicability beyond obsessive-compulsive disorder, its best hope lay in Tourette's syndrome. Recent evidence indicates that this disease strikes about 5 people per 1,000. Although its precise cause remains to be worked out, Tourette's has a strong genetic component. Drs. James Leckman and David Pauls of Yale University had shown in 1986 that there is a biological link between OCD and Tourette's, such that the presence of Tourette's in a family puts relatives at risk of OCD. But I was interested in a different common characteristic of the two diseases. The defining symptoms of Tourette's are sudden stereotypical outbursts, called tics. They include vocalizations such as grunting or barking or spouting profanities, as well as muscle movements such as twitches and jerks of the face, head, or shoulders. These echo the compulsions of OCD, but there is more: the motor and vocal tics that characterize Tourette's usually have a harbinger, a vague discomfort that patients often describe as an irresistible urge to perform the head jerk, to utter the profanity. The more the patient suppresses the tic, the more insistent the urge becomes, until the inevitable surrender brings immediate (albeit temporary) relief. The similarities to OCD are obvious: Tourette's patients suffer a bothersome urge to twitch their face, blink their eyes, purse their lips, sniffle, grunt, or clear their throat in much the same way as OCD patients feel compelled to count, organize, wash, or check.

The two diseases also seem to share a neural component. The symptoms of Tourette's apparently arise from impaired inhibition in the circuit linking the cortex and the basal ganglia—a circuit that is also impaired in OCD. The basal ganglia, you'll recall from Chapter 2, play a central role in switching from one behavior to another. Impairment there could account for the perseveration of obsessions and compulsions, as well as the tics characteristic of Tourette's.

The first case study of this disease appeared in 1825, with a description of one Marquise de Dampierre. "In the midst of a conversation that interests her extremely, all of a sudden, without being able to prevent it, she interrupts what she is saying or what she is listening to with bizarre shouts and with words that are more extraordinary and which make a deplorable contrast with her intellect and her distinguished manners," reads the translation of the account in *Archives Générales de Médecine*. "The words are for the most part gross swear words and obscene epithets and, something that is no less embarrassing for her than for the listeners, an extremely crude expression of a judgment or of an unfavorable opinion of someone in the group." Sixty years later, Georges Gilles de la Tourette took up "the case of the cursing marquise," identifying it as the prototypical example of what he called *maladie des tics*. The disease was given its current name in 1968 by the psychiatrist Arthur Shapiro and his wife, the psychologist Elaine Shapiro. In the search for a cause, physicians suspected patients' families of inflicting early psychological trauma, and patients themselves were blamed for a failure of will.

The disease's mysterious cause inspired a wide range of treatments, from leeches to hypnosis and even lobotomies. More recently, physicians, suspecting that dopamine transmission in the basal ganglia circuit causes the disease, have tried to treat Tourette's with haloperidol and pimozide, drugs that block the neurotransmitter dopamine. (Uncertainty remains, however, over whether an excess of dopamine, sensitivity of dopamine receptors, basal ganglia malfunction, or some combination of these is at fault. In any case, it is likely that the genetic basis of the disease manifests itself somewhere in the dopamine system.) Drugs typically reduce tic symptoms 50 to 60 percent in the 80 percent of patients who respond at all. But dopamine blockers do not work for every patient. Worse, the drugs have big drawbacks, often producing serious side effects, even leaving some patients in a zombielike state. In fact, side effects lead up to 90 percent of patients to discontinue the drugs. Of those who stick with the regimen, inconsistent compliance is common. Many parents, concerned about the lack of information on the long-term effects of the medications on children, are understandably reluctant to keep their kids drugged. No wonder drugs have fallen from their perch as the treatment of choice.

That leaves behavioral treatment. But behavioral approaches have been hampered by—and there is no way to put this politely—lousy science. Over several decades, almost a hundred studies have investigated half a dozen behavioral therapies. They looked into *massed practice,* in which the patient performs his worst tic for perhaps five minutes at a time, with a minute of rest, before repeating the process for a total of about thirty minutes. Other studies investigated *operant conditioning,* in which parents and others are taught to praise and encourage a child when he is not performing a tic, and punish the performance of a tic. Others tried *anxiety management* (since tics seem to get worse with stress) and *relaxation training,* with deep breathing and imagery; although many patients managed to control tics during the therapy session, the improvement seldom carried over into the real world. Some studies explored awareness training, using videotapes and mirrors to make the patient realize how bad his tics were. But few of the studies included more than nine subjects, few included children, most failed to use standard measures of tics, and many piled so many behavioral interventions on top of one another that it was impossible to tease out which therapy was responsible for any observed effects. Follow-up was poor (a real problem since tics wax and wane naturally). In other words, the scientific underpinnings of this generation of behavioral therapies were so seriously flawed as to compromise their credibility.

Suspecting that Tourette's might be amenable to a mindfulness-based approach like the one that was succeeding with OCD, I began (in a nice way) jawboning researchers who might be sympathetic. In 1989, at the annual meeting of the American College of Neuropsychopharmacology (ACNP), I struck up a conversation with Jim Leckman. Jim is arguably the country's leading expert on the cause of Tourette's. On top of that, he is trained as a psychoanalyst and so is keenly interested in the mind-brain interface. After 1989 we became good friends, but it was not until the mid-1990s, at another ACNP meeting, that I began telling Jim about the broader implications of the PET data we had collected on OCD patients—implications about the power of the mind to shape the brain. Although he kept an open mind, he was decidedly skeptical. Because Jim is perhaps the most polite member of the baby boom generation, it took me almost a decade to realize just how much he had been graciously humoring me. Eventually, he became convinced that there might be clinical advantages to giving patients an active role in therapy—in the case of Tourette's, by using mindfulness to modulate the physical expression of the tics. Only then did he start to believe that my arguments were more than a lot of hot air.

Even after Jim started to come around, his boss at the Yale Child Study Center remained less than a true believer in this idea. When I visited Jim's lab in July 1998, he and his department chair, Donald Cohen, arranged for me to meet with an adolescent boy with OCD in what's called an observed interview. As Jim, Cohen, and a group of the clinical staff looked on, I did a brief, interactive, and animated overview of the Four Steps with this bright kid. Afterward, as we reviewed the clinical interaction, Cohen looked at me with an amused expression and said, "So, it seems like you managed to sell that young man on your shtick." "Well, it's not really a shtick . . ." I began. "It sounds like a shtick to me," he shot back. I tried to point out that it's not a gimmick to teach patients suffering with OCD that their intrusive thoughts and urges are caused by brain imbalances, and that we now know they can physically alter those imbalances through mindfulness and self-directed behavioral therapy techniques. Although to psychiatry professors the Four Steps of Relabel, Reattribute, Refocus, Revalue may initially seem like a shtick, we had strong scientific evidence that this approach can bring about changes in brain function. (Of course Cohen already knew all this, or I wouldn't have been at Yale in the first place.)

This seemed to calm matters down a bit. In any event, we all went out for a nice dinner. And besides, the Yale Child Study group had already done a major brain imaging study relevant to this point. Just three months before, in April 1998, in a study based on reasoning quite similar to the Four Steps approach to changing the brain circuits underlying OCD, Brad Peterson, Jim Leckman, and their Yale colleagues published important data on what happens when Tourette's patients use willful effort to suppress tic expression. The Yale group had patients undergo fMRIs while they alternated forty seconds of letting their tics be expressed with forty seconds of volitionally suppressing them. Hearing the word *now* told the volunteers when to switch—particularly, when to call up whatever reserves of will they could muster to prevent the tics from causing bodily movements. The investigators noted how brain activity changed during tic suppression compared to when tics are given free rein. The most relevant regions seemed to house the circuit involving our old friends the prefrontal cortex, anterior cingulate gyrus, basal ganglia, and thalamus. Activity in this circuit—the very one involved in OCD and in the formation of habits—was noticeably altered (activity in the caudate increased and activity in the putamen decreased) when patients willfully prevented themselves from moving in response to the intrusive bothersome urges of Tourette's. The study also found that the worse the tics, the less the basal ganglia and thalamus activity change when the tics are suppressed.

This finding is quite consistent with the notion of a "brain lock" in Tourette's, which may be similar to that in OCD patients. You'll remember from Chapter 2 that prior to cognitive-behavioral treatment the brain structures of the OCD circuit—the orbital frontal cortex, anterior cingulate gyrus, caudate, and thalamus—showed such high correlations in their activity they seemed to be functioning in lockstep. This same circuit also seems to be "locked up" in patients with Tourette's. As the Yale researchers put it, "A failure to inhibit tics may result from an impaired ability to alter subcortical neuronal activity." Thus in Tourette's, as in OCD, the gearshift of the basal ganglia seems to be locked. As it happened, my UCLA colleague John Piacentini was in the middle of a study designed to gauge whether mindful awareness and directed mental force could help unfreeze this jammed transmission.

In August 2000 Jim Leckman was in Los Angeles attending to family matters, so he, John Piacentini, and I got together. Piacentini had just put together the data from his ongoing study using a cognitive-behavioral approach incorporating mindfulness to treat children with Tourette's. This new approach had been designed with an eye toward combining classical behavioral techniques for treating tics with the mindfulness component of the Four Steps. The key was to make patients understand that tics are an expression of a biological brain malfunction, much as the Four Steps makes OCD patients aware that their obsessions and compulsions originate in an overactive brain circuit. This new treatment aims to teach the patient that the behavioral response to the tic urge can be modified so that both the functional impairments (social and otherwise) and the physical damage to joints and muscles are reduced. After all, tics can be painful.

As Piacentini explained it to Jim Leckman and me, he asks each child to describe his or her tic occurrences in detail and reenact them in front of a mirror. Piacentini points out a tic if one occurs during a session. He also teaches the patient to identify the situations when a tic is most likely to recur, to recognize the very first glimmerings of the urge to tic, and to enhance that awareness by labeling it with the verbal or mental note *t:* as soon as the child feels a tic coming on, he says *t* to himself.

But the distinctive ingredient is training patients to develop what are called *competing responses.* Then, every time the urge to tic arises, that urge is paired with a behavior designed to modify the expression of the urge in order to control it better. If it is a verbal tic, John teaches the patient to breathe through the nose, slowly; that makes it physically impossible to bark out a curse. If it is a motor tic, John coaches him to hold his arm close to his body, tense the neck muscles, or slowly open and close his eyes—activities that preclude wild arm swings, head jerks, or fast blinking, respectively. In the really creative part of the therapy John teaches the patient attenuated behavior, such as moving the arm slowly and wiping his brow to make the movement more volitional and controlled. The strategy has a good deal in common with directing OCD patients to Refocus attention away from a pathological compulsion and onto a healthy behavior. "What you want to do is substitute voluntary, controlled movement for the involuntary tic," says Piacentini. "You need to be able to recognize the onset of the urge to tic, pay attention, and be motivated. It's similar to the Four Steps approach to OCD. Patients are trained to recognize and label tic urges consciously and then either try to resist these urges, or else respond in a controlled and attenuated way. Most youngsters are eventually able to significantly diminish and/or eliminate targeted tics."

When Leckman, Piacentini, and I sat down to look at John's preliminary data, it was clear he had something. Twenty-four Tourette's children, aged seven to seventeen, had enrolled in the study. Piacentini had divided the kids into two groups. In one, the children practiced recognizing when an urge to tic arose. In this stage, analogous to the Relabel and Reattribute parts of the Four Steps, they would realize that the urge to tic had arisen and give it that label *t*. "This is the urge to tic." This is called *awareness training*. In the other group, Piacentini combined awareness training with habit modulation, in which the child is taught to respond to the urge to tic in a safe way by, for instance, executing a less intense movement. This is analogous to the Refocus step. Seventeen of the children completed the eight-week program. The assessors, gauging the children's tic severity, were blind to which treatment group each child had been assigned to.

The results were striking. Patients receiving awareness training alone had an approximately 10 percent improvement in tic severity. But those also receiving habit modulation training had a 30 percent reduction in tic severity and a 56 percent improvement in tic-related impairment. "Now it is being accepted that you can use behavioral intervention to treat a biologically-mediated disease," Piacentini says. Although John has not yet done before-and-after brain scans to ascertain whether the children's clinical improvement is accompanied by brain changes of the kind we detected in OCD patients, it is quite likely that brain changes analogous to those we found in OCD were occuring.

Innumerable studies have now shown that the mind can affect the body: mere thoughts can set hearts racing and hormones surging. Although mind-body medicine is usually understood as the mind's effect on the body from the neck down, the power of the Four Steps to remodel neuronal connections—strengthening those underlying healthy habits and inhibiting those between the frontal cortex and basal ganglia (the OCD circuit) underlying pathological ones—strongly suggests that the mind can also affect the brain. In 1997, colleagues who knew of my interest in mindfulness told me about the work of John Teasdale at the Medical Research Council Cognition and Brain Sciences Unit in Cambridge, England. They said that Teasdale, working with Mark Williams and Zindel Segal, seemed to be harnessing exactly this power of the mind, but to treat depression: he proposed that patients would lower their risk of falling back into clinical depression if they learned to experience their depressive thoughts "simply as events in the mind." That, of course, is a hallmark of the Impartial Spectator and mindful awareness. Teasdale and colleagues suspected that this perspective would diminish the power of so-called triggering cues to tip someone into a depressive episode. Just as my OCD patients learned to recognize intrusive thoughts as the manifestation of their brain's misbehavior, so Teasdale's depressives, the researchers thought, could learn to prevent a relapse by processing emotional material in a new way. By 1995, they were was boldly using *mindfulness* in the titles of their research papers, and in 2000 Teasdale named his approach *mindfulness-based cognitive therapy.*

Depression is often a chronic disorder, characterized by frequent relapses. Over a lifetime, a patient has an 80 percent chance of suffering a recurrence and, on average, experiences four major depressive episodes lasting twenty weeks each. Antidepressants are the most widely used treatment. But in the 1990s, studies had begun to suggest that cognitive therapy, too, had the power to prevent relapses. In one study of 158 patients who had been treated with only partial success by antidepressants, some received cognitive therapy as well as drugs for their remaining symptoms. The rest received only the medication. The difference in relapse rates over the sixty-eight-week period of the study was significant: patients undergoing cognitive therapy experienced a 40 percent reduction in their rate of relapse compared to the drugs-only group.

Clearly, cognitive therapy helps prevent depressive relapse, and Teasdale thought he knew why. Other studies were showing that people are at greatest risk for becoming depressed when a sad, dysphoric, or "blue" mood produces patterns of negative, hopeless thinking. What distinguishes people who get depressed from those who don't, Teasdale suspected, may be this: in the first group, dysphoria, that down-in-the-dumps feeling that most of us experience at least once in a while, triggers "patterns of depressogenic thinking" powerful enough to trigger full-blown depression. In these patients, dysphoria that may make a healthy person feel kind of blue plunges depression-prone patients into a well of despair. They ruminate on their perceived deficiencies and on the hopelessness of life. Although in a healthy person a bad day at work or a disastrous date may induce some passing sadness, in someone susceptible to depression it can readily escalate to "I'm completely incompetent and life is pointless." In these vulnerable people, a dysphoric thought or experience seems to trigger the onslaught of depression much as a burning ember kindles a brushfire: the sad thought ignites a conviction that one is pathetic and worthless, or that one's current problems are irreparable and eternal. In relapsing depressives, this connection becomes so habitual and automatic that they "keep the system 'stuck' in repetitively generating [depressogenic thoughts]." It follows, then, that the risk of relapse can depend on how easily sadness tips someone into the self-perpetuating biological imbalances that characterize major depression.

To Teasdale, the corollary to this hypothesis was clear. To prevent the recurrence of depression, it may be sufficient for a patient to process her emotions in a new way, a way that does not trigger the thoughts and mood states characteristic of a depressive episode. That is, it may be enough to find a way to disrupt the automatic segue from sadness to sickness—and the pathological brain states associated with it. With the right therapy, thoughts and feelings that once tipped the person into full-blown depression would instead become "short-lived and self-limiting," Teasdale suggested. His proposed therapy would change the very way patients think about their thoughts.

If Teasdale's hunch seems reminiscent of my work with OCD patients—in which the urge to perform some compulsive act still intrudes but is thwarted when patients think about their thoughts and feelings differently—well, so it seemed to me, too. Studies at a handful of labs were already demonstrating the power of cognitive therapy over depression. When patients were presented with a scenario such as "You go out with someone and it goes badly" and were then asked how much they agreed or disagreed with such statements as "My value as a person depends greatly on what others think of me," patients receiving only drugs were more likely to sink into dysfunctional thinking than those who also received cognitive-behavioral therapy. (The extent to which dysfunctional thinking follows dysphoria predicts the likelihood that a patient will suffer a relapse of depression.) That patients receiving cognitive-behavioral therapy are better able to resist being plunged into despair by sad thoughts suggests that this therapy changes emotional processing—the way people think about their feelings—in ways that prevent dysphoria from triggering full-blown depression. Such research, Teasdale concludes, "suggests that emotional processing should focus primarily on changing emotional responses to internal affective events and thoughts, so that these responses are short-lived and self-limiting, rather than the first stages of an escalating process."

Teasdale then began to construct a therapy to achieve this. Through mindfulness-based cognitive therapy, he wanted to make patients more aware of their thoughts. In particular, he wanted them to recognize that sadness can (through a brain-based biological mechanism) escalate into depression. To prevent it from doing so, they would learn to meet the onset of dysphoria with such responses as "Thoughts are not facts" and "I am not my thoughts." Or as Teasdale puts it, they would learn to prevent what he calls *depressive interlock* (again, reminiscent of my "brain lock"): the strong, physical connection between unhappy thoughts and the memories, associations, and modes of thought that inflate sadness into depression. To do that, the therapist needs to help patients encode in memory alternative thought patterns that can be activated by the very same cues that otherwise tap into the despairing ones.

When I read this, in 1999, I was thrilled. Finally, I thought, I've found a kindred spirit in this profession. This guy is actually using mindfulness to help patients to see the passing, ephemeral nature of their depressive thoughts. Teasdale's proposed treatment also gave me a sense of déjà vu. Healthy emotional processing prevents the dysphoria from triggering global thoughts of hopelessness and self-worthlessness. It instead activates alternative memories and associations, so that the next time the patient encounters something that makes her sad she reacts not with despair but by calling up other, healthier associations. To me, this was reminiscent of OCD patients learning to respond to the compulsive urge to wash by deciding instead to crochet or garden—that is, by Refocusing. As Teasdale put it, "The new schematic models rather than the old will be accessed and these new models will determine emotional response." Much as I had, he was proposing that patients could learn to weaken the physical connections to the old, pathological schema—habitual way of thinking—and strengthen those to a new, healthier one. And as with the Four Steps approach, mindfulness was to be the key.

How, then, to apply mindfulness to depression? Teasdale identified three ways that depressives can process emotion-laden thoughts. They can mindlessly emote, or allow themselves to be engulfed by their feelings with little self-awareness or reflection. Patients who respond this way typically have poor outcomes to psychotherapy. Alternatively, patients can engage in "conceptualizing/doing." By this, Teasdale means having impersonal, even detached thoughts about the self, about depression, and about its causes and consequences. Conceptualizing/doing lacks the introspection inherent in mindfulness. Depressed patients who think this way also tend not to do well in therapy.

The third option is what Teasdale named "mindful experiencing/being." In this way of thinking about your emotions, you sense feelings, sensations, and thoughts from the perspective of the Impartial Spectator. You regard your thoughts and feelings as passing, ephemeral "mental events" rather than as accurate reflections of reality. Instead of reacting to negative thoughts and feelings as "these are me," you come to regard them as "events in the mind that can be considered and examined." You recognize that thoughts are not facts (just as my OCD patients learned that their obsessions are only their brain's causing their mind to misbehave) but are instead "events that come and go through the mind," as Teasdale explains it. Mindfulness gives patients the attentional skills that allow them to disengage from, and focus instead on alternatives to, the dysfunctional ways of thinking that trigger a relapse of their depression. Teasdale had independently constructed, and taken the first steps toward proving, a model of depression much like my model of OCD.

In a landmark paper in August 2000, Teasdale and his colleagues reported the results of his yearlong study on using mindfulness to prevent the relapse of depression, offering strong support for the findings in OCD patients that mindfulness can alter brain circuits. Using an approach pioneered by the American psychologist Jon Kabat-Zinn, Teasdale had his patients participate in two-hour group sessions once a week for eight weeks, receiving mindfulness training through tape-recorded instructions that taught them to direct their attention to specific regions of the body in succession. The goal was to become acutely aware of whatever sensations an arm, a cheek, a knee was experiencing at the moment. The patients then learned to focus on their breathing. If the mind wandered, patients acknowledged the distractions with "friendly awareness"—that is, not with frustration or anger—and learned to return calmly to a focus on the breath. Repeating this process over and over, patients learned to use their inhalations and exhalations as an anchor to pull them back to a mindful awareness of the present moment. The patients also had homework, including exercises designed to increase their moment-by-moment awareness of feelings, thoughts, and sensations and to allow them to view thoughts and feelings (particularly negative ones) as merely passing events in the mind and brain.

The results were impressive. Of the 145 patients from ages eighteen to sixty-five, who had suffered at least two episodes of major depression within the last five years, about half were randomly assigned to receive the standard treatment and half to receive the mindfulness training, too. All had been off antidepressants for at least the previous twelve weeks, long enough to clear the drugs from their system. Over the sixty-week study period (eight weeks of treatment then fifty-two weeks of follow-up), among patients who had suffered at least three episodes of major depression there was a 44 percent reduction in the rate of relapse among those who received mindfulness therapy compared to those receiving standard therapy. Adding mindfulness, then, cut the rate of relapse by almost half. This was the first demonstration that a mindfulness-based psychological intervention can reduce the rate of relapse in depression.

The will, it was becoming clear, has the power to change the brain—in OCD, in stroke, in Tourette's, and now in depression—by activating adaptive circuitry. That a mental process alters circuits involved in these disorders offers dramatic examples of how the ways someone thinks about thoughts can effect plastic changes in the brain. Jordan Grafman, chief of cognitive neuroscience at the National Institute of Neurological Disorders and Stroke, calls this top-down plasticity, because it originates in the brain's higher-order functions. "Bottom-up" plasticity, in contrast, is induced by changes in sensory stimuli such as the loss of input after amputation. Merzenich's and Tallal's work shows the power of this bottom-up plasticity to resculpt the brain. The OCD work hints at the power of top-down plasticity, the power of the mind to alter brain circuitry. I suspect that when the requisite brain imaging is done with Teasdale's depressives, that research will also show the power of mind to change the brain. In fact, recent studies using a somewhat different form of psychotherapy called interpersonal therapy already have.

Sitting somewhere between purely mental events and purely sensory ones is this vast sea of life called experience. Research into how experience affects the brain is only in its infancy, but one of my favorite examples suggests where we may be heading.

One wag called the study "taxicology." When researchers at University College London decided to study how navigation expertise might change the brain, they didn't have to look far for subjects. London cabbies are renowned for their detailed knowledge of the capital's streets: to get their license, they have to pass a stringent police test assessing how well they know the fastest way from point A to point B and what streets are where. Drivers call it "being on The Knowledge," and it takes them an average of two years to learn it.

Earlier studies in small mammals, monkeys, birds, and humans had established that the right half of an unassuming little structure near the center of the brain called the hippocampus is involved in the formation of directional memories; in fact, the back of the right hippocampus seems to store a mental map of the environment. Eleanor Maguire and her colleagues at the university therefore decided to examine the hippocampi of London taxi drivers, using magnetic resonance imaging, and compare them to the hippocampi of Londoners who hadn't the faintest notion of the best way to get from Fleet and Chancery to Gresham and Noble.

Maguire scanned the brains of sixteen cabbies, aged thirty-two to sixty-two, and fifty ordinary right-handed men of the same age. Everyone's brain structures looked about the same, in both size and shape—except the hippocampus. In the taxi drivers, the back was significantly larger than it was in the other men, and the front was smaller. That might simply reflect the fact that if you're born with a big rear hippocampus, you are a navigational ace, and hence are more likely to take up hacking than if you can't tell east even at sunrise. So to see if the brain differences reflected experience, Maguire plotted the differences in the volume of the hippocampus against how experienced a driver was. There it was: the more years a man had been a taxi driver, the smaller the front of his hippocampus and the larger the posterior. "Length of time spent as a taxi driver correlated positively with volume in . . . the right posterior hippocampus," found the scientists. Acquiring navigational skills causes a "redistribution of gray matter in the hippocampus" as a driver's mental map of London grows larger and more detailed with experience.

What cabbies might be sacrificing in the front part of their hippocampus for an enlarged posterior part remains unknown, as does the mechanism for the volume changes. Although neurogenesis might explain the enlargement of the rear of the hippocampus, the London scientists have their money on an overall reorganization of the hippocampus's circuitry "in response to a need to store an increasingly detailed spatial representation." One thing, however, is clear: a key brain structure can change in response to your experience as an adult. Published in 2000, this was the first demonstration that the basic anatomy of the adult brain, not just the details of its wiring, can be altered by the demands its owner places on it.

The study of neuroplasticity began with scientists' cataloguing the changes in sensory input that induce cortical remapping and rewiring. Now, even as they add to the list of examples of neuroplasticity, researchers are also exploring the cellular and molecular mechanisms that underlie it. We know that the formation of new synapses, as a result of the growth of existing axons or dendrites, is involved in both the remodeling of circuits and cortical remapping. A change in the quantity of available neurotransmitters, or the enzymes that regulate them, can also foster plasticity. But now researchers are examining a mechanism that had long been dismissed as an avenue to plasticity: the actual creation of new neurons. Although a slew of animal experiments had demonstrated that new synapses can form when the animal is exposed to an "enriched" environment, that was one step short of showing that new neurons, as opposed to new connections between neurons, were being born.

That changed in 1997. Fred Gage and colleagues at the Salk Institute in La Jolla, California, placed adult mice in an "enriched" environment (one that resembles the complex surroundings of the wild more than the near-empty cages of the rats in the "nonenriched" environment do). By the end of the experiment, the formation and survival of new neurons had increased 15 percent in a part of the hippocampus called the dentate gyrus. These animals also learned to navigate a maze better. In 1999 Elizabeth Gould of Princeton University used similar techniques in adult rats to demonstrate that the creation of new neurons, called neurogenesis, was not a talent lost in infancy: the increased neurogenesis, she found, is directly related to learning tasks that involve the hippocampus. Also in 1999, Gage showed again that new neurons grow in the hippocampus of adult mice as a result of exercising on a wheel, and in 2001 Gould and colleagues demonstrated that newly generated neurons are "associated with the ability to acquire . . . memories."

"Neurogenesis was a hard thing for scientists to come to grips with," said Gage. But by the new millennium it was clear that new neurons arise from stem cells, immature cells capable of differentiating into virtually any type of cell. There is now abundant evidence that neural stem cells persist in the adult brain and support ongoing neurogenesis. And the evidence is no longer confined to mice. In 1998, Peter Eriksson of Goteborg, Sweden, working with Gage, demonstrated that neurogenesis occurs in the adult human hippocampus. Thus Gage's and Gould's discoveries suggest that the possibilities for neuroplasticity are greater than even diehard believers thought: the brain may not be limited to working with existing neurons, fitting them together in new networks. It may, in addition, add fresh neurons to the mix. The neural electrician is not restricted to working with existing wiring, we now know: he can run whole new cables through the brain.

Neuroplasticity has come a long way since Nobel laureates ridiculed Mike Merzenich for his audacity in claiming to have shown that the mature central nervous system has the capacity to change. Even in the early 1990s neuroplasticity was viewed as, at best, an interesting little field. By the middle of the decade, however, it had become one of the hottest topics in neuroscience, and as the decade ended, hundreds of researchers had made it the focus of their studies. "If you had taken a poll of neuroscientists in the early 1990s, I bet only 10 to 15 percent would have said that neuroplasticity exists in the adult," Merzenich says. "Even by the middle of the decade the split would have been 50–50. What changed that was the human experiments" like Taub's. Now there is no question that the brain remodels itself throughout life, and that it retains the capacity to change itself as the result not only of passively experienced factors such as enriched environments, but also of changes in the ways we behave (taking up the violin) and the ways we think ("That's just my OCD acting up"). Nor is there any question that every treatment that exploits the power of the mind to change the brain involves an arduous effort—by patients afflicted by stroke or depression, by Tourette or OCD—to improve both their functional capacity and their brain function.

We began our discussion of neuroplasticity by quoting from the Spanish neuroanatomist Santiago Ramón y Cajal's description of the "nerve paths" in the adult brain as "fixed" and "immutable." It seems only right, then, to close with another passage from Cajal, who despite his pessimism about the seeming lack of malleability in the adult brain nevertheless saw a glimmer of hope: "It is for the science of the future to change, if possible, this harsh decree. Inspired with high ideals, it must work to impede or moderate the gradual decay of the neurones, to overcome the almost inevitable rigidity of their connections." The science of the future has arrived. And in what may be the most remarkable form of neuroplasticity, scientists are seeing glimpses that internal mental states can shape the structure and hence the function of the brain. Faced with examples of how the brain can be changed—from Taub's stroke patients and Piacentini's Tourette's patients to Teasdale's depressives and Merzenich's dyslexics—I had become more convinced than ever that such self-directed neuroplasticity is real. It was time to explore how attention in general, and wise attention—mindfulness—in particular, wields its power.

{EIGHT}

THE QUANTUM BRAIN

Anyone who is not shocked by quantum theory has not understood it.
—*Niels Bohr*

I am now convinced that theoretical physics is actual philosophy.
—*Max Born,* My Life and My Views

Science no longer is in the position of observer of nature, but rather recognizes itself as part of the interplay between man and nature. The scientific method . . . changes and transforms its object: the procedure can no longer keep its distance from the object.
—*Werner Heisenberg, 1958*

In the autumn of 1997, I was laboring to rework my second book. One of the joys of my life, *A Return to Innocence* is based on a series of letters I exchanged with a sixteen-year-old named Patrick on how to use mindfulness to cope with the raging changes in body, mind, and brain that accompany the transition from adolescence to adulthood. As respite, I drove up to Santa Cruz that October to visit the forty acres of redwood forest I had been fortunate enough to acquire in the nearby town of Boulder Creek several years before. During the drive north, I tried to work through what I considered a key reason for the disagreement between the materialist-reductionist view I abhorred and the view of those who, in the philosopher David Chalmers's words, "take consciousness seriously." That disagreement, I was more and more convinced, largely turned on the differing perspectives the two sides took on a deceptively simple question: what counts as primary, or irreducible, data? To those who take consciousness seriously, it is not only permissible but even vital to count subjective phenomena—phenomena such as what consciousness feels like from the inside. Such subjective phenomena have not been reduced to phenomena of a different type. The feeling of pain, the feeling of red (as discussed in Chapter 1), the subjective experience of sadness—none has in any way been convincingly explained away as the mere by-product of the firings of neurons and the release of neurotransmitters. In 1997, no less than now, it was beginning to look as if maybe they never would be. From my perspective, they never could.

In a book published the year before, *The Conscious Mind: In Search of a Fundamental Theory*, Chalmers had advanced a thesis consistent with this view. He argued that contemporary neuroscience doesn't even begin to explain how subjective experience arises from the electrochemical chatter of brain neurons. Such reductionist explanations fall so far short of the goal, Chalmers maintained, that the whole enterprise of seeking an explanation of consciousness solely in the stuff of the brain was fatally flawed, and therefore doomed from the start. Instead, Chalmers suggested that conscious experience is inherently and forever irreducible to anything more "basic"—including anything material. Perhaps it should instead be understood in an entirely new way, he said: perhaps conscious experience is an irreducible entity, like space, or time, or mass. No member of that triad can be explained, much less understood, as a manifestation of either of the other two. Perhaps conscious experience, too, is such a fundamental.

Yet Chalmers also believed in *causal closure,* the idea that only the physical can act on the physical: if a phenomenon is nonphysical (like consciousness), then it is powerless to affect anything made out of the tissues, molecules, and atoms of the physical (like a brain). Chalmers, good epiphenomenalist that he was, accepted that consciousness is real and nonreducible but believed that it is caused by the brain and cannot act back on anything material. (Within a couple of years, as we saw in Chapter 1, Chalmers had modified this last part.) I, of course, disagreed vehemently with the idea of causal closure. My OCD data, not to mention twenty-five years of practice in mindfulness meditation, had persuaded me that the nonphysical entity we call the mind has the power to change the brain. Just before I left L.A., as I was struggling to find a way to explain my notions of mindfulness in *A Return to Innocence,* I hit on the idea of *mental force*. It was little more than a literary device, really, a way to convey the notion that through intense effort we can resist our baser appetites. And although I believed in the concept in a spiritual sense, I wasn't yet thinking about whether such a mental force might have a physical reality.

That Chalmers did not subscribe to diehard materialism—and was quite a nice guy—had led to a budding philosophical friendship over the past year. A few weeks before, I had sent the first e-mail of my life—to Chalmers. "I'm really just checking to see if there is any consciousness on the other end of this," I typed, full of doubts about this newfangled toy. Almost immediately Dave replied, "Consider it affirmed: consciousness exists." So during my trip to Santa Cruz I arranged to drop in, with no firmer plans than to shoot the breeze and catch up with Dave. Which is how I came to be sitting with him on his porch that Sunday afternoon, overlooking Santa Cruz, and discussing all things mind-brain. Over a couple of beers, I began lamenting the terrible social consequences of materialism and my view that the less-than-laudable moral condition of America in general and Santa Cruz in particular (I was grumpy from overwork and have never been particularly enamored of the moral condition of Santa Cruz in any event) could be laid at the feet of nearly three centuries of materialist ascendancy. The reigning belief that the thoughts we think and the choices we make reflect the deterministic workings of neurons and, ultimately, subatomic particles seemed to me to have subverted mankind's sense of morality. The view that people are mere machines and that the mind is just another (not particularly special) manifestation of a clockwork physical universe had infiltrated all our thinking, whether or not someone knew a synapse from an axon. Do you know what the most addressable cause of all this moral decrepitude is?, I asked Dave. Materialism! Not the materialism of Rodeo Drive, SUVs, and second homes in Telluride, but materialism as a worldview, a view that holds that the physical is all that exists, and that transcendent human mental experiences and emotions, no matter what grandeur they seem—from within—to possess, are in reality nothing but the expressions of electrical pulses zipping along neurons. Chalmers wouldn't be the first (or the last) to express incredulity that I was blaming the moral morass of the late twentieth century on a school of philosophy that most people had never heard of. Still, there was a hint of sympathy in Chalmers's voice as he asked, "Do you really believe that?"

I did. Chalmers and I then tossed around the names of scholars who might be positing tenable, scientifically based alternatives to materialism. One was David Hodgson. Like Chalmers an Aussie, and a justice on the Australian Supreme Court, he had written *The Mind Matters: Consciousness and Choice in a Quantum World*, published in 1991. Although it may seem odd for a jurist to focus on something as abstruse as materialism and consciousness, materialism clearly poses a bit of a problem for a central tenet of the justice system—namely, that people exert free will in their actions, including their criminal actions. If actions are merely the inevitable consequences of hard-wired brain circuitry—or, pushing the chain of causation back a step, of the genes we inherit from our parents—then the concept of genuine moral culpability becomes untenable. The second researcher whose work suggested an alternative to materialist reductionism, said Chalmers, was a physicist. His name was Henry Pierce Stapp.

When I returned to Los Angeles, I got hold of a copy of Hodgson's book, an imposing volume with a very long section on quantum physics smack in the middle. In a key passage on free will, Hodgson mentioned this same Stapp. A quick search on Amazon.com turned up a collection of Stapp's papers, which I borrowed from the UCLA physics library. After spending a night that stretched into dawn with the collection, I purchased Stapp's 1993 book *Mind, Matter and Quantum Mechanics* that February. The physics of mind-brain relations expounded in his book, I was startled to see, echoes William James's theories, especially James's belief in the reality of will and the efficacy of mental effort. Although hugely influential in the late nineteenth and early twentieth centuries, James's ideas on the power of mind and the existence of free will fell into disrepute with the rise of behaviorism. Why? Well, who can nail down exactly what forces conspire to nourish a radically new scientific proposal or smother it at birth? James himself despaired of the possibility of scientifically demonstrating the efficacy of will. "The utmost that a believer in free-will can ever do will be to show that the deterministic arguments are not coercive," he wrote. He knew that his theories of psychology—in particular his idea that "the feeling of effort" is an "active element which . . . contributes energy" to bodily action—failed to mesh with classical physics. In contrast, because the behaviorists' theories rested on the classical, deterministic physics of the late nineteenth and early twentieth centuries, they easily trumped James's. The behaviorist paradigm of John Watson held out the promise of a science of psychology in which researchers would discover the rules that govern why humans act, think, and emote as they do, much as they discovered the rules of electricity or hydrology. Behaviorism denies the reality of thoughts and emotions—indeed, of any sort of inner life. Instead of being afraid of something, it claims, we exhibit "a conditioned fear response"; instead of loving someone, we show "conditioned love responses."

Stapp suspected that James's theories on mind and brain had been ignored largely because the physics of James's time—the classical, deterministic physics of Newton—not only failed to support his argument but even undermined it, casting it as so much mysticism. For James's ideas to gain traction, then, they had to await not developments in neuroscience or psychology, but a revolution in physics. For as James himself realized, what he was saying about mind in general and will in particular—namely, that "the brain is an instrument of possibilities, not certainties," and that "consciousness . . . will, if endowed with causal efficacy, reinforce the favorable possibilities and repress the unfavorable or indifferent ones"—directly contradicted the materialist perspective of the late nineteenth century.

In fact, James's perspective on mind and brain is thoroughly modern, Stapp had observed in his 1993 book. James's theories, Stapp argued, are actually more modern than those of the psychologists and philosophers who dominated the field in the decades after James's death in 1910. Indeed, the consistency between James's perspective on attention and will on one hand and the orthodox interpretation of quantum mechanics on the other was almost eerie. It was as if a ghost from psychology past were whispering into the ear of physics present. For once we recognize that classical Newtonian physics does not accurately describe the world and replace it with the quantum physics that does, it emerges naturally and inevitably that the mind has the power to act back on the brain just as James suggested. That makes the notion that mind is strictly determined by the movements of atoms and electrons seem as dated as Newton's powdered wig.

Classical physics held that the reality of the physical world is constituted of infinitesimal particles in a sea of space. Causation, in this scheme, reflects, at bottom, one particle's acting on its immediate neighbor, which in turn acts on its neighbor, until—well, until something happens. Wholly deterministic natural laws govern the behavior of matter. Furthermore, reality consists of material objects forever separated, by the chasm of Cartesian dualism, from the immaterial mind. This mechanistic view—stimulus in, behavior out—evolved into today's neurobiological model of how the mind works: neurotransmitter in, behavior, thought, or emotion out.

But physics in the years since James had undergone a revolution. The development of quantum physics, in the opening decades of the twentieth century, gave James's conclusions about the relationship between attention and will a grounding in physical science that they lacked during his lifetime. Although classical physics had failed to validate—had even undermined—James's theories, it had not had the last word. The very origin of the mind-brain problem lies in a physics that has been outdated for almost a century. Although biologists, as well as many philosophers, cite "the ordinary laws of physics and chemistry" as an explanation for all of the events we gather under the tent labeled "mind," the laws to which they refer "are a matter of the past," as the Nobel physicist Eugene Wigner wrote way back in 1969. "[They] were replaced, quite some time ago, by new laws"—the laws of quantum mechanics.

It has been a century since the German physicist Max Planck fired the opening shot in what would become the quantum revolution. On October 19, 1900, he submitted to the Berlin Physical Society a proposal that electromagnetic radiation (visible light, infrared radiation, ultraviolet radiation, and the rest of the electromagnetic spectrum) exists as tiny, indivisible packets of energy rather than as a continuous stream. He later christened these packets *quanta.* Planck, a new professor at the University of Berlin, had no false modesty about the significance of his new radiation formula: he told his son, Erwin, during a walk that day, "Today I have made a discovery as important as that of Newton." In a lecture to the German Physical Society on December 14, he made his proposal public. Planck viewed his quanta as mere mathematical devices, something he invoked in "an act of desperation" to explain why heated, glowing objects emit the frequencies of energy that they do (an exasperating puzzle known as the *black-body radiation* problem). He did not seriously entertain the possibility that they corresponded to physical entities. It was just that if you treated light and other electromagnetic energy as traveling in quanta, the equations all came out right. But "nobody, including himself, realized that he was opening the door to a completely new theoretical description of nature," said Anton Zeilinger, one of today's leading quantum experimentalists, on the one hundredth anniversary of Planck's talk.

The notion that electromagnetic energy exists as discrete packets of energy rather than a continuous stream became the foundation on which physicists erected what is inarguably the most successful (and strangest) theory in the history of science. The laws of quantum physics not only replicate all the successes of the classical theory they supplanted (that is, a quantum calculation produces an answer at least as accurate as a classical one in problems ranging from the fall of an apple to the flight of a spaceship). They also succeed where the laws of classical physics fail. It is quantum physics, not classical physics, that explains the burning of stars, accounts for the structure of elementary particles, predicts the order of elements in the periodic table, and describes the physics of the newborn universe. Although devised to explain atomic and electromagnetic phenomena, quantum physics has "yielded a deep understanding of chemistry and the solid state," noted the physicist Daniel Greenberger, a leading quantum theorist: quantum physics spawned quantum technologies, including transistors, lasers, semiconductors, light-emitting diodes, scans, PET scans, and MRI machines. "[T]he extent of the success of quantum theory," concluded Greenberger, "comes rather as an undeserved gift from the gods."

Yet gifts from gods, no less than gifts from crafty Greeks, can conceal unwelcome surprises. For quantum physics, in addition to predicting and explaining phenomena that range over fifteen orders of magnitude in energy, has done something else: it has triggered a radical upheaval in our understanding of the world. In place of the tidy cause-and-effect universe of classical physics, quantum physics describes a world of uncertainties, or indeterminism: of limits to our knowledge. It describes a world that often seems to have parted company with common sense, a world at odds with some of our strongest intuitive notions about how things work. In the quantum world, subatomic particles have no definite position until they are measured: the electron orbiting the nucleus of an atom is not the pointlike particle we usually imagine but instead a cloud swathing the nucleus. In the quantum world, a beam of light can behave as a wave or a barrage of particles, depending on how you observe it. Quantities such as the location, momentum, and other characteristics of particles can be described only by probabilities; nothing is certain. "It is often stated that of all the theories proposed in this century, the silliest is quantum theory," the physicist Michio Kaku wrote in his 1995 book *Hyperspace*. "In fact, some say that the only thing that quantum theory has going for it is that it is unquestionably correct."

Correct it may be, but at its core quantum physics departs from classical physics in a very discomfiting way. Integral to quantum physics is the fundamental role played by the observer in choosing which of a plenitude of possible realities will leave the realm of the possible and become actual. For at its core, quantum physics challenges the ontology that permeated the scientific enterprise for centuries, the premise that a real world—independent of human choice and interference—is out there, uninfluenced by our observation of it. Quantum physics makes the seemingly preposterous claim (actually, more than claim, since it has been upheld in countless experiments) that there is no "is" until an observer makes an observation. Quantum phenomena seem to be called into existence by the very questions we ask nature, existing until then in an undefined fuzzy state. This feature of the quantum world led the American physicist John Archibald Wheeler to say that the world comes into being through our knowledge of it—or, as Wheeler put it, we get "its from bits" (bits of knowledge). The Danish physicist Niels Bohr captured the counterintuitive world that physicists were now playing in when he told a colleague, "We all agreed that your theory is crazy.

The question which divides us is whether it is crazy enough to have a chance of being correct."

The role of observation in quantum physics cannot be emphasized too strongly. In classical physics, observed systems have an existence independent of the mind that observes and probes them. In quantum physics, however, only through an act of observation does a physical quantity come to have an actual value. Many are the experiments that physicists have dreamed up to reveal the depths of quantum weirdness. I'll follow the lead of the late physicist Richard Feynman, who contributed as much as any scientist after the founders to the development of quantum mechanics. Feynman thought that "the only mystery" of quantum physics resides in a contemporary version of an experiment first performed in 1801. It is called the *double-slit experiment,* and Feynman found it so revealing of the quantum enigmas that he described it as having in it "the heart of quantum mechanics." (I am using the terms *quantum physics*, *quantum mechanics*, and *quantum theory* interchangeably.) In his book *The Character of Physical Law,* Feynman declared, "Any other situation in quantum mechanics, it turns out, can always be explained by saying, 'You remember the case of the experiment with the two holes? It's the same thing.' "

In 1801, the English polymath Thomas Young rigged up the test that has been known forever after as the two-slit experiment. At the time, physicists were locked in debate over whether light consisted of particles (minuscule corpuscles of energy) or waves (regular undulations of a medium, like water waves in a pond). In an attempt to settle the question, Young made two closely spaced vertical slits in a black curtain. He allowed monochromatic light to strike the curtain, passing through the slits and hitting a screen on the opposite wall. Now if we were to do a comparable experiment with something we know to be corpuscular rather than wavelike—marbles, say—there is no doubt about the outcome. Most marbles we fired at, for instance, a fence missing two slats would hit the fence and drop on this side. But a few marbles would pass through

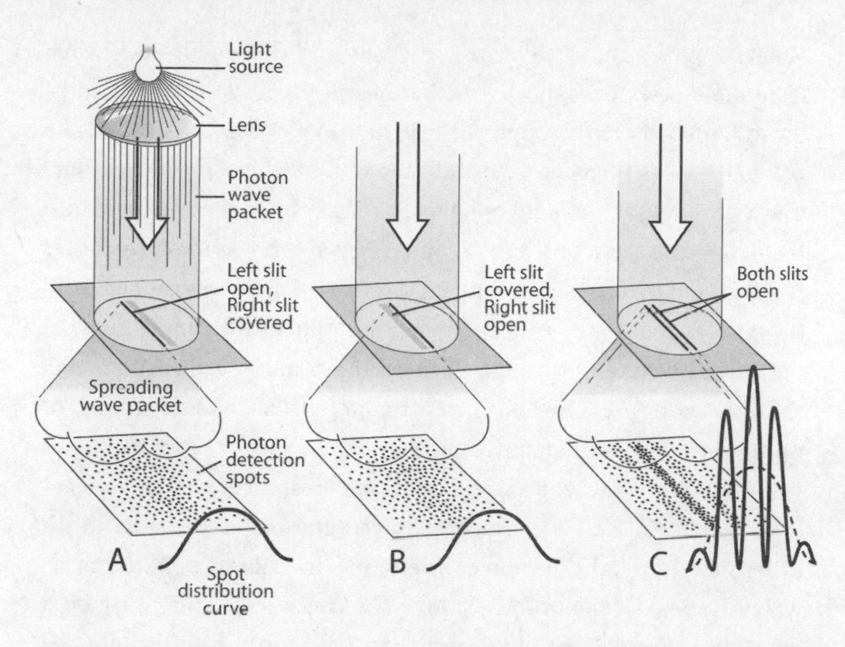

Figure 7: The Double-Slit Experiment. Monochromatic light passes through a two-slit grating. In Experiment A, only one narrow slit is open. The narrowness of the slit, coupled with the quantum Uncertainty Principle, causes the beam that passes through the slit to spread out and cover a wide area of the photographic plate. But each photon is observed to land in a tiny spot. The bell curve shows the distribution of spots, or photons. In B, only the right slit is open, and again the beam is spread out over a wide area. In C, both slits are open, but the result is not the sum of the single-slit results (dotted curve). Instead, the photons are observed in narrow bands that resemble the interference pattern formed when water waves pass through two openings in a sea wall: as semicircular waves from the two openings ripple outward, they combine where crest meets crest (in the photon experiment, the bands where many photons are found) and cancel when crest meets trough (where photons are scarce). Opening the second slit makes it clear that light behaves like a wave, since interference is a wave phenomenon. Yet each photon is found to land, whole and undivided, in a tiny region, like a particle would. Even when photons are emitted one at a time, they still form the double-slit interference pattern. Does a single photon interfere with itself?

the gaps and, if we had coated the marbles with fresh white paint, leave two bright blotches on the wall beyond, corresponding to the positions of the two openings.

This is not what Young observed.

Instead, the light created, on a screen beyond the slitted curtain, a pattern of zebra stripes, alternating dark and light vertical bands. It's called an *interference pattern.* Its genesis was clear: where crests of light waves from one slit met crests of waves from the other, the waves reinforced each other, producing the bright bands.

Where the crest of a wave from one slit met the trough of a wave from the other, they canceled each other, producing the dark bands. Since the crests and troughs of a light wave are not visible to the naked eye, this is easier to visualize with water waves. Place a barrier with two openings in a pool of water. Drop a heavy object into the pool—watermelons work—and observe the waves on the other side of the barrier. As they radiate out from the watermelon splash, the ripples form nice concentric circles. When any ripple reaches the barrier, it passes through both openings and, on the other side, resumes radiating, now as concentric half-circles. Where a crest of ripples from the left opening meets a crest of ripples from the right, you get a double-height wave. But where crest meet trough, you get a zone of calm. Hence Young's interpretation of his double-slit experiment: if light produces the same interference patterns as water waves, which we know to be waves, then light must be a wave, too. For if light were particulate, it would produce not the zebra stripes he saw but, rather, the sum of the patterns emerging from the two slits when they are opened separately—two splotches of light, perhaps, like the marbles thrown at our broken fence.

So far, so understandable. But, for the next trick, turn the light source way, way down so that it emits but a single photon, or light particle, at a time. (Today's photodetectors can literally count photons.) Put a photographic plate on the other side, beyond the slits. Now we have a situation more analogous, it would seem, to the marbles going through the fence: zip goes one photon, perhaps making it through a slit. Zip goes the next, doing the same. Surely the pattern produced would be the sum of the patterns produced by opening each slit separately—again, perhaps two intermingled splotches of light, one centered behind the left slit and the other centered behind the right.

But no.

As hundreds and then thousands of photons make the journey (this experiment was conducted by physicists in Paris in the mid-

1980s), the pattern they create is a wonder to behold. Instead of the two broad patches of light, after enough photons have made the trip you see the zebra stripes. The interference pattern has struck again. But what interfered with what? This time the photons were clearly particles—the scientists counted each one as it left the gate—and our apparatus allowed only a single photon to make the journey at a time. Even if you run out for coffee between photons, the result is eventually the same interference pattern. Is it possible that the photon departed the light source as a particle and arrived on the photographic plate as a particle (for we can see each arrive, making a white dot on the plate as it lands)—but in between became a wave, able to go through both slits at once and interfere with itself just as a water wave from our watermelon goes through the two openings in the barrier? Even weirder, each photon—and remember, we can release them at any interval—manages to land at precisely the right spot on the plate to contribute its part to the interference pattern.

So, to recap, we seem to have light vacillating between a particlelike existence and a wavelike one. As a particle, the light is emitted and detected. As a wave, it goes through both slits at once. Lest you discount this as just some weird property of light and not of matter, consider this: the identical experiment can be done with electrons. They, too, depart the source (an electron microscope, in work by a team at Hitachi research labs and Gakushuin University in Tokyo) as particles. They land on the detector—a scintillation plate, like the front of a television screen, which records each electron arrival as a minuscule dot—as particles. But in between they act as waves, producing an interference pattern almost identical to that drawn by the photons. Dark stripes alternate with bright ones. Again, the only way single electrons can produce an interference pattern is by acting as waves, passing through both slits at once just as the photons apparently did. Electrons—a form of matter—can behave as waves. A single electron can take two different paths from source to detector and interfere with itself: during its travels it can be in two places at once. The same experiments have been performed with larger particles, such as ions, with the identical results. And ions, as we saw back in Chapter 3, are the currency of the brain, the particles whose movements are the basis for the action potential by which neurons communicate. They are also, in the case of calcium ions, the key to triggering neurotransmitter release. This is a crucial point: ions are subject to all of the counterintuitive rules of quantum physics.

The behavior of material particles means that these particles have associated waves. Like a wave in water, the electron wave goes through both slits and interferes with itself on the other side. The interference pattern at the detector defines the zebra pattern. Particles travel through the hole as waves but arrive as particles. How can we reconcile these disparate properties of such bits of matter as electrons? A key to understanding the whole bizarre situation is that we actually measure the photon or electron at only two points in the experiment: when we release it (in which case a photodetector counts it) and when we note its arrival at the end. The conventional explanation is that the act of measurement makes a spread-out, fuzzy wave (at the slits) collapse into a discrete, definite particle (on the scintillation plate or other detector). According to quantum theory, what in fact passes through the slits is a wave of probability. In fact, quantum physics describes the behavior of a particle by something called the *Schrödinger wave equation* (after Erwin Schrödinger, who conceived it in 1926). Just as Newton's second law describes the behavior of particles, so Schrödinger's wave equation specifies the continuous and smooth evolution of the wave function at all times when it is not being observed. The wave function encodes the entire range of possibilities for that particle's behavior—where the particle is, when. It contains all the information needed to compute the probabilities of finding the particle in any particular place, any time. These many possibilities are called superpositions. The element of chance is key, for rather than specifying the location, or the energy, or any other trait of a particle, the equation modestly settles for describing the probability that those traits will have particular values. (In precise terms, the square of the amplitude of the wave function at any given position gives the probability that the particle will be found in some region near that position.) In this sense the Schrödinger wave can be considered a probability wave.

When a quantum particle or collection of particles is left alone to go its merry way unobserved, its properties evolve in time and space according to the deterministic wave equation. At this point (that is, before the electron or photon is observed), the quantum particle has no definite location. It exists instead as a fog of probabilities: there are certain odds (pretty good ones) that, in the appropriate experiment, it will be in the bright bands on the plate, other odds (lower) that it will land in the dark bands, and other odds (lower still, but nonzero) that it will be in the Starbucks across the street. But as soon as an observer performs a measurement—detecting an electron landing on a plate, say—the wave function seems to undergo an abrupt change: the location of the particle it describes is now almost definite. The particle is no longer the old amalgam of probabilities spread over a large region. Instead, if the observer sees the electron in *this* tiny region, then only that part of the wave function representing the small region where observation has found it survives. Every other probability for the electron's position has vanished. Before the observation, the system had a range of possibilities; afterward, it has a single actuality. This is the infamous *collapse of the wave function.*

There is, to put it mildly, something deeply puzzling about the collapse of the wave function. The Schrödinger equation itself contains no explanation of how observation causes it; as far as that equation is concerned, the wave function goes on evolving forever with no collapse at all. And yet that does not seem to be what happens. All that we know from experiment and hard-nosed mathematical calculations is that the Schrödinger wave equation, describing a microworld of superposed wave functions, somehow becomes a macroworld of definite states. In sharp contrast to the unqualified success of quantum physics in predicting the outcome of experiments stands the mess of diverse opinion that lies under the umbrella "interpretations of quantum mechanics": what happens to turn the Schrödinger wave equation into a single observed state, and what does that process tell us about the nature of reality?

There are at least three ways to account for the shift from a microworld of probabilities defined by the Schrödinger wave equations to a macroworld of definite states that we measure. Each interpretation implies a different view of the essential nature of the world. One view, preferred by Einstein, holds that the world is governed by what are called *hidden variables.* Although so-far undiscovered and perhaps even undiscoverable (hence the *hidden* part), they are supposed to be the certainties of which the wave function of quantum physics describes the probabilities. This view can be thought of as the way a tank of goldfish might think about the arrival every day of food flakes drifting through their water. It seems to be random, and without cause. There is (say) a fifty-fifty chance that food will arrive before the shadow of a little plastic skin diver reaches the little plastic mermaid, and a fifty-fifty chance that the food will land later. If only our little finned friends knew more about the world, they would understand that the arrival of the food is completely causal (the human walks over and sprinkles flakes on the water's surface). The hidden variables view, in other words, says that things look probabilistic only because we are too stupid to identify the forces that produce determinism. If we were more clever, we would see that determinism rules. Einstein's beliefs tended in this direction, leading him to his famous pronouncement "God does not play dice" with the universe. Einstein notwithstanding, however, hidden variables have been out of favor since the 1960s, for reasons too technical to get into. Suffice it to say that the physicist John Bell showed that hidden variables require instantaneous action at a distance—that is, causal influences that break the Einsteinian speed limit and travel faster than the speed of light.

A second interpretation of quantum physics holds that superposed waves exist for quantum phenomena, all right, but never really collapse. This *many-worlds view* is the brainchild of the late physicist Hugh Everett III, a student of John Wheeler, who suggested it at Wheeler's urging in a 1957 paper. Instead of attempting to answer how the act of observation induces the wave function to collapse into a single possibility, the many-worlds view holds that no single possibility is ever selected. Rather, the wave function continues evolving, never collapsing at all. How, then, do we manage to see not superpositions—electrons that are a little bit here and a little bit there—but discrete states? Every one of the experiential possibilities inherent in the wave function is realized in some superrealm, Everett proposed. If the wave function gives a fifty-fifty probability that a radioactive atom will decay after thirty minutes, then in one world the atom has decayed and in another it has not. Correspondingly, the mind of the observer has two different branches, or states: one perceiving an intact atom and the other perceiving a decayed one. The result is two coexisting parallel mental realities, the *many-minds view.* Every time you make an observation or a choice your conscious mind splits so that, over time, countless different copies of your mind are created. This, needless to say, is the ultimate in having your cake and eating it, too: sure, you may have uttered that career-ending epithet in this branch of reality, but in some other branch you kept your mouth shut.

From its inception, this theory caused discomfort. "For at least thirteen years after Everett's paper was published, there was a deafening silence from the physics community," recalls Bryce DeWitt, a physicist who championed the many-worlds view even into his eighties. "Only John Wheeler, Everett, and I supported it. I thought Everett was getting a raw deal."

As a psychiatrist I was certainly familiar with the idea of a split personality, but here was an interpretation of quantum mechanics that took the concept absolutely literally. Many scholars investigating the ontological implications of quantum mechanics squirm at the notion that all possible experiences occur, and that a new world and a new version of each observer's mind are born every time a quantum brain reaches a choice point. But many others prefer this bizarre scenario to the idea of a sudden collapse of the wave function. They like, too, that many-worlds allows quantum mechanics to operate without requiring an observer to put questions to nature—that is, without human consciousness and free choice rearing their unwelcome heads, and without the possibility that the human mind can affect the physical world. In fact, at a 1999 quantum conference in England, of ninety physicists polled about which interpretation of quantum mechanics they leaned toward, thirty chose many-worlds or another interpretation that includes no collapse. Only eight said they believed that the wave function collapses. But another fifty chose none of the above. (And of course, if they were honest, the ones who chose many-worlds would have to believe they had simultaneously made a near-infinity of different choices on their other branches.)

A third view of the change from superpositions to a single definite state is the one advanced by Niels Bohr. In this case, the abrupt change from superpositions to single state arises from the act of observation. This is the interpretation that emerged in the field's earliest days. During a period of feverishly intense creativity in the 1920s, the greatest minds in physics, from Paul Dirac and Niels Bohr to Albert Einstein and Werner Heisenberg, struggled to explain the results of quantum experiments. Finally, at the fifth Solvay Congress of physics in Brussels 1927, one group—Bohr, Max Born, Paul Dirac, Werner Heisenberg, and Wolfgang Pauli—described an accord that would become known as the Copenhagen Interpretation of quantum mechanics, after the city where Bohr, its chief exponent, worked. Bohr insisted that quantum theory is about our knowledge of a system and about predictions based on that knowledge; it is not about reality "out there." That is, it does not address what had, since before Aristotle, been the primary subject of physicists' curiosity—namely, the "real" world. The physicists threw in their lot with this view, agreeing that the quantum state represents our knowledge of a physical system.

Before the act of observation, it is impossible to know which of the many probabilities inherent in the Schrödinger wave function will become actualized. Who, or what, chooses which of the probabilities to make real? Who, or what, chooses how the wave function "collapses"? Is the choice made by nature, or by the observer? According to the Copenhagen Interpretation, it is the observer who both decides which aspect of nature is to be probed and reads the answer nature gives. The mind of the observer helps choose which of an uncountable number of possible realities comes into being in the form of observations. A specific question (Is the electron here or there?) has been asked, and an observation has been performed (Aha! the electron is there!), corralling an unruly wave of probability into a well-behaved quantum of certainty. Bohr was silent on how observation performs this magic. It seems, though, as if registering the observation in the mind of the observer somehow turns the trick: the mental event collapses the wave function. Bohr, squirming under the implications of his own work, resisted the idea that an observer, through observation, is actually influencing the course of physical events outside his body. Others had no such qualms. As the late physicist Heinz Pagels wrote in his wonderful 1982 book *The Cosmic Code*, "There is no meaning to the objective existence of an electron at some point in space . . . independent of any actual observation. The electron seems to spring into existence as a real object only when we observe it!"

Physical theory thus underwent a tectonic shift, from a theory about physical reality to a theory about our knowledge. Science is what we know, and what we know is only what our observations tell us. It is unscientific to ask what is "really" out there, what lies behind the observations. Physical laws as embodied in the equations of quantum physics, then, ceased describing the physical world itself. They described, instead, our knowledge of that world. Physics shifted from an ontological goal—learning what is—to an epistemological one: determining what is known, or knowable. As John Archibald Wheeler cracked, "No phenomenon is a phenomenon until it is an observed phenomenon." The notion that the wave function collapses when the mind of an observer registers a new bit of knowledge was developed by the physicist Eugene Wigner, who proposed a model of how consciousness might collapse the wave function—something we will return to. But why human consciousness should be thus privileged has remained an enigma and a source of deep division in physics right down to today.

It is impossible to exaggerate what a violent break this represented. Quantum physics had abandoned the millennia-old quest to understand what exists in favor of our knowledge of what exists. As Jacob Bronowski wrote in *The Ascent of Man*, "One aim of the physical sciences has been to give an exact picture of the material world. One achievement of physics in the twentieth century has been to prove that that aim is unattainable." The Copenhagen Interpretation drew the experiences of human observers into the basic theory of the physical world—and, even more, made *them* the basic realities. As Bohr explained, "In our description of nature the purpose is not to disclose the real essence of phenomena but only to track down as far as possible relations between the multifold aspects of our experience." With this shift, Heisenberg said, the concept of objective reality "has thus evaporated." Writing in 1958, he admitted that "the laws of nature which we formulate mathematically in quantum theory deal no longer with the particles themselves but with our knowledge of the elementary particles." "It is wrong," Bohr once said, "to think that the task of physics is to find out how nature is. Physics concerns what we can say about nature."

To many, this surrender was nothing short of heresy. Einstein, perhaps the most passionate opponent of the abandonment of efforts to understand nature, objected that this shift from what exists to what we know about what exists violated what he considered to be "the programmatic aim of all physics," namely, a complete description of a situation independent of any act of observation. But Einstein lost. With the triumph of quantum, physics stopped being about nature itself and instead became about our knowledge of nature.

Right about here every discussion of quantum epistemology invokes Schrödinger's cat, a thought experiment that Schrödinger proposed in 1935 to illustrate the bewilderments of quantum superpositions. Put a pellet inside a box, he said, along with a radioactive atom. Arrange things so that the pellet releases poison gas if and only if the atom decays. Radioactive decay is a quantum phenomenon, and hence probabilistic: a radioactive atom has a finite probability of decaying in a certain window of time. In thirty minutes, an atom may have a 50 percent chance of decaying—not 70 percent, not 20 percent, but precisely 50 percent. Now put a cat in the box, and seal it in what Schrödinger called a "diabolical device." Wait a while. Wait, in fact, a length of time equal to when the atom has a fifty-fifty chance of decaying. Is the cat alive or dead?

Quantum mechanics says that the creature is both alive and dead, since the probability of radioactive decay and hence release of poison gas is 50 percent, and the possibility of no decay and a safe atmosphere is also 50 percent. Yet it seems absurd to say that the cat is part alive and part dead. Surely a physical entity must have a real physical property (such as life or death)? If we peek inside the box, we find that the cat is alive or dead, not some crazy superposition of the two states. Yet surely the act of peeking should not be enough to turn probability into actuality? According to Bohr's Copenhagen Interpretation, however, this is precisely the case. The wave function of the whole system, consisting of kitty and all the rest, collapses when an observer looks inside. Until then, we have a superposition of states, a mixture of atomic decay and atomic intactness, death and life.

Observations, to put it mildly, seem to have a special status in quantum physics. So long as the cat remains unobserved, its wave function encodes equal probabilities of life and death. But then an observation comes along, and *bam*—the cat's wave function jumps from a superposition of states to a single observed state. Observation lops off part of the wave function. The part corresponding to living or deceased, but not the other, survives.

If the power of the observer to coax a certain value out of a probability wave sounds like the wrong answer to the question of whether a tree falls in an empty wood even if no one hears it, take heart: physicists are just as puzzled about how the world can work in such a bizarre way. As Eugene Wigner stated in 1964, "This last step [the entering of an observation into consciousness] is . . . shrouded in mystery and no explanation has been given for it so far." The collapse of the wave function, which gives rise to the centrality of consciousness in physical theory, "enters quantum mechanical theory as a deus ex machina, without any relation to the other laws of this theory."

One of the things I admired about Stapp's *Mind, Matter and Quantum Mechanics* was his willingness to address the ethical implications of the change from Newtonian physics to quantum physics. In particular, Stapp made the point that there is no stronger influence on human values than man's belief about his relationship to the power that shapes the universe. When medieval science connected man directly to his Creator, man saw himself as a child of the divine imbued with a will free to choose between good and evil. When the scientific revolution converted human beings from the sparks of divine creation into not particularly special cogs in a giant impersonal machine, it eroded any rational basis for the notion of responsibility for one's actions. We became a mechanical extension of what preceded us, over which we have no control; if everything we do emerges preordained by the conditions that prevail, then we can have no responsibility for our own actions. "Given this conception of man," Stapp argued, "the collapse of moral philosophy is inevitable." But just as Newtonian physics undermines moral philosophy, Stapp thought, so quantum physics might rescue it. For quantum physics describes a world in which human consciousness is intimately tied into the causal structure of nature, a world purged of determinism.

Impressed by what seemed a kindred mind, I therefore e-mailed Stapp on March 2, 1998, introducing myself as a friend of Chalmers (whom Stapp had met several times) and telling him that I "had started reading your book Mind Matter and QM—I'm still working on it and finding it of great importance." Stapp responded on March 10, with requests for some of my reprints. In a phone call soon after, we discussed, among other things, how Newtonian approaches had evolved so as to stifle morality, and why science was therefore vastly overrated as a force for good. Stapp, a courtly man, seemed sympathetic (if not as passionate as I about all this), and we arranged to get together. On June 9, I drove my old used 1988 copper Mercedes up to Berkeley and met Henry for the first time and spent the afternoon in his office. At a long dinner that evening at a restaurant hard by the railroad tracks on the outskirts of Berkeley, the conversation ranged from quantum mechanics to phantom pain, from statistical tests of the paranormal to the attempts by some theologians to find spiritual messages in the discoveries of cosmologists. (That week, Berkeley was hosting a symposium, "Science and the Spiritual Quest.")

Back home, I was juggling a couple of tasks. Although I had only a rudimentary understanding of quantum physics at the time I tackled Stapp's 1993 book, its relevance to the mind-brain question, and to my interpretation of the OCD brain data, made it clear that I needed to learn a whole lot more. Quantum mechanics seems to contain a role for consciousness in the physical world. Fortunately, Stapp's book, as well as papers for nonphysicists that he posted on his web site, addressed those questions in accessible, if not entirely elementary, ways. Even better, Stapp himself is the soul of patience, who proved to be extraordinarily open to question-and-answer sessions by phone and e-mail. And so, over the next two and a half years, I slowly got a handle on key physics concepts supporting the causal efficacy of volition and attention. In a September 7, 1998, phone conversation, Stapp told me, "In quantum theory, experience is the essential reality, and matter is viewed as a representation of the primary reality, which is experience." I wrote it down verbatim and tacked it on my office wall, where it still hangs today.

As I worked to deepen my understanding of quantum mechanics, I was also trying to apply the basic structure of Stapp's physics reasoning to a philosophical paper I was writing about the OCD work. In April, I had presented a paper at a conference that Dave Chalmers had helped organize at the University of Arizona's Tucson campus, "Toward a Science of Consciousness." I had spoken about my OCD work, particularly the "mind can change brain" aspects. Now, in June, I was in the midst of turning the oral presentation into a paper for the conference proceedings. In addition, *A Return to Innocence* was in galleys at this point, on track for a September 1998 publication. The combination of the book, in which I used "mental force" as a literary device, plus my philosophically grounded reanalysis of the PET scans of my OCD patients for the Tucson paper, created a powerful alchemy. As I turned over in my mind yet again the four steps I taught my OCD patients to go through when in the throes of a compulsion—Relabel, Reattribute, Refocus, Revalue—it occurred to me that mental force might be more than just a way to help readers of *Return to Innocence* understand how mindfulness and directed effort can help reshape the way they think and behave. To the contrary: there was nothing about mental force as I conceived it that condemns it to be just a metaphor. Whether it had any genuine scientific import, I had no idea—not yet, anyway.

But Stapp might. My manuscript contained the first use of "mental force" in more than a metaphorical sense and marked the first time I had used it in a scientific paper. On June 21, I e-mailed Stapp that I was writing something he might consider beyond the pale: "I know I took some serious risks with my use of the concepts of energy, power and force, . . . so I'll definitely need your feedback on that—hopefully you'll still be willing to talk to me after you see it." Three days later, I prepared to e-mail the paper to Stapp to get his reaction. I stared at the computer screen for what seemed like ages before I could screw up my courage to hit "send." I think this is pretty good, I thought, but considering what I know about physics and forces I might be about to embarrass myself royally.

Stapp replied by e-mail on June 25: "Your draft was masterful," he wrote.

> *It should act to focus the attention of workers in this field on the fact that THE fundamental issue before us, namely the question of the efficacy of consciousness, need not be walked away from as something totally intractable. Your work makes it clear that "will" involves different levels or aspects of consciousness, with higher-level experiences presiding over lower-level experiences in much the way that the "subjective I" controls bodily actions. . . . In this connection, it may be important that the key issue in the interpretation of quantum theory is "at what level do the collapses occur?" . . . The quality of the experience of "will" strongly suggests that [it] . . . acts on the physical level. . . . Contemporary quantum theory is built on experience, and makes no sense without experience. . . . Within the richer theoretical framework "will" will almost certainly be efficacious.*

Henry Stapp had been interested, since his student days, in what has come to be called the interpretation of quantum theory. When Wolfgang Pauli visited UC Berkeley to deliver a series of lectures in 1958, the physics department, as was customary, assigned a postdoc to take lecture notes. Stapp got the nod. That put him in frequent and close contact with Pauli, who invited Stapp to go to Zurich to work with him. Stapp arrived in the fall of 1958, but Pauli died that December. Since Stapp's fellowship was to last six months, he found himself with unexpected time on his hands. He used it to delve into the work of the mathematician John von Neumann, in particular his book on the foundations of quantum theory. This work raised in Stapp's mind questions about the role of mind in physics. In 1959, still in Zurich, he foreshadowed his later book by writing an essay, "Mind, Matter and Quantum Mechanics," in which he discussed the notion that reality comes into being only when an observer observes it. But he also recognized serious problems with this idea. In 1965, when the United States sent the unmanned *Mariner 4* probe to pass by Mars, Stapp asked, Are we to believe that a mountain on Mars springs into existence only when some guy at mission control calls up *Mariner*'s imaging data on the console screen? Like most others, Stapp resisted von Neumann's suggestion that mind had anything to do with creating reality. But he continued to ponder the mystery of what turned all of the potentials embodied in the Schrödinger wave function into a single observed reality.

Back in Berkeley, this was the challenge Stapp took up. "I worked long and hard trying to figure out what led to the collapse of the wave function. In the end, I became more convinced that conscious experiences needed to be taken seriously," he recalls. In 1968, Stapp went to Munich, where he became engrossed in discussions with Heisenberg about his and Bohr's more philosophical papers. "Then as now, physicists pay lip service to these writings, but quantum physics is taught as engineering," says Stapp. "This is how you apply it and these are the mathematical rules. The philosophy is brushed under the rug; you don't try to think what's really happening." Or as physicists sometimes put it: "Don't think. Calculate."

But Stapp was deeply curious about the philosophy implied by quantum physics, and whether in fact the act of observation has a hand in bringing about one possible reality rather than another. "When I came to Munich I was filled with lots of questions," he recalls.

> *I had quite a few discussions with Heisenberg, and came to realize that his and Bohr's positions were not the same. Heisenberg talked in terms of propensities or tendencies for an event to occur, which would happen even if an observer were not there. This is the common understanding of quantum theory by most practitioners, who almost to a man do not believe that human observers have much to do with this. Heisenberg separated himself from [Bohr's] interpretation by being willing to discuss what is actually happening, in spite of the fact that the official doctrine says you are not supposed to talk about that. He acknowledged that, but said this is nevertheless "what I think."*

Heisenberg believed that the infamous cat was indeed either alive or dead, even before an observer looked and collapsed the wave function; it is nature herself who collapses the wave function. "It was very useful for me to hear right from him that there was not total agreement" on the role of the observer, Stapp says. "I came to realize that the interpretation of quantum physics, particularly the underlying ontology, was not totally worked out."

Shortly after his discussions with Heisenberg, Stapp was returning from Europe and had to overnight in London. He walked to a park and settled in on a bench with William James's *The Meaning of Truth*. What he read produced an epiphany. "That was when it all came together," says Stapp. "James argues that we'll never know for sure the absolute truth, and that science is therefore provisional. In the end all you can do is say how well your theories are working. Once I read James's idea, it allowed me to understand what Bohr was saying," with his conclusion that we cannot know what really happens, but only what we observe to happen. Was the Danish physicist (who died in 1962) familiar with the work of the American psychologist? The science historian Thomas Kuhn once asked Bohr whether there was any connection between his ideas and James's. Bohr responded, "Yes, it's all coming back; we'll talk about that tomorrow." He died that very night, November 18, taking the answer with him.

As Stapp interpreted quantum mechanics, the observer plays two roles. He experiences the output of the measuring devices, of course—the clicks of the Geiger counter in our radioactive atom experiment, for instance. What he records depends on which choice nature has made: the atom decays or it doesn't. This is known as the Dirac choice after P. A. M. Dirac, the English physicist who, at the fifth Solvay Congress of physics in Brussels in 1927, conceptualized this random event as a choice by nature (but is better known for predicting the existence of antimatter). It is, as far as physicists know, a truly random choice. But the observer plays another role: he chooses which questions to pose to nature. Stapp named this the Heisenberg choice, because Heisenberg stressed it at the 1927 congress. "In quantum theory," Stapp says, "the observer must decide which question to put to nature, which aspect of nature his inquiry will probe. A person's conscious thoughts can and . . . must . . . play a role that is not reducible to the combination of the Schrödinger and Dirac processes." Until a question is posed, nothing happens. Without some way of specifying what the question is, the quantum process seizes up like a stuck gear and grinds to a halt. There is, then, a three-part process: the evolution of the wave equation as described by the Schrödinger equation, the choice of which question to pose (the Heisenberg choice), and nature's statistical choice of which answer to give (the Dirac choice).

This three-part description of quantum mechanics had never been presented publicly in any detail when, in July 1999, Stapp and I, along with Dave Chalmers and a host of renowned physicists, neuroscientists, and philosophers, ascended into the cool clear mountain air of Flagstaff, Arizona, for a conference, "Quantum Approaches to Consciousness." I eagerly looked forward to this meeting, both because of its lovely Grand Canyonesque location and because I knew that here, before a solemn (well, at least during the daytime sessions) gathering, Stapp would tackle the thorny question of how the probabilities described by the Schrödinger equation collapse into the actualities we observe and measure. Eugene Wigner, as I hinted earlier, followed the new realizations to their inevitable conclusion. "The laws of quantum mechanics cannot be formulated . . . without recourse to the concept of consciousness," he wrote in 1961. Matter has become intrinsically connected to subjective experiences. And that leads to a profound implication. It makes little sense, Wigner argued, to describe the mind and consciousness in terms of the positions of atoms, for one simple reason: the latter are derived from the former and have no fixed and non-probabilistic existence outside the former. "It seems inconsistent," Wigner said in 1969, "to explain the state of mind of [an] observer . . . in terms of concepts, such as positions of atoms, which have to be explained, then, in terms of the content of consciousness." If the positions of atoms (and thus, for our purposes, the state and arrangement of neurons, since neurons are only collections of zillions of atoms) have no unambiguous existence independent of the consciousness of an observer, Wigner asked, then how can that very consciousness depend on those same atoms? "The extreme materialistic point of view . . . is clearly absurd and . . . is also in conflict with the tenets of quantum mechanics," he concluded.

Classical physics had no way to account for consciousness; Copenhagen brought in consciousness, all right, but at the ghastly price of substituting it for objective reality. The von Neumann/ Wigner theory that Stapp referred to in his first e-mail to me, in 1998, seemed to offer a way out. Eugene Wigner and John von Neumann had joined the wave of refugees fleeing Hitler and had wound up at Princeton University. In 1932, von Neumann formulated a new version of quantum mechanics. Its main point of departure from the Copenhagen Interpretation is this: Copenhagen describes measuring devices (things like Geiger counters and scintillation counters as well as the human brain that registers the results of those measurements) in Newtonian rather than quantum terms. This makes the theory inherently inconsistent, since one part of the physical world (subatomic particles) gets the quantum treatment but the rest of the physical world (lab equipment and brains) stays Newtonian. And yet the stuff in the second group is made of the same atoms and subatomic particles as the stuff in the first. Von Neumann realized that this made no sense: a measuring device is not intrinsically different from the atoms that make it up. So he fixed the problem. In the mathematical rules of quantum theory he worked out, he first incorporated measuring devices, so that when physicists did a calculation they would have to apply quantum rules to these devices. And then he incorporated everything made of atoms and their constituents—in particular, the human brain. In von Neumann's formulation, every experiential event, such as reading a measuring device or otherwise making an observation, has a corresponding brain event. No surprise there. But von Neumann went further: the brain, he argued, operates according to the rules of quantum mechanics.

Applying quantum theory to the brain means recognizing that the behaviors of atoms and subatomic particles that constitute the brain, in particular the behavior of ions whose movements create electrical signals along axons and of neurotransmitters that are released into synapses, are all described by Schrödinger wave equations. Thanks to superpositions of possibilities, calcium ions might or might not diffuse to sites that trigger the emptying of synaptic vesicles, and thus a drop of neurotransmitter might or might not be released. The result is a whole slew of quantum superpositions of possible brain events. When such superpositions describe whether a radioactive atom has disintegrated, we say that those superpositions of possibilities collapse into a single actuality at the moment we observe the state of that previously ambiguous atom. The resulting increment in the observer's knowledge of the quantum system (the newly acquired knowledge that the atom has decayed or not) entails a collapse of the wave functions describing his brain. This point is key: once the brains of observers are included in the quantum system, the wave function describing the state of the brain of any observer collapses to the form corresponding to his new knowledge. The quantum state of the brain must collapse when an observer experiences the outcome of a measurement. The collapse occurs in conjunction with the conscious act of experiencing the outcome of the observation. And it occurs in the brain of the observer—the observer who has learned something about the system.

What do we mean by collapsing the quantum state of the brain? Like the atom threatening Schrödinger's cat, the entire brain of an observer can be described by a quantum state that represents all of the various possibilities of all of its material constituents. That brain state evolves deterministically until a conscious observation occurs. Just before an observation, both the observed quantum system (let's stick with the radioactive atom) and the brain that observes it exist as a profusion of possible states. Think of each possible state as a branch on a tree. Each branch corresponds to some possible state of knowledge, or course of action. But when an observation registers in the mind of the observer, the branches are brutally pruned: only the branches compatible with the observer's experience remain. If, say, the observation is that the sun is shining, then the associated physical event is the updating of the brain's representation of the weather. Branches corresponding to "the sky is overcast" are chopped off. An increase in knowledge is accompanied by an associated reduction of the quantum state of the brain. And with that, the quantum brain changes, too.

Because the observer's only freedom is the choice of which question to pose (Shall I look up at the sky?), it is here that the mind of the observer has a chance to affect the dynamics of the brain. An examination of the mathematics, Stapp argued, shows that "the conscious intentions of a human being [reflected in the choices he makes about what question to put to nature] can influence the activities of his brain. . . . Each conscious event picks out from the multitude of . . . possibilities that comprise the quantum brain the subensemble that is compatible with the conscious experience." The physical event reduces the state of the brain to that branch of it that is compatible with the particular experience or observation.

Each of the principal interpretations of quantum theory—hidden variables, many-worlds, the dynamical role of consciousness, von Neumann's application of quantum rules to the brain—has its passionate partisans. For many physicists, unfortunately, which interpretation they subscribe to seems to be more a matter of intellectual fashion and personal taste than rigorous analysis. So is whether they bother with interpretation at all. Just as many neuroscientists are perfectly happy to sweep the question of consciousness under the rug and stick to something they can measure and manipulate—the brain—so a similar attitude prevails among physicists (though physicists may not be quite so oblivious as neuroscientists. "It is surprising," Wigner dryly noted, "that biologists are more prone to succumb to the error of disregarding the obvious than are physicists"). Engineers who design or use transistors, which exploit quantum phenomena, rarely think about the ontological implications of quantum mechanics and whether the mind shapes reality; neither do high-energy physicists, as they work out the chain of reactions in a particle accelerator. For every hundred scientists who use quantum mechanics, applying the standard equations like recipes, probably no more than one ponders the philosophy of it. They don't have to. You can do perfectly good physics if you just "shut up and calculate," as the physicist Max Tegmark puts it. Physicists can safely continue to believe in classical epistemology and ontology, whether consciously or not, and stash the epistemology and ontology demanded by quantum mechanics in a rarely opened room of their mind like an ugly lamp exiled to Grandma's attic. "The reason I went into physics was my fascination with the fundamental issues raised by quantum mechanics, but I quickly realized the subject was taboo," recalls Tegmark.

"Real physicists just didn't spend time on these questions, and you realize pretty quickly that you can't get a job doing this. So what I would do is write quantum papers secretly, when no one knew what I was doing. When I was a grad student at Berkeley, I would make sure my adviser was far away from the printer when I printed them out."

The reluctance of most physicists to face, let alone explore, the interpretation and philosophical implications of quantum mechanics has had an unfortunate consequence. Scientists in other fields, many of whom consider physics the most basic of all sciences and the one with whose tenets their own findings must accord (an attitude sometimes denigrated as "physics envy"), have remained, for the most part, painfully naïve about the revolutionary implications of quantum theory. For neuroscientists, this ignorance exacts a price: the view of reality demanded by quantum physics challenges the validity of the Cartesian separation of mind and material world, for in the quantum universe "there is no radical separation between mind and world." As Wolfgang Pauli stated in a letter to Niels Bohr in 1955, "In quantum mechanics . . . an observation here and now changes in general the 'state' of the observed system. . . . I consider the unpredictable change of the state by a single observation . . . to be an abandonment of the idea of the isolation of the observer from the course of physical events outside himself." This is the textbook position on quantum mechanics and the nature of reality: that the Cartesian separation of mind and matter into two intrinsically different "substances" is false.

Ignoring quantum physics thus deprives both philosophers and neuroscientists of an avenue into, if not a way out of, the mystery of mind's relationship to matter. The unfortunate result, as we're seeing, is the belief that interactions among large assemblies of neurons are causally sufficient to account for every aspect of mind. As the philosopher Daniel Dennett put it, "A brain was always going to do what it was caused to do by local mechanical disturbances." In this view, mind is indeed nothing more than billions of interacting neurons—in short, nothing more than brain processes. There is no power we ascribe to mind—even what we experience as the power to choose, to exert the will in a way that has measurable consequences—that is not completely accounted for by electrochemistry. Most attempts to resolve the mind-matter problem, derived as they are from a Newtonian worldview, dismiss both consciousness and will as illusions, products of human fallibility or hubris. And yet such conclusions, built as they are on an outdated theory of the physical world, are built on a foundation of sand. The classical formulations are wrong "at the crucial point of the role of human consciousness in the dynamics of human brains," Stapp argues. If the mind-brain problem has resisted resolution for three centuries, it is because the physical theory that scientists and philosophers have wielded is fundamentally incorrect. If we are foundering in our attempts to resolve the mind-matter problem, the fault lies with the physics more than with the philosophy or the neuroscience. In other words, we are not doing all that badly in our efforts to understand the mind side of the equation; it's our understanding of the role of matter that is seriously off. For this, we can thank the materialist view that grew to predominance over the last three centuries.

The more I talked with Stapp throughout the summer of 1998, the more I became convinced that quantum physics would provide the underpinning for the nascent idea of mental force. The fact that the collapse of the wave function so elegantly allows an active role for consciousness—which is required for an intuitively meaningful understanding of the effects of effort on brain function—is itself strong support for using a collapse-based interpretation in any scientific analysis of mental influences on brain action. In my discussions with Stapp, it became clear that a genuine scientific synergy was possible. It would not be just my OCD patients and their PET scans, or any other data from neuroscience alone, that would drive the final nail in the coffin of materialism. It would be the integration of those data with physics. If there is to be a resolution to the mystery of how mind relates to matter, it will emerge from explaining the data of the human brain in terms of these laws—laws capable of giving rise to a very different view of the causal efficacy of human consciousness. Quantum mechanics makes it feasible to describe a mind capable of exerting effects that neurons alone cannot.

Historically, the great advances in physics have occurred when scientists united two seemingly disparate entities into a coherent, logical whole. Newton connected celestial motions with terrestrial motion. Maxwell unified light and electromagnetism. Einstein did it for space and time. Quantum theory makes exactly this kind of connection, between the objective physical world and subjective experiences. It thus offers a way out of the morass that the mind-brain debate has become, because it departs most profoundly from classical physics at a crucial point: on the nature of the dynamical interplay between minds and physical states, between physical states and consciousness. It ushers the observer into the dynamics of the system in a powerful way. Following quantum theory into the thicket of the mind-matter problem actually leads to a clearing, to a theory of mind and brain that accords quite well with our intuitive sense of how our mind works. In Stapp's formulation, quantum theory creates a causal opening for the mind, a point of entry by which mind can affect matter, a mechanism by which mind can shape brain. That opening arises because quantum theory allows intention, and attention, to exert real, physical effects on the brain, as we will now explore.

{NINE}

FREE WILL, AND FREE WON'T

If the atoms never swerve so as to originate some new movement that will snap the bonds of fate, the everlasting sequence of cause and effect—what is the source of the free will possessed by living things throughout the earth?

—*Lucretius,* On the Nature of the Universe, Book 2

The question is whether such a technique can really make a man good. Greatness comes from within, 6655321. Goodness is something chosen. When a man cannot choose he ceases to be a man.

—*Anthony Burgess,* A Clockwork Orange

Attending the Tucson III conference, "Toward a Science of Consciousness," in April 1998 was both a great learning experience and a lot of fun. Dave Chalmers had encouraged me to present a talk on how my OCD work provided evidence for the power of the mind over the physical stuff of the brain. That alone would have made the meeting worthwhile, but the gathering also turned out to be a great place to make friends and (at least to some degree) influence people. At the very first session, I attended a presentation that immediately made me realize I wasn't alone in denying that the mind is a mere appendage of the brain. The paper was by someone who was about to have a significant impact on my life: Jonathan Shear. A professor of philosophy at Virginia Commonwealth University and managing editor of the *Journal of Consciousness Stud-*

ies, Shear is also a serious student (and practitioner) of Transcendental Meditation. He was an early adopter, as they say in the world of technology: by 1963 he was already deeply involved in the study of where meditation meets science, and he knew about the maharishi before he was The Maharishi (that is, before the Beatles worked with him in India). Fittingly, Shear's talk was on Eastern philosophies and their views of consciousness—and he attracted quite a crowd.

The next day Shear and I ran into each other outside one of the meeting rooms and started talking. We quickly realized we had important interests in common, especially the use of meditation to investigate consciousness. After about fifteen minutes, we slipped out for a long lunch at the hotel restaurant. There, over the buffet (he seemed relieved that I wasn't one of those "tofu-and-veggies-or-die" meditators), Shear peppered me with questions about Buddhism. My answers were long and technical, and so were his replies. We vowed to keep in touch, and after returning to Virginia, Shear asked whether I might contribute a long theoretical article to a single-topic issue being planned by the *Journal of Consciousness Studies* (*JCS*) to be called "The Volitional Brain." The guest editor would be the renowned neurophysiologist Benjamin Libet of the University of California, San Francisco. I was eager to take it on, since it offered a chance to develop further my ideas on the philosophical implications of the OCD work.

On May 31, 1998, I sent Shear the abstract of the paper I had presented the month before in Tucson. In it, as I've mentioned, I first used the term *mental force* in a scientific sense, as I explored the importance of volition to my OCD patients in changing their neural activity. As my title posed the question, "A Causal Role for Consciousness in the Brain?" I described how PET studies of patients with obsessive-compulsive disorder had demonstrated systematic alterations in cerebral activity in those who were successfully treated with a drug-free cognitive-behavioral therapy. I outlined the Four Step method and explained how it teaches patients to regard the intrusion of OCD symptoms into consciousness as the manifestation of a "false brain message," training them to willfully select alternative actions when experiencing obsessions and compulsions. Although such willful behavioral change is difficult, I went on, it both relieves OCD symptoms and brings about systematic changes in metabolic activity in the OCD circuit. It turns out that the key predictor of whether the Four Steps will help an OCD patient is whether he learns to recognize that a pathological urge to perform a compulsive behavior reflects a faulty brain message—in other words, to Revalue it.

This work seemed appropriate for an issue on the volitional brain because it flew in the face of the widespread notion, dating back to at least the time of Descartes, that mind is incapable of acting on and changing matter. As noted in Chapter 1, this philosophical position, known nowadays as *epiphenomenalism,* views conscious experience as nothing more special than the result of physical activity in the brain, as rain is the result of air pressure, wind, and cloud conditions in the atmosphere. Epiphenomenalism is a perfectly respectable, mainstream neurobiological stance. But it denies that the awareness of a conscious experience can alter the physical brain activity that gives rise to it. As a result, it seemed to me, epiphenomenalism fails woefully to account for the results I was getting: namely, that a change in the valuation a person ascribes to a bunch of those electrochemical signals can not only alter them in the moment but lead to such enduring changes in cerebral metabolic activity that the brain's circuits are essentially remodeled. That, of course, is what PET scans of OCD patients showed.

On June 3, Shear responded to the abstract I had sent him. Two of the *JCS* editors he had shown it to, he said, had reacted "quite positively." One of them, Keith Sutherland, answered Shear's query about whether to include something along those lines in the *JCS* volume with a succinct "Yes—go for it!" Sutherland remembered an article on my work that appeared in *New Scientist*, a popular British science weekly, the previous summer, and asked, "Does he touch on any similarities between cognitive therapy and Buddhist practice?" This was the first time a fellow scientist had independently suggested tying the OCD results, and implicitly my Four Step therapy, to Buddhist philosophy and meditation. Another editor, Bob Forman, called it "a counter punch, long overdue, to the meaning-ignoring epiphenomenalist types."

Working that summer to refine my theory of mental force, I spent many long nights sweating bullets over the paper. I also spent hours discussing the details with Stapp, who, as it happened, had also been invited to contribute a paper to the *JCS* issue. As soon as I learned this, it struck me that this would be a great opportunity to integrate the OCD work with Stapp's interpretation of quantum mechanics to create something like a grand synthesis. He and I discussed the possibility of writing back-to-back papers and decided to give it a shot. So one Sunday in late July, when I had to be in Berkeley for the opening of a film a friend had just produced, I drove up early that morning and took the opportunity to visit Stapp at home. Sitting beside the pool in his backyard, with its breathtaking view of San Francisco Bay, we started talking about quantum physics, and how the philosophy that it supports seems quite Jamesian in implying that the willful expression of consciousness has causal efficacy in the material realm. What struck us both was how close William James had come to formulating a persuasive, scientifically based theory of how attention reifies intention. He lacked only a mechanism, but that was because only quantum physics, and not the classical physics of his day, provided one. We talked, too, about how both quantum physics and classical Buddhism give volition and choice a central role in the workings of the cosmos. For quantum physics, until and unless a choice is made about what aspect of nature to investigate, nothing definite occurs; the superposition of possibilities described by the Schrödinger wave equation never collapses into a single actuality, as discussed in the previous chapter. As Stapp puts it, "For the quantum process to operate, a question must be addressed to Nature." Formulating that question requires a choice about which aspect of nature is to be probed, about what sort of information one wishes to know. Critically, in quantum physics, this choice is free: in other words, no physical law prescribes which facet of nature is to be observed. The situation in Buddhist philosophy is quite analogous. Volition, or Karma, is the force that provides the causal efficacy that keeps the cosmos running. According to the Buddha's timeless law of Dependent Origination, it is because of volition that consciousness keeps arising throughout endless world cycles. And it is certainly true that in Buddhist philosophy one's choice is not determined by anything in the physical, material world. Volition is, instead, determined by such ineffable qualia as the state of one's mind and the quality of one's attention: wise or unwise, mindful or unmindful. So in both quantum physics and Buddhist philosophy, volition plays a special, unique role.

In neuroscience, on the other hand, to take an interest in the role of volition and the mental effort behind it, and further to wonder whether volition plays a critical role in brain function, is virtually unheard of. Piles of brain imaging studies have shown that volitional processes are associated with increases in energy use in the frontal lobes: "right here," you can say while pointing to the bright spots on the PET scan, volition originates. But the research is mute on the chicken-and-egg question of what's causing what. Does activity in the frontal lobes cause volition, or does volition trigger activity in the frontal lobes? If the former, does the activity occur unbidden, as a mere mechanical resultant, or is it in any sense free? Generally, neuroscientists assume that the brain causes everything in the mind, period—further inquiry into causality is most unwelcome.

In the final version of my "Volitional Brain" paper, I was trying to do better than this glib dismissal. The feel of OCD symptoms and the feeling of mental effort that accompanies the Four Steps make this disease and its treatment a perfect fit for a volume examining phenomena at the nexus of mind and brain, I argued to Stapp on that summer morning. The intrusive thoughts that plague patients feel like extraneous intrusions into consciousness, as if they were invaders from another brain. Experiencing OCD symptoms is a purely passive process. In contrast, Relabeling the OCD symptoms and Refocusing attention on healthy circuitry are wholly active processes. The difference between the two "feels" makes genuine choice and the causal efficacy of that choice possible. Going further, I argued that the undeniable role of effort and the possibility of an associated mental force to explain the observed changes in the OCD circuit suggest a mechanism by which the mind might affect—indeed, in a very real sense, reclaim—the brain. That mechanism would allow volition to be real and causally efficacious, not the "user illusion" that determinists call it; it would allow volition to act on the material brain by means of an active and purposeful choice about how to react to the conscious experience of OCD symptoms. As I laid all this out, Stapp expressed confidence that it was all consistent with quantum physics.

The mechanism that allows volition to be physically efficacious is the one I called mental force. Similarly to what has been called "mind as a force field," mental force also echoes what Ben Libet, a pioneer in the study of the neurobiology of volition, has named the "conscious mental field." I proposed in the final version of my *JCS* paper that mental force is a physical force generated by mental effort. It is the physical expression of will. And it is physically efficacious. At the moment an OCD patient actively changes how he responds to the obsessive thoughts and compulsions that besiege him, the volitional effort and refocusing of attention away from the passively experienced symptoms of OCD and toward alternative thoughts and behaviors generate mental force. Mental force acts on the physical brain by amplifying the newly emerging brain circuitry responsible for healthy behavior and quieting the OCD circuit. We know that directed mental effort causes measurable changes in brain function, the self-directed neuroplasticity discussed earlier. And we know that mental effort is not reducible to brain action: hence the need for a new actor—mental force.

This notion of mental force fit an idea about free will that Libet had long propounded, one known as the "free won't" version of volition. In a nutshell, "free won't" refers to the mind's veto power over brain-generated urges—exactly what happens when OCD patients follow the Four Steps. Since Libet served as a guest editor for the *JCS* volume, it didn't hurt that I was able to acknowledge my intellectual debt to him. But it was hardly a stretch to make the connection to his work: OCD symptoms can be viewed as high-powered, painfully persistent versions of the desultory mental events that pop into consciousness countless times each day. Most of these thoughts do not insist on action, or demand attention, because the will can ignore them rather easily, Libet had argued. But in OCD patients the thoughts aren't nearly this well mannered: they are as insistent and intrusive as a nagging toddler. The discomfort they cause demands attention. Making that attention mindful and wise requires effort of the highest degree. That effort, I suspected, becomes causally efficacious on brain action through the mechanism of mental force. At the 1999 Quantum Brain conference in Flagstaff, I had discussed this possibility with Libet, and now it became part of my argument.

The fact that willful refocusing of attention caused brain changes in patients with OCD had exciting implications for the physics of mind-brain. "Ideas that I had long been working on, but which seemed to have no practical application, tied in very well with Jeff's discovery of the power of mental effort to keep attention focused," Stapp recalled. "That gave me the impetus to pursue this." In his own *JCS* paper, Stapp argued that neither scientists nor philosophers who adhered to the ideas of classical Newtonian physics would ever resolve the mind-brain mystery until they acknowledged that their underlying model of the physical world was fundamentally flawed. For three centuries classical physics has proved incapable of resolving the mind-body problem, Stapp noted. And although quantum physics supplanted classical physics a century ago, the implications of the quantum revolution have yet to penetrate biology and, in particular, neuroscience. And that's a problem, for the key difference between classical and quantum physics is the connection they make between physical states and consciousness. Quantum theory "allows for mind—pure conscious experience—to interact with the 'physical' aspect of nature. . . . [I]t is [therefore] completely in line with contemporary science to hold our thoughts to be causally efficacious," Stapp argued. He ended his *JCS* paper with a discussion of my OCD therapy, calling it "in line with the quantum-mechanical understanding of mind-brain dynamics." According to that understanding, mental events influence brain activity through effort and intentions that in turn affect attention. "The presumption about the mind-brain that is the basis of Schwartz's successful clinical treatment," Stapp concluded, "is that willful redirection of attention is efficacious. His success constitute[s] prima facie evidence" that "will is efficacious."

This statement was tremendously gratifying because it stated, from a physicist's perspective, what seemed to me the essential core of all my OCD work: that effort itself is the key to altering one's brain function. Stapp's insight was that quantum theory naturally allows for the direct influence of mental effort on the function of the brain. It thus makes mental effort and its effect on attention a primary causal agent.

In addition to our individual papers for the *JCS* issue, Stapp and I wrote an "appendix" that appeared between them. It became our strongest argument yet of the power of quantum physics to support the causal efficacy of mental force: "The basic principles of physics, as they are now understood, are not the deterministic laws of classical physics," we wrote. The basic physical laws are, rather, those of quantum physics, which allow mental effort to "keep in focus a stream of consciousness that would otherwise become quickly defocused as a consequence of the Heisenberg uncertainty principle, and keep it focused in a way that tends to actualize potentialities that are in accord with consciously selected ends. Mental effort can, within contemporary physical theory, have, via the effects of willful focus of attention, large dynamical consequences that are not automatic consequences of physically describable brain mechanisms acting alone."

Stapp's and my contributions stood apart from the rest of the "Volitional Brain" papers in arguing that modern physics provides a basis for volition and mental effort to alter brain function. Other contributions, taken together, constituted a grand tour of what neuroscience at the end of the twentieth century knew about volition. Better known as free will, volition has had a tough time of it lately. The very notion of "willpower" now carries a whiff of the Victorian, like the smell rising from a musty old hatbox. Invoking "a failure of willpower" to explain someone's succumbing to the temptations of alcohol or illegal drugs or shopping until the credit card maxes out seems—at least to science sophisticates—as outdated and discredited as applying leeches to the sick. "There is no magical stuff inside you called willpower that should somehow override nature," James Rosen, a professor of psychology at the University of Vermont, told a reporter. "It's a metaphor." "Willpower as an independent cause of behavior is a myth," said Michael Lowe, professor of clinical psychology at M. C. P. Hahnemann University in Philadelphia.

How did we arrive at this pass? The confusion is nothing new. No less an eminence than Kant threw up his hands in the face of the problem, identifying "freedom of the will" as one of three metaphysical mysteries beyond the reach of the human intellect (the other two are immortality and the existence of God). Kant, in fact, succumbed to the same temptation as others who have grappled with free will: in order to reconcile the discoveries of a universe governed by natural law and the felt experience of freedom of action, he concluded that the world simply must have room (albeit a hidden room) for free moral choices—even if physical determinism rules the world of which we have sensory knowledge. For

Kant, the fact that he could not disprove this notion sufficed to sustain it; the fact that he could not prove it did not deter him from believing it. This leitmotif recurs throughout modern attempts to come to grips with free will: free will seems to violate all we know of how the world works, but as long as we cannot construct a logical proof of its nonexistence we cling to it tenaciously, even desperately.

With attempts to find scientific support for free will failing badly, it is no surprise that the twentieth century saw the slow decline of free will as a scientifically tenable concept. In 1931, Einstein had declared it "man's illusion that he [is] acting according to his own free will." In 1964 the great humanist Carl Rogers wrote that "modern psychological science, and many other forces in modern life as well, hold the view that man is unfree, that he is controlled, that words such as purpose, choice and commitment have no significant meaning." In 1971, B. F. Skinner offered what may be the definitive statement of this view, arguing in *Beyond Freedom and Dignity* that our behavior reflects nothing more noble than conditioned responses to stimuli.

The scientific and philosophic basis for this perspective, of course, goes back to Descartes's clockwork universe and is a primary feature of all radical materialist perspectives. But materialist determinism truly gained ascendancy in biology and psychology more recently. It is hard to date precisely the moment when biological determinism turned free will into a "myth" or a mere "metaphor." Perhaps it was in 1996, with the discovery of the first gene associated with a common behavior—risk taking. Perhaps it was in 1995, with the discovery of leptin, the hormone associated with a loss of appetite control. Or perhaps it was even earlier, with the avalanche of discoveries in neuroscience linking a serotonin deficit with depression, and dopamine imbalances with addiction. Each connection that neuroscientists forged between a neurochemical and a behavior, or at least a propensity toward a behavior, seemed to deal another blow to the notion of an efficacious will.

Even if historians will never agree on the precise turning point, what is clear is that the cascade of discoveries in neuroscience and genetics has created an image of individuals as automata, slaves to their genes or their neurotransmitters, with no more free will than a child's windup toy. As Stapp has observed, "The chief philosophies of our time proclaim, in the name of science, that we are mechanical systems governed, fundamentally, entirely by impersonal laws that operate at the level of our microscopic constituents." This scientific determinism holds that every happenstance has a causally sufficient antecedent in the physical world. Given those antecedents, only the happenstance in question could have occurred. Determinism professes, as James put it, that "the future has no ambiguous possibilities hidden in its womb. . . . Any other future complement than the one fixed from eternity is impossible." That which is not necessary is impossible; though we may conceive of an alternate future as possible, that is an illusion. That which fails to come about was never a real possibility at all. In ancient times, determinism rested on a belief in an omniscient God. Today, it is not old-time religion but, rather, our culture's newfound faith—science—that challenges the belief in free will. "The self . . . is not imagined to be ultimately responsible for itself, or its ends and purposes. Rather, the self is entirely a function of environment and genetics," as one explanation of this view states it. Or, more bluntly, "My genes (or my neurotransmitters) made me do it." In this view it is never the "I" who acts, but always the neurochemicals, or the genes, or the neuronal circuits that determine our choices and our course of action. Behavior, in this view, "is solely the consequence of the past history of the system, that has brought it to a state where various neuronal populations form an excitatory consortium that organizes and ineluctably triggers the correlated synaptic volleys needed for a particular movement," as the neuroscientist Robert Doty described it. The sense that one is exercising free will when one orders the cheesecake or moves the cursor on the computer screen to another game of hearts rather than to the spreadsheet program with your overdue taxes—is an illusion, an artifact of a prescientific era, says the prevailing paradigm. The idea that we might choose cantaloupe over cheesecake is as illusory as the apparent underwater "bending" of an oar dipped into a river.

Before we explore the reality of will, it's worth noting that, for a quality whose reality most people wish dearly to believe in, will is hardly something most of us go around exercising every waking minute. For instance, most of our movements are nonmindful and occur without direct conscious control; we generally don't need to will the right foot to lift off the ground and swing forward when the left foot has finished its step. Rather, habitual patterns of action such as those controlled by the basal ganglia and cerebellum, and stimulus-response pairings explain more of our behavior than we perhaps care to admit. The only time volition enters into that walk may be in inspiring us to set out in the first place. But when you reach the last word on the right-hand page of a book, you probably do not (unless reading a mindfulness meditation tract) pause in profound deliberation over turning the page. James called these "effortless volitions," which "are mechanically determined by the structure of that physical mass, [the] brain." But it is effort*ful* volitions that concern us here. It is no exaggeration to call the question of the causal efficacy of will the most critical issue that any mature science of human beings must confront.

In contrast to determinism, indeterminism holds that there exist some actions whose antecedents in the material world are causally insufficient to produce them; given those same antecedents, the agent could have acted differently. It holds that the world of possibilities exceeds the number of actualities, in that the existence (or the coming into existence) of one thing does not strictly determine what other things shall be. When we conceive of alternative futures, more than one is indeed truly possible. "Actualities"—James again—"seem to float in a wider sea of possibilities from out of which they are chosen; and somewhere, indeterminism says, such possibilities exist, and form part of truth." It is obvious from this why the question of free will excites our passions: it seems to be the quality of mental life that, more than any other, holds the key to who we are and why we act. To believe in free will, or to deny it, is to imply a position, too, on such profound questions as the reality of fate and the relation of mind to matter, as well as on such practical ones as the locus and source of moral responsibility and the power all of us hold to shape our destiny. To assert a belief in free will is to accept responsibility for our actions and to recognize the mind as "more or less a first cause, an unmoved mover," as the theorist Thomas Clark says: it is to hold the view that "we could have willed otherwise in the radical sense that the will is not the explicable or predictable result of any set of conditions that held at the moment of choice."

More often than not, to believe that we have such freedom is also to believe that, without it, the moral order is in danger of collapse. If the human mind is not in some sense an unmoved mover, one cannot reasonably assign personal responsibility, or ground a system of true justice. In this sort of world, the person who kills or robs or steals is in the grip of an inexorable mechanical process, and there is no rational basis for belief in taking responsibility for one's actions and choices. If consciousness and its handmaiden, will, are "a benign user illusion," as the philosopher Daniel Dennett argued in 1991 in *Consciousness Explained*, then we come face to face with what he calls "the Spectre of Creeping Exculpation." This is a world most people find abhorrent, in a way the American justice system reflects. Although the law allows for an insanity defense, "insanity" is understood as an inability to understand that one's actions were wrong. Insanity, to the courts, is not an inability to choose to act otherwise. True, occasionally a defendant walks on the basis of the so-called Twinkie defense ("The sugary food I ate made me crazy"). But in the vast majority of cases a defense based on a brain abnormality, or a genetic one, fails. Carried to its logical limits, a system in which no one has a choice about what action to take is unworkable. Despite the messages from genetics and neuroscience, most Americans greatly prefer to believe that we can choose freely—that Adam truly had a choice about whether to eat from the Tree of Knowledge. A Buddhist way of putting this is that you alone are responsible for the motives you choose to act on. In Gotama's words, you are "the owner" of the state of your will and "heir" to the results of your actions. The essence of the Buddhist perspective is that you are free to choose the way in which you exert effort and strive.

In this atmosphere of skepticism about the existence of free will, the *Journal of Consciousness Studies* brought out its 298-page volume, "The Volitional Brain: Towards a Neuroscience of Free Will," in the summer of 1999. The *towards* in the title signaled that we were not there yet. But the pairing of *neuroscience* and *free will* signaled a sea change in attitude about whether free will is even a valid subject for scientific, as distinct from philosophical, inquiry. The scientist who, more than any other, put free will on the neurobiology radar screen was Ben Libet. His experiments have incited as much controversy and as many battling interpretations as any in the field of neuroscience.

Libet was inspired by work reported in 1964 by the German neurophysiologists Hans Kornhuber and Luder Deecke. Using an electroencephalograph (EEG), these researchers discovered that the pattern of electrical activity in the cerebral cortex shifts just before you consciously initiate a movement. It's sort of like the whine of an idling jet engine shifting in pitch just before the plane takes off. The scientists also used a then-new technique that allowed them to analyze stored EEG data and thereby explore the chronological relationship between a voluntary movement (of the hand or foot) and brain activity. What they found was that, between 0.4 and 4 seconds before the initiation of a voluntary movement, there appears a slow, electrically negative brain wave termed the *Bereitschaftpotential*, or "readiness potential." Detectable at the surface of the scalp, the electrical activity was interpreted as being related to the process of preparing to make a movement. But no scientist was prepared to take the next step, investigating whether that electrical activity might have anything to do with the will to make a movement. "Their work just sat there for almost twenty years," Libet said over lunch at a Japanese restaurant in New York in late 2000. "John Eccles, with whom I studied, said to me one day that Kornhuber and Deecke's experiment made the case that conscious will starts almost a second before you act to express that will. I myself thought that was quite unlikely, and in any case I thought it would be hopeless to try to time things accurately enough to fix the moment when conscious will arose. But finally I got this idea."

That idea was to find a way to pinpoint the moment when a person became aware of the conscious desire to act. In experiments he reported in 1982 and 1985, Libet asked volunteers to decide to flick or flex their wrist whenever they chose. These movements were to be performed, as Libet put it, "capriciously, free of any external limitations or restrictions." Devices on the subjects' scalps detected the readiness potential that marks neuronal events associated with preparation for movement. Libet found that this readiness potential began, on average, 550 milliseconds before the activation of the muscles moving the wrist. But not all readiness potentials were followed by movements. "The brain was evidently beginning the volitional process in this voluntary act well before the activation of the muscle that produced the movement," Libet noted in 1999. That is, the readiness potential he was detecting appeared too long before muscle activation to correspond directly with a motor command to the muscle.

What, then, was this odd cerebral signal, which seemed to be acting as a sort of advance scout blazing a trail for the motor command? Libet had instructed his subjects to move the wrist any time they had an urge to do so. His next—and key—question became, When does the conscious intention to execute some movement arise? According to the traditional view of will as something that initiates action, this sense of volition would have to appear before the onset of the readiness potential, or at worst coincidently with it; otherwise the neuronal train would have left the station, as it were, before the will could get into the act. In that case, will would be pretty wimpish, merely assenting to an action that was already under way. But 550 milliseconds is, neuronally speaking, an eternity: "An appearance of conscious will 550 msec or more before the act seemed intuitively unlikely," Libet thought, preceding the act by way too long an interval to make sense. Was it possible, instead, that conscious will *followed* the onset of the readiness potential? If so, "that would have a fundamental impact on how we could view free will."

In his next experiments, Libet therefore tried to establish when will showed up. At first, measuring the onset of will "seemed to me an impossible goal," he recalls. But after giving the matter some thought, he decided to ask subjects, sitting in chairs, to note the position of the second hand on a clock at precisely the moment when they first became aware of the intention to move. Because he was dealing in intervals of less than a second, Libet knew that an ordinary sweep second hand would not suffice. He needed something faster. He came up with the idea of using a spot of light on the face of an oscilloscope. The spot swept around like a second hand, but twenty-five times faster. Each marked-off "second" on the oscilloscope's face therefore amounted to 40 milliseconds. Although this might seem to present a stiff challenge to anyone trying to pinpoint the position of the spot of light, in a dry run Libet found that subjects (including him) were pretty accurate in their readings: when he gave them a weak electrical jolt to the skin and asked them what time the spotlight indicated, the subjects got it right to within 50 milliseconds. "We were ready to go," he says.

Following Libet's instructions, all of the five subjects flicked their wrist whenever the spirit (or something) moved them. They also reported where the oscilloscope spot was when they first became aware of the will to move. Libet compared that self-report with concurrent measurements of the onset of the readiness potential. The results of forty trials—which have since been replicated by other researchers—are straightforward to relate, if difficult to interpret. The readiness potential again appeared roughly 550 milliseconds before the muscle moved. Awareness of the decision to act occurred about 100 to 200 milliseconds before the muscle moved. Simple subtraction gives a fascinating result: the slowly building readiness potential appears some 350 milliseconds before the subject becomes consciously aware of his decision to move. This observation, which held for all of the five subjects in each of the six sessions of forty trials, made it seem for all the world as if the initial cerebral activity (the readiness potential) associated with a willed act was unconscious. The readiness potential precedes a voluntary act by some 550 milliseconds. Consciousness of the intention to move appears some 100 to 200 milliseconds before the muscle is activated—and about 350 milliseconds after the onset of the readiness potential.

Libet thus produced the first experimental support for the version of free will that Richard Gregory famously called "free won't." At first glance, the detection of a readiness potential before consciousness of the wish to act appears to bury free will: after all, cortical activity leading to a movement is well under way before the subject makes what he thinks is a conscious decision to act. The neuronal train has indeed left the station. If free will exists, it seems to be like a late passenger running beside the tracks and ineffectually calling, "Wait! Wait!" Yet Libet does not interpret his work as proving free will a convenient fiction. For one thing, the 150 or so milliseconds between the conscious appearance of will and the muscle movement "allow[s] enough time in which the conscious function might affect the final outcome of the volitional process," he notes. Although his results have been widely and vigorously debated, one interpretation with significant experimental support is this: there exists conscious cerebral activity whose role may be "blocking or vetoing the volitional process so that no actual motor action occurs," as Libet wrote in 1998. "Veto of an urge to act is a common experience for individuals generally." It is also, of course, the essence of mindfulness-based OCD treatment and reaffirms Sherrington's insight that "to refrain from an act is no less an act than to commit one": thus, "free won't."

Experiments published in 1983 clearly showed that subjects could choose not to perform a movement that was on the cusp of occurring (that is, that their brain was preparing to make) and that was preceded by a large readiness potential. In this view, although the physical sensation of an urge to move is initiated unconsciously, will can still control the outcome by vetoing the action. Later researchers, in fact, reported readiness potentials that precede a planned foot movement not by mere milliseconds but by almost two full seconds, leaving free won't an even larger window of opportunity. "Conscious will could thus affect the outcome of the volitional process even though the latter was initiated by unconscious cerebral processes," Libet says. "Conscious will might block or veto the process, so that no act occurs." Everyone, Libet continues, has had the experience of "vetoing a spontaneous urge to perform some act. This often occurs when the urge to act involves some socially unacceptable consequence, like an urge to shout some obscenity at the professor." Volunteers report something quite consistent with this view of the will as wielding veto power. Sometimes, they told Libet, a conscious urge to move seemed to bubble up from somewhere, but they suppressed it. Although the possibility of moving gets under way some 350 milliseconds before the subject experiences the will to move, that sense of will nevertheless kicks in 150 to 200 milliseconds before the muscle moves—and with it the power to call a halt to the proceedings. Libet's findings suggest that free will operates not to initiate a voluntary act but to allow or suppress it. "We may view the unconscious initiatives for voluntary actions as 'bubbling up' in the brain," he explains. "The conscious will then selects which of these initiatives may go forward to an action or which ones to veto and abort. . . . This kind of role for free will is actually in accord with religious and ethical strictures. These commonly advocate that you 'control yourself.' Most of the Ten Commandments are 'do not' orders." And all five of the basic moral precepts of Buddhism are restraints: refraining from killing, from lying, from stealing, from sexual misconduct, from intoxicants. In the Buddha's famous dictum, "Restraint everywhere is excellent."

The evolution of Libet's thoughts about his own experiments mirrors that of neuroscience as a whole about the reality of volition. Libet had long shied from associating his findings with free will. For years he refused even to include the words in his papers and resisted drawing any deeper conclusions from his results. At the 1994 "Toward a Scientific Basis of Consciousness" conference (Tucson I), Libet was asked whether his results could be used to support the existence of free will. "I've always been able to avoid that question," he demurred. But in later years he embraced the notion that free will serves as the gatekeeper for thoughts bubbling up from the brain and did not duck the moral implications of that. "Our experimental work in voluntary action led to inferences about responsibility and free will," he explained in late 2000. "Since the volitional process is initiated in the brain unconsciously, one cannot be held to feel guilty or sinful for simply having an urge or wish to do something asocial. But conscious control over the possible act is available, making people responsible for their actions. The unconscious initiation of a voluntary act provides direct evidence for the brain's role in unconscious mental processes. I, as an experimental scientist, am led to suggest that true free will is a [more accurate scientific description] than determinism."

This may seem an enfeebled sort of free will, if it does not initiate actions but only censors them. And yet the common notion of free will assumes the possibility of acting otherwise in the same circumstances, of choosing not to perform actions that tempt us each and every day. By "the possibility of acting otherwise," I mean not that possibility as judged by an outside observer, one who might sneer that, well, you didn't have to scream at me. I mean, instead, that the possibility of an alternative action is one that you feel as more than theoretical. It must be one that you consider, even if only briefly, before acting. As a matter of fact, William James believed that will seized the moment after the first thought about an intended action, but before the actual action. Consistent with his feeling that "volition is nothing but attention," James argued that the ability to "emphasize, reinforce or protract" certain thoughts at the expense of others percolating into consciousness—an ability he identified with attention—manifests itself as will. So for James, too, will derives not from the freedom to initiate thoughts, but to focus on and select some while stifling, blocking—or vetoing—others. For Buddhist mindfulness practice, it is the moment of restraint that allows mindful awareness to take hold and deepen. The essence of directed mental force is first to stop the grinding machine–like automaticity of the urge to act. Only then can the wisdom of the prefrontal cortex be actively engaged.

Free will as gatekeeper raises a deeper question: how does the gatekeeper decide which thoughts to let pass and which to turn back, which to allow to be expressed and which to veto unceremoniously? Libet himself concedes that although his discovery of the 550-millisecond gap offers a hint of how free will operates, it does not address whether our willed actions are strictly determined by the prior history and state of the neurons in our brain, or whether the will to action is truly free—by which I mean not reducible to and predictable from material processes. The initiatives that bubble up in the brain are, he suspects, based on the person's past—memories, experiences, the values transmitted by the larger society—as well as on present circumstances. If willed actions are strictly determined, and if the brain's veto of possible actions is strictly determined by neural antecedents, then we are right back where we started, with (presumably) unconscious neural states' calling all the shots. Such "free" will seems hardly worth having, and we find ourselves once again in a purgatory where our brains or our genes control our actions as a puppeteer controls a marionette. But Libet insists that such is not the case. "I propose . . . that our conscious veto may not require or be the direct result of preceding unconscious processes," he declared. "There is no logical imperative in any mind-brain theory, even identity theory, that requires specific neural activity to precede and determine the nature of a conscious control function. And there is no experimental evidence against the possibility that the control process may appear without development by prior unconscious processes."

Libet turned eighty-five in 2001, and he had lost none of his fire. He seemed resigned, though, to remain a voice in the wilderness. "Most neuroscientists shy away from my argument invoking free will and a mental field that are not encompassed by existing physical law," he says with a hint of a grin.

> *It violates determinism, which makes them very uncomfortable. But physical laws were discovered as a result of the study of physical objects, not of subjective experience. Even if we had perfect knowledge of all the trillions of synaptic connections in a brain, of all the circuits that comprise it—even with all this, as we have learned from the Heisenberg Uncertainty Principle as well as chaos theory, you cannot predict what that brain will do.*

Both Buddhist and William James's philosophy are quite consistent with this interpretation of volition. In Buddhism, the quality of awareness or attention determines the nature of the consciousness that arises, and thus the action (karma) that takes place. The only willful choice one has is the quality of attention one gives to a thought at any moment. Similarly, James believed that "th[e] strain of attention is the fundamental act of will." And in the Four Steps, of course, to Refocus mindfully away from a destructive obsession or compulsive urge and onto a benign object of attention is the core volitional act, as I will describe further in the next chapter.

Libet's discovery of the 550-millisecond gap in the mid-1980s launched a thousand symposia and inspired a neuroscience of volition. Typically, considering how enamored brain scientists are of mapping the regions that correspond to mental events, they have had a field day (or decades) recording cerebral activity during willed acts. As early as 1977, for instance, researchers led by the Swedish physiologist David Ingvar had volunteers first automatically and rhythmically clench their hand, and then imagine doing the same act without moving their hand. Measuring cerebral blood flow, which serves as a proxy of neuronal activity, they found activation of motor cortex during automatic hand clenching. In addition, and quite markedly, the prefrontal cortex was activated during the willful mental activity. Many subsequent studies have similarly found that willed mental effort results in prefrontal cortex activation. In schizophrenics who show symptoms of a "sick will," which is marked by autistic behavior and inactivity, the dorsolateral prefrontal cortex shows lower-than-normal activity. In depression, one symptom of which is a lack of initiative (will?), three decades of brain mapping have shown consistently low activity in the prefrontal cortex. This has led to the suspicion that the prefrontal cortex houses, at minimum, what Ingvar calls "action programs for willed acts."

Study after study has indeed found a primary role for the prefrontal cortex in freely performed volitional activity. "That aspect of free will which is concerned with the voluntary selection of one action rather than another critically depends upon the normal functioning of the dorsolateral prefrontal cortex and associated brain regions," Sean Spence and Chris Frith concluded in "The Volitional Brain." Damage to this region, which lies just behind the forehead and temples and is the most evolutionarily advanced brain area, seems to diminish one's ability to initiate spontaneous activity and to remain focused on one task rather than be distracted by something else. These symptoms are what one would predict in someone unable to choose a particular course of action. Large lesions of this region turn people into virtual automatons whose actions are reflexive responses to environmental cues: such patients typically don spectacles simply because they are laid before them, or eat food presented to them, mindlessly and automatically. (This is what those who have had prefrontal lobotomy do.) And studies in the 1990s found that when subjects are told they are free to make a particular movement at the time of their own choosing—in an experimental protocol much like Libet's—the decision to act is accompanied by activity in the dorsolateral prefrontal cortex. Without inflating the philosophical implications of this and similar findings, it seems safe to conclude that the prefrontal cortex plays a central role in the seemingly free selection of behaviors, choosing from a number of possible actions by inhibiting all but one and focusing attention on the chosen one. It makes sense, then, that when this region is damaged patients become unable to stifle inappropriate responses to their environment: a slew of possible responses bubbles up, as it does in all of us, but brain damage robs patients of the cerebral equipment required to choose the appropriate one.

Typical of the new breed of neuroscientists intrepid enough to investigate the existence and efficacy of will is Dr. David Silbersweig. As a philosophy major at Dartmouth College, in 1980 he wrote his senior thesis on the philosophy of mind. A slight man with an intense manner (perhaps a side effect of the caffeine he was substituting for sleep, thanks to his newborn, on the summer day in 2000 when we met), Silbersweig chose not to hole up in a garret and think deep thoughts about Mind. Instead, he enrolled in medical school. But after training at Cornell Medical Center and working at the Medical Research Council in London, Silbersweig returned to the passion of his youth. At the functional neuroimaging lab at Cornell that he runs with his wife, Emily Stern, he looks for the neural footprints that volition leaves as it darts through the brain. As he puts it, "We are now in an era where you can address questions of volition through neuroscience."

Silbersweig and Stern do that with PET scans, testing how volition affects conscious perception. Sensory input, of course, does not necessarily produce conscious sensory awareness: if it did, people would be aware of every sight that their visual system takes in, and we plainly aren't. (To test this, ask yourself what lies at the extreme right of your peripheral vision this very second, but without concentrating on it.) In such a case, sensory information is clearly being processed—one can trace the sequence of activation along the visual pathway—yet conscious perception of it is absent. To investigate the role of volition in sensory disturbances, Silbersweig and Stern study, among other conditions, schizophrenia. When schizophrenics hear voices, the brain is constructing a simulacrum of reality. The patient is conscious of sounds that are not there, suggesting that a brain state underlying the mental state ("I hear voices!") is sufficient for conscious awareness. But no volition is involved; the patient does not wish to hear voices. So here we have conscious sensory perception in the absence of both sensory stimuli and volition.

Volition can coexist with conscious perception, but in the absence of sensory stimuli. This is the well-known case of mental imagery. One can voluntarily (volitionally) evoke a sensory experience, calling up the image of a giraffe or the voice of Dr. Martin Luther King, Jr., delivering his "I Have a Dream" speech. If you just did either of these, then your visual association cortex almost certainly became active in the first case, your auditory association cortex in the second. Imagery thus presents a neat comparison to schizophrenic hallucinations: the same lack of sensory input, a similar albeit internally experienced conscious percept—but with volition.

For another example of how volition can affect sensation, fortune sent Stern and Silbersweig a young man known as S. B., who was eighteen when they began studying him in 1992. S. B. had suffered two strokes of the middle cerebral artery, in 1990 and 1991. The strokes had produced lesions that left him cortically deaf: although his ear and the rest of the peripheral components of his auditory system are fine, S. B. fails to hear environmental sounds—a door closing, a car backfiring. But cortical deafness is more nuanced than this. When S. B. concentrates hard, he can make out simple sounds—when they begin and when they cease, as well as their volume. So in experiments with S. B., volition alone is responsible for conscious perception.

PET scans pick out striking differences in brains receiving or not receiving external auditory stimuli, with or without awareness, with or without volition. In one unmedicated schizophrenic's brain, Stern and Silbersweig found, auditory-language association cortices in the temporal lobe became more active at the very moment he reported hearing voices. There was, as expected, no activity in the primary auditory cortex, the region that processes input from the ear. Among five schizophrenic patients who were on medication but still heard voices, the active regions included those generally believed to retrieve contextual information (the hippocampus), integrate emotional experience with perception (ventral striatum), and help maintain conscious awareness (thalamus). These regions, together, probably generate complex, emotionally charged hallucinations. But just as Sherlock Holmes solved a mystery by noting that a dog had failed to bark, it was the brain region that remained dark that offered the tantalizing clue to volition, which is absent during schizophrenic hallucinations. The frontal cortex remained quiet.

Silbersweig and Stern compared this pattern to that in healthy patients who were asked to imagine sounds. "There was a preponderance of activity in the frontal lobes," Silbersweig said. When S. B. became aware of sounds to which he was otherwise deaf—sounds that he could hear only if he willed himself to do so—the same frontal regions lit up. What they were seeing, Silbersweig believes, "was an effect of volition and attention on perceptual function, a top down modulation of activity." The PET results support the hypothesis that these prefrontal systems play a role in "volitional control of conscious sensory experience," Silbersweig and Stern conclude.

One of the more striking hints of the reality, and power, of will came from experiments in the late 1990s on patients with "locked-in syndrome." In this terrifying condition, a patient's cognitive and emotional capacities are wholly intact, but he is completely paralyzed. He cannot point, nod, whisper, smile, or perform any other voluntary motor function. (Occasionally some muscles around the eye are spared, allowing the patient to blink voluntarily and so achieve a rudimentary form of communication.) Such a patient's muscles are deaf to the wishes of his mind. Locked-in syndrome is generally caused by stroke or other brain injury; it can also result from Lou Gehrig's disease, amyotrophic lateral sclerosis (ALS). The damage blocks the pathways by which the brain initiates voluntary movement. For decades scientists despaired of ever helping these patients. But then a few groups began investigating a long shot: might they somehow bypass the muscles and enable the patients to communicate through computers controlled by the brain alone?

Johnny Ray had been locked-in ever since a brainstem stroke in December 1997. His powers of reason, cognition, emotion all remained intact. But his brain could no longer communicate with his body, for those messages run through the brainstem, where the neuronal wires were no more functional than electric utility lines after Hurricane Andrew. He could no longer move or talk. So in a twelve-hour operation the following March, Johnny, a Vietnam veteran, had electrodes implanted into the region of his motor cortex that controlled the movement of his left hand. The electrodes, encased in glass cones, contained growth-promoting substances that caused some of the patient's functioning brain cells to grow into the cones. When they did, an electric signal passing along an axon in the part of the cortex controlling the left hand excited the minuscule gold contacts in the electrodes, which amplified and transmitted the signal through a gold wire to a receiver in Johnny's pillow at the Veterans Affairs Medical Center in Decatur, Georgia, and from there to a computer. Soon, Johnny was imagining moving his left hand, causing a wave of action potentials to sweep through his motor cortex. By altering the frequency of the signals, he managed to move a cursor to various icons ("help," "pain"). He stared at the computer monitor, focusing on the imagined movement of his paralyzed hand, willing the cursor to move. He eventually learned to control the cursor well enough to spell, choosing letters from the screen one at a time. In a few months, he got up to three characters per minute. And then he skipped the middle step: rather than imagine moving his hand, he simply concentrated on moving the cursor. And it moved. He had willed the cursor to move.

The herculean mental effort required to operate the cursor system provides strong evidence that what is involved here is real, volitional effort. As such, it mirrors the tremendous mental effort that OCD patients must make to veto the urge to execute some compulsion. In the absence of effort the OCD pathology drives the brain's circuitry, and compulsive behaviors result. But mental effort, I believe, generates a directed mental force that produces real physical effects: the brain changes that follow cognitive-behavioral therapy for OCD. The heroic mental effort required underlines the power of active mental processes like attention and will to redirect thoughts and actions in a way that is detectable on brain scans. Let me be clear about where mental effort enters the picture. The OCD patient is faced with two competing systems of brain circuitry. One underlies the passively experienced, pathological intrusions into consciousness. The other encodes information like the fact that the intrusions originate in faulty basal ganglia circuits. At first the pathological circuitry dominates, so the OCD patient succumbs to the insistent obsessions and carries out the compulsions. With practice, however, the conscious choice to exert effort to resist the pathological messages, and attend instead to the healthy ones, activates functional circuitry. Over the course of several weeks, that regular activation produces systematic changes in the very neural systems that generate those pathological messages—namely, a quieting of the OCD circuit. Again quoting James, "*Volitional effort is effort of attention. . . . Effort of attention is thus the essential phenomenon of will.*"

I propose, then, that "mental force" is a force of nature generated by volitional effort, such as the effort required to refocus attention away from the obsessions of OCD and onto an actively chosen healthy behavior. Directed mental force, I suggest, accounts for the observed changes in brain function that accompany clinical improvement among OCD patients who have been successfully treated with the Four Steps. The volitional effort required for Refocusing can, through the generation of mental force, amplify and strengthen alternative circuitry that is just beginning to develop in the patient's brain. The results are a quieting of the OCD circuit and an activation of healthy circuits. Through directed mental force, what begin as fragile, undependable processes—shifting attention away from the OCD obsessions and onto less pathological behaviors—gradually become stronger. This is precisely the goal of the therapy: to make the once-frail circuits prevail in the struggle against the OCD intruder. The goal is to become, as Gotama Buddha termed it, "a master of the course of thought." Volitional effort and attentional Refocusing generate a mental force that changes brain circuitry, thus resulting in a lessening of OCD symptoms—and, over time, produces a willfully induced change in the very circuitry of the brain.

One should not, needless to say, posit the existence of a new force of nature lightly. The known forces—gravity, electromagnetism, and the strong and weak forces that, respectively, hold atomic nuclei together and cause radioactive decay—do a pretty good job of explaining a dizzying range of natural phenomena, from the explosion of a supernova to the photosynthesis of a leaf, from the flight of a falling autumn leaf to the detonation of the Hiroshima bomb. But mental force, its name notwithstanding, is not strictly analogous to the four known forces. Instead, I am using *force* to imply the ability to affect matter. The matter in question is the brain. Mental force affects the brain by altering the wave functions of the atoms that make up the brain's ions, neurotransmitters, and synaptic vesicles. By a direct action of mind, the brain is thus made to behave differently. It is in this sense of a direct action of mind on brain that I use the term *mental force*. It remains, for now, a hypothetical entity. But explaining phenomena like the self-directed neuroplasticity observed in OCD patients undergoing Four Steps therapy, like the brain changes detected in those of Alvaro Pascual-Leone's piano players who only imagined practicing a keyboard exercise, like the brain changes in Michael Merzenich's monkeys who paid attention to incoming sensory stimuli—explaining all of these phenomena and more requires a natural force of this kind. Mental force is the causal bridge between conscious effort and the observed metabolic and neuronal changes.

Let me anticipate an objection. Materialists may argue that although the experience of effort is caused by the brain's activity (as are all mental experiences, in this view), it has no effect on the brain. If the brain changes, according to this argument, it is because the same brain events that generate the feeling of mental effort also act back on (other parts of) the brain; this intervening thing called "the feeling of mental effort," they might argue, is a mere side effect with no causal power of its own. But this sort of reasoning is inconsistent with evolutionary theory. The felt experience of willful effort would have no survival value if it didn't actually *do something*. Therefore, positing that the feeling is the mere empty residue of neuronal action is antibiological reasoning and an unnecessary concession to the once-unquestioned but now outdated tenet that all causation must reside in the material realm. Moreover, the "brain changes itself" hypothesis fails to account for the observed clinical data, in which OCD patients describe making a concerted mental effort to master the task that changes their brain. Denying the causal efficacy of mental effort, then, means ignoring the testimony of individuals who describe the enormous exertion of will required to wrestle their obsessions into submission. (Of course, psychology has a long history of dismissing such verbal reports as a misleading source of data. But as James pointed out in 1890, that dismissal reflects the "strange arrogance with which the wildest materialist speculations persist in calling themselves 'science.'") To those of us without a constitutional aversion to the idea of an active role for the mind in the real world, the facts speak loud and clear: there are no rational grounds for denying that conscious mental effort plays a *causal role* in the cerebral changes observed in these OCD patients.

In contrast to classical physics, with its exclusive focus on material causation, quantum physics offers a mechanism that validates the intuitive sense that our conscious thoughts have the power to affect our actions. Quantum theory, in the von Neumann–Wigner formulation as developed by Henry Stapp, offers a mathematically rigorous alternative to the impotence of conscious states: it allows conscious experience to act back on the physical brain by influencing its activities. It describes a way in which our conscious thoughts and volitions enter into the causal structure of nature and focus our thoughts, choose from among competing possible courses of action, and even override the mechanical aspects of cerebral processes. The quantum laws allow mental effort to influence the course of cerebral processes in just the way our subjective feeling tells us it does. How? By keeping in focus a stream of consciousness that would otherwise diffuse like mist at daybreak. Quantum theory demonstrates how mental effort can have, through the process of willfully focusing attention, dynamical consequences that cannot be deduced or predicted from, and that are not the automatic results of, cerebral mechanisms acting alone. In a world described by quantum physics, an insistence on causal closure of the physical world amounts to a quasi-religious faith in the absolute powers of matter, a belief that is no more than a commitment to brute, and outmoded, materialism.

An obvious question is how far one can extend the reach of the hypothesized mental force. As the Decade of the Brain ended, neuroscientists had mapped out the neural circuits that underlie myriad states and behaviors, from depression to aggression to suicidal impulses. Does the existence of mental force imply that with enough attention and volition the violent teen can will himself the brain circuits that make a civilized adult of him? That the suicidal widow can will herself the neural circuits correlated with a love of life, or at least spiritual acceptance? That the schizophrenic can will his voices to be silent, and his visions to disappear? The power of cognitive-behavioral therapy to alter brain circuits in people with either depression or OCD implies that similar therapy, drawing on mental force, should be able to change other circuitry that underlies an aspect of personality, behavior, even thought. And that, of course, encompasses approximately everything, from the mundane to the profound: addiction or temperance, a bad temper or a forgiving nature, impatience or patience, love of learning or antipathy to it, generosity or niggardliness, prejudice or tolerance.

There is a danger to this way of thinking: it treads close to the position that anyone with a mental illness remains sick because of a failure of will, anyone with an undesirable personality trait retains it because she has failed to exert sufficient mental effort. Even those of us who distrust the "My genes (or my neurochemicals) made me do it" school of human behavior back away from the implication that will alone can bring into being the neural circuitry capable of supporting any temperament or behavioral tendency—indeed, any state of mental health. But to frame the issue in this all-or-nothing way is to create a simplistic, and false, choice. The distinction between active and passive mental events offers us some flexibility as we search for where free will and mental force might exhaust their powers. That the passive side of the picture is largely determined, no one can deny. That the intensity of that passive conscious content can at times be overwhelming, no one with an ounce of empathy can fail to realize. Sometimes the power of those passive, unbidden, and unwanted brain processes—the voices the schizophrenic hears, the despair a depressive feels—is simply too great for mental force to overcome. And although directed mental force allows will to change the brain in both the stroke patients Edward Taub has treated and my own OCD patients, of course it is not will alone. It is knowledge, training, support from the community and loved ones, and appropriate medical input.

Twenty-five hundred years ago, a culture very distant from our own in both time and place produced an astonishing prescient insight. In the Pali texts, Gotama says, "It is volition, monks, that I declare to be Karma (Action). Having willed, one perfoms an action by body, speech or mind." By these words the Buddha meant that it is the state of one's will that determines the nature of one's actions (karma), and so profoundly influences one's future states of consciousness. This is the Law of Karma. As the Buddhist scholar Ledi Sayadaw explains, "Volition becomes the chief and supreme leader in the sense that it informs all the rest. Volition, as such, brings other psychical activities to tend in one direction." In addition, Gotama vividly described how the quality of attention that one places on a mental or physical object determines the type of conscious state that arises in response to that object. As the next few months of my collaboration with Henry Stapp were to show, Gotama wasn't a bad neuroscientist—or physicist either, for that matter. By the time Stapp wrote his paper for "Volitional Brain," we were well on the way toward identifying a quantum-based mechanism by which the mental effort that generates "willful focus of attention" would bring about brain changes like those detailed in the OCD brain imaging work. Attention was key.

The implication of the preceding chapters—particularly the power of mental effort and mindfulness to alter neuronal connections—is that will is neither myth nor metaphor. It is, or at least exerts, a real physical force. The research on OCD makes clear that will involves different levels of consciousness, with high-order mental functions potentially presiding over lower-level ones. As 1999 passed, with fireworks and laser shows, into 2000, Stapp and I worked to connect the seemingly disparate threads of a nascent theory: William James's observations about will and attention, my work showing the power of mental effort to change the brain patterns of OCD patients, and quantum physics. James foreshadowed the mechanism by which, according to Stapp, volition acts through quantum processes: "At various points," James wrote, "ambiguous possibilities shall be left open, either of which, at a given instant, may become actual. [One] branch of these bifurcations become[s] real."

{TEN}

ATTENTION MUST BE PAID

The task is . . . not so much to see what no one has yet seen;
but to think what nobody has yet thought,
about that which everybody sees.
—*Erwin Schrödinger*

On Christmas Eve day 1999, I flew up to San Francisco, taking advantage of a seventy-nine-dollar round-trip fare so that I could touch base with Henry Stapp. Henry picked me up at the airport, and after lunch at a trattoria in North Beach we drove across the bridge to Berkeley. I had recently been rereading William James's work on attention, I told Henry, and realized how uncannily the perspective of this nineteenth-century psychologist foreshadowed the work Henry was doing. I hadn't taken my copy of James's *Psychology: A Briefer Course* with me, so Henry and I set out to find one.

After parking near Telegraph Avenue, we walked past street people slumped against doorways and sprawled across sidewalks, the gaiety of stores' Christmas decorations forming an incongruous backdrop. We ducked into Cody's bookstore and split up, looking for James everywhere from self-help to religion. No luck. But at Moe's used bookstore down the block, we hit paydirt. There in the psychology section, amid what seemed like oceans of Jung, was a single slim volume of James. I opened it to this passage:

> *I have spoken as if our attention were wholly determined by neural conditions. I believe that the array of* things *we can attend to is so determined. No object can* catch *our attention except by the neural machinery. But the* amount *of the attention which an object receives after it has caught our mental eye is another question. It often takes effort to keep the mind upon it. We feel that we can make more or less of the effort as we choose. If this feeling be not deceptive, if our effort be a spiritual force, then of course it contributes coequally with the cerebral conditions to the result. Though it* introduce *no new idea, it will deepen and prolong the stay in consciousness of innumerable ideas which else would fade more quickly away. . . . [I]t is often a matter of but a second more or less of attention at the outset, whether one system shall gain force to occupy the field and develop itself, and exclude the other, or be excluded itself by the other. . . . [T]he whole drama of the voluntary life hinges on the amount of attention, slightly more or slightly less, which rival motor ideas may receive. . . . Effort may be an original force and not a mere effect, and it may be indeterminate in amount.*

As we stood at the counter paying for our find, I could tell by the change in Henry's usually impassive demeanor that I had piqued his interest. "See—I told you it was uncanny how relevant James was to the physics of attention!" I said. Even the guy behind the cash register seemed interested.

Walking down the street to Henry's car, I continued reading aloud. (This was Berkeley on Christmas Eve: no one looked twice at us.) But there was more, I told Henry. Riffling through the book, I opened it to a passage several chapters later: "Volitional effort is effort of attention." And this: "The function of the effort is . . . to keep affirming and adopting a thought which, if left to itself, would slip away." And, "Effort of attention is thus the essential phenomenon of will." And finally, "To sustain a representation, to think, is, in short, the only moral act." Here we got to the nub of it, the conviction that the act of focusing attention so that one thought, one possible action, prevails over all the other possible ones competing for dominance in consciousness—this is the true moral act, the point where volition enters into what James had just called "the cerebral conditions" and, moreover, "contribute[s] coequally" to them in determining which of those competing thoughts and actions will be chosen. It is this power of attention—to select one possibility over all others—that invests us with an efficacious will.

"It's uncanny," I repeated. "It's unbelievable," Henry said. A man of the nineteenth century had described in detail the connection between the quantum-based theory of attention and volition that we described in our "Volitional Brain" papers. The causal efficacy of will, James had intuited more than one hundred years ago, is a higher-level manifestation of the causal efficacy of attention. To focus attention on one idea, on one possible course of action among the many bubbling up inchoate in our consciousness, is precisely what we mean by an act of volition, James was saying; volition acts through attention, which magnifies, stabilizes, clarifies, and otherwise makes predominant one thought out of many. The essential achievement of the will is to attend to one object and hold it clear and strong before the mind, letting all others—its rivals for attention and subsequent action—fade away like starlight swamped by the radiance of the Sun. That was just the idea that had emerged from the quantum approach. I handed the book to Henry and said, "Merry Christmas, and happy New Millennium."

Once settled in Henry's car, we drove back across the Bay Bridge, talking animatedly about how James had come to a scientific understanding of the origin and efficacy of volition that was exactly in line with what quantum theory was telling us a century later. We were just picking up where James left off, I felt; it was as if we'd encountered a time warp that bypassed the entire twentieth century and took us directly from the late nineteenth century to the year 2000.

Given James's strong philosophical bent, it's hardly surprising these twin concepts, attention and will, were of such tremendous importance to him. He was well aware, especially given his goal of placing psychology squarely within natural science, that thickets of controversy awaited anyone willing to tackle the question of free will. But on the key point of the causal efficacy of attention, and its relation to will, James held fast to his belief—one he suspected could not be proved conclusively on scientific grounds, but to which he clung tenaciously on ethical grounds—that the effort to focus attention is an active, primary, and causal force, and not solely the result of properties of a stimulus that acts on a passive brain. Between his 1,300-plus-page *Principles* and the 443-page *Briefer Course* published fifteen months later, he did not budge from (indeed, he elaborated on) the statement that effortful attention "would deepen and prolong the stay in consciousness of innumerable ideas which else would fade more quickly away." If we can but understand the effort of attention, James believed, we will have gone a very long way toward understanding the nature of will.

What particularly struck me was James's recognition of the high stakes involved. The question of whether attention (and therefore will) follows deterministically upon the predictable response of brain cells to stimuli, or whether the amount of attention can be (at least sometimes) freely chosen and causally efficacious, "is in fact the pivotal question of metaphysics, the very hinge on which our picture of the world shall swing from materialism, fatalism, monism, towards spiritualism, freedom, pluralism,—or else the other way." James was scrupulously fair in giving equal time to the view that attention is a fully determined result of brain function rather than a causally efficacious force. As he notes, it is entirely plausible that attention may be "fatally predetermined" by purely material laws. In this view, the amount of attention we pay a stimulus, be it one from the world outside or an internally generated thought or image, is determined solely by the properties of that stimulus and their interaction with our brain's circuits. If the words you hear or the images you see are associated with a poignant memory, for instance, then they trigger—automatically and without any active effort by you—more attention than stimuli that lack such associations. In this case, "attention only fixes and retains what the ordinary laws of association bring 'before the footlights' of consciousness," as James put it. That is, the stimuli themselves provoke neural mechanisms that cause them to be attended to and fixed on. This is the attention-as-effect school of thinking.

But James did not think that attention was always and only a fully determined effect of the stimuli that are its object. On the flight back to Los Angeles, I went over in my own mind what we knew about attention, and why it mattered.

We go through our lives "seeing" countless objects that we do not pay attention to. Without attention, the image (or the sound, or the feel—attention plays a role in every sense) does not register in the mind and may not be stored even briefly in memory. I can guarantee that if you were to scan every square centimeter of a crowd scene in a photograph, visual information about every person depicted would reach your visual cortex. But if I asked you, after you had scanned the photo of the crowd, where the man in the fedora and vest was, you would doubtless be flummoxed. Our minds have a limited ability to process information about multiple objects at any given time. "Because of limited processing resources," as the neuroscientists Sabine Kastner and Leslie Ungerleider of NIH wrote in a 2000 review of attention, "multiple objects present at the same time in the visual field compete for neural representation. . . . Two stimuli present at the same time within a neuron's receptive field are not processed independently. [R]ather, . . . they interact with each other in a mutually suppressive way." They compete for neural representation. The key question for attention is, What determines the winner?

Let's say I asked you, before you looked at the picture, to find the man in fedora and vest. With your mind thus primed, you would probably find him in seconds. You would have selected the relevant stimulus and filtered out the extraneous ones. How? According to our best understanding, the images of scores of people (to continue with our example of the crowd photo) sped from your retina and into your visual cortex, in parallel. But then competition set in. The winner was determined by the strength of the stimulus (perhaps the man in the fedora is brighter than the other images), by its novelty (we tend to pick out, say, the tuxedoed monkey at a black-tie dinner before we notice the humans), by its strong associations (if you scan a crowd scene for someone you know, you can generally pick her out before a stranger), or, most interestingly, by the demand of the task—in this case, looking for "the man in fedora and vest." Selectively focusing attention on target images significantly enhances neuronal responses to them.

This is especially true when nearby stimuli, if not for the power of attention, would distract us. In general, when two images are presented simultaneously, each suppresses the neuronal activity that the other triggers. But selective focusing of attention can override this effect and thereby filter out distractions. How do we know? When physiologists record electrical activity in the brains of monkeys doing tasks that require selective attention, they find that the firing of neurons activated by a target image becomes significantly enhanced when the monkeys selectively focus attention on it, effectively eliminating the suppressive influence of nearby images. Human brains act the same way, according to functional magnetic resonance imaging (fMRI): neurons that respond to a target (the image attracting your attention) fire more strongly than neurons that respond to a distraction. The act of paying attention, then, physically counteracts the suppressive influences of nearby distractions. Robert Desimone of the National Institute of Mental Health, one of the country's leading researchers into the physiology of attention, explains it this way: "Attention seems to work by biasing the brain circuit for the important stimuli. When you attend to a stimulus, the suppression that distracters otherwise cause is reduced."

In other words, selective attention can strengthen or weaken neural processing in the visual cortex. This seems to happen in at least two ways. In the first, the neural response to the object of attention becomes stronger. In one fascinating series of experiments, monkeys were trained to look for the color of an object that flashed on a screen. When they did, neurons that respond to color became more active. Similarly, when the monkeys were trained to keep an eagle eye on the direction an object was moving, or on its orientation, neurons that perform those tasks became more active. Attention to shape and color pumps up the volume of neuronal activity in the region of the visual cortex that processes information about shape and color; attention to speed turns up the activity of neurons in the region that processes information about speed. In people, paying attention to faces turns up activity in the region whose job it is to scan and analyze faces.

If this seems somewhat self-evident, it's worth another look: the visual information reaching the brain hasn't changed. What has changed—what is under the observer's control—is the brain's response to that information. Just as visual information about the color of this book's cover reached your brain as you opened it, so every aspect of the objects on the screen (their shape, color, movements, etc.) reached the monkey's brain. The aspect of the image that monkey pays attention to determines the way its brain responds. Hard-wired mechanisms in different brain areas get activated, or not, depending on what the monkey is interested in observing. An activity usually deemed to be a property of the mind—paying attention—determines the activity of the brain.

Attention can do more than enhance the responses of selected neurons. It can also turn down the volume in competing regions. Ordinarily—that is, in the absence of attention—distractions suppress the processing of a target stimulus (which is why it's tough to concentrate on a difficult bit of prose when people are screaming on the other side of a thin wall). It's all well and good for a bunch of neurons to take in sounds at a boisterous party, but you can't make out a damn thing until you pay attention. Paying attention to one conversation can suppress the distracting ones. Neurons that used to vibrate with the noise of those other conversations are literally damped down and no longer suppress the response of neurons trying to hear the conversation you're interested in. Anyone who has ever had the bad luck to search for a dropped contact lens has also had the experience of paying attention to one object (the lens) and thus suppressing neuronal responses to other objects (bits of lint in a rug). If you are searching for a contact lens on a Persian rug, you can thank this effect for hushing the neurons that respond to those flowers and colors, and turning up the responses of neurons that respond to the glimmer of light reflecting off little clear disks. Specifically, it is the activity of neurons deep in the brain's visual pathway, rather than in the primary visual cortex, that is damped down or turned up by attention.

It often takes real effort to maintain the appropriate focus, which is why it takes so much concentration to get into the proper exit lane at a complicated freeway interchange. But once you muster the appropriate focus, you can literally direct your brain to filter out the suppressive effects of distracting signals. Willfully directed attention can filter out unwanted information—another example of how directed mental force, generated by the effort of directed attention, can modulate neuronal function.

When it comes to determining what the brain will process, the mind (through the mechanism of selective attention) is at least as strong as the novelty or relevance of the stimulus itself. In fact, attention can even work its magic in the total absence of sensory stimuli. If an experimenter teaches a monkey to pay attention to a certain quadrant of a video screen, then single-cell recordings find that neurons responsible for that area will fire 30 to 40 percent more often than otherwise, even when there is no there there—even, that is, when that quadrant is empty. So here again we have the mental act of paying attention acting on the activity of brain circuits, in this case turning them up before the appearance of a stimulus. fMRIs find that activity spikes in human brains, too, when volunteers wait expectantly for an object to appear in a portion of a video monitor. Even before an object appears, attention has already stacked the neuronal deck, activating the visual cortex and, even more strongly, the frontal and parietal lobes—the regions of the brain where attention seems to originate. As a result, when the stimulus finally shows up it evokes an even greater response in the visual cortex than if attention had not primed the brain. This, says Robert Desimone (who happens to also be Leslie Ungerleider's husband), "is the most interesting finding. In attention without a visual stimulus, you get activation in the same cells that would respond to that stimulus, as if the cells are primed. You also get activation in the prefrontal cortex and parietal lobes. That seems like strong evidence that these lobes exert top-down control on what the sensory system processes." To summarize, then, selective attention—reflecting willful activation of one circuit over another—can nudge the brain into processing one signal and not another.

Much of what neuroscientists have learned about attention lately has come from brain imaging. As in so many other areas of neurobiology, imaging beckons with the siren call of finding "the neural correlates of . . .": that is, pinpointing activity in some part of the brain that corresponds to a mental activity. And although I am the last person to equate brain states, or areas of neuronal activity, with attention or any other mental act or experience, it is worth exploring the results of imaging for what they tell us about what is happening in the brain (and where it's taking place) when we pay attention. Briefly, these imaging studies have shown that there is no single attention center in the brain. Rather, there are multiple distributed systems, including those in the prefrontal cortex (involved in task-related memory and planning), parietal cortex (bodily and environmental awareness), and anterior cingulate (motivation). Also activated are the underlying cerebellum and basal ganglia (habit formation and coordination of movement). That's all very nice, but it doesn't really tell us much about how attention works (that's the trouble with the neural-correlates approach). Fortunately some brain imaging studies have gone beyond this, to reveal some truly interesting things about attention.

In 1990, researchers led by Maurizio Corbetta at Washington University went beyond the monkey work to study attention in humans, showing that when you pay attention to something, the part of your brain that processes that something becomes more active. The scientists' subjects watched a computer screen while an array of a dozen identical little boxes appeared for 400 milliseconds. After a 200-millisecond pause, another screen, also filled with geometric shapes, appeared. Half the time, the first and second frames were identical; half the time they differed in one feature or more, such as color or shape or motion of the elements. The volunteers were sometimes told to determine whether the two succeeding images differed at all, and sometimes told to determine whether the images differed specifically in color, in shape, or in motion. Looking for any old difference is an example of "divided attention," in that subjects have to pay attention to more than a single attribute in their visual field, searching and scanning to find a difference. Focusing on a specific attribute, on the other hand, requires "selective attention."

As you might expect, when the volunteers focused attention on a single attribute ("Are the colors of these objects different from the ones you just saw?"), they did much better at identifying how the second screen differed from the first than when they divided their attention among several attributes ("What's different here?"). But then the study turned up what has become a key finding in the science of attention. Active, focused attention to a specific attribute such as color, they discovered, ramps up the activity of brain regions that process color. In other words, the parts of the brain that process color in an automatic, "hard-wired" way are significantly and specifically activated by the willful act of focusing on color. Activity in brain areas that passively process motion are amplified when volunteers focus attention on motion; areas that passively process shape get ramped up when the volunteers focus on shape. Brain activity in a circuit that is physiologically dedicated to a particular task is markedly amplified by the mental act of focusing attention on the feature that the circuit is hard-wired to process. In addition, during the directing of such selective attention, the prefrontal cortex is activated. As we saw in Chapter 9, this is also the brain region implicated in volition or, as we are seeing, in directing and focusing attention's beam.

The following year, another team of neuroscientists confirmed that attention exerts real, physical effects. This time, they looked not for increased neuronal activity but for something that often goes along with it: blood flow. After all, blood carries oxygen to neurons just as it does to every other cell in the body. Just as a muscle engaged in strenuous aerobic activity is a glutton for oxygen, so a neuron that's firing away needs a voluminous supply of the stuff. In the 1991 experiment, some subjects were instructed to pay attention to vibrations applied to the tips of their fingers, while others were not. The researchers found that, in the subjects paying attention to the vibrations, activation in the somatosensory cortex region representing the fingertips increased 13 percent compared to activation in subjects receiving the identical stimulation but not paying attention. It was another early hint that paying attention to some attribute of the sensed world—colors, movements, shapes, faces, feels, or anything else—affects the regions of the brain that passively process that stimulus. Attention, then, is not some fuzzy, ethereal concept. It acts back on the physical structure and activity of the brain.

Attending to one sense, such as vision, does not simply kick up the activity in the region of the brain in charge of that sense. It also reduces activity in regions responsible for other senses. If you are really concentrating on the little black lines and curves on this white page, you are less likely to feel someone brush against you, or to hear someone speaking in the background. When you watch a ballet, if you're focusing on the choreography, you don't hear the music so well. If you're deep in conversation at a noisy party and your partner in dialogue has a deep baritone voice, it is probable that those parts of your auditory cortex that are tuned to low frequency will get an extra activation boost; at the same time, regions of the auditory cortex that process sopranos are likely turned down, with the result that you may literally not hear (that is, be conscious of) a high-pitched voice across the room. Attention, as the neuroscientist Ian Robertson of Trinity College Dublin says, "can sculpt brain activity by turning up or down the rate at which particular sets of synapses fire. And since we know that firing a set of synapses again and again makes [them] grow . . . stronger, it follows that attention is an important ingredient" for neuroplasticity, a point we will return to later. For now, it is enough simply to emphasize that paying attention to a particular mode of sensation increases cerebral activity in the brain region that registers that sensation. More generally, the way an individual willfully focuses attention has systematic effects on brain function, amplifying activity in particular brain circuits.

A growing body of evidence demonstrates that mindfulness itself may be a key factor in the activating process. In one fascinating experiment, Dick Passingham of Oxford University and colleagues at London's Institute of Neurology compared the brain activity of a young man as he tried to figure out a mystery sequence on a keypad, to the brain activity after he had mastered it. All the man was told was that he had to figure out which sequence of eight keys was correct. He did it by trial and error: when he pressed an incorrect key, a low-pitched tone sounded, much as hearing a sour note tells you that you have hit the wrong key on a piano. When he pressed a correct one, a high-pitched tone sounded. Now he both had to remember the correct key and figure out the next one, and the next six after that. Throughout his trial-and-error ordeal, PET scans showed, the man's brain was ablaze with activity. In particular, the prefrontal cortex, parietal cortex, anterior cingulate, caudate, and cerebellum were very active; all are involved in planning, thinking, and moving.

When the young man finally worked out the correct sequence, he was instructed to keep tapping it out until he could do so effortlessly and without error. After an hour, though he was beginning to rebel at the boredom of it all, his fingers could fly over the keypad as if on automatic pilot. In fact, they were: he could tap out the sequence flawlessly while verbally repeating strings of six digits, or even while generating lists of verbs. The effortless automaticity was reflected in a marked change in his brain: according to the PET scan, the man's brain had shut off the lights in numerous regions as if they were offices at quitting time. Although his brain was still remembering the eight keys in order, and signaling the fingers how to move, the mental and cerebral activity behind that output had diminished dramatically. Only motor regions, which command the fingers to move, remained active.

Passingham then took the experimental step that really caught my eye because of its implications for my own nascent theory of directed mental force. What happens in the brain, he asked, if the person carrying out an automatic task suddenly makes a special effort to pay attention to that task? The PET scan kicked out the answer. When the young man again focused on the now-automatic keypad movements, his prefrontal cortex and anterior cingulate jerked awake, becoming metabolically active once again. This is a finding of tremendous importance, for it shows that mindful awareness has an activating effect on the brain, lighting it up. The take-home message of Passingham's studies is that willfully engaging in mindful awareness while performing an automatic task activates the action-monitoring circuitry of the prefrontal cortex. It is this activation that can transform us from automatons to members in good standing of the species *Homo sapiens* (from Latin *sapere*, "to be wise"). Given the strong evidence for the involvement of the prefrontal cortex in the willful selection of self-initiated responses, the importance of knowing we can modulate the brain activity in that very area with a healthy dose of mindfulness can't be overstated.

More evidence for the capacity of willfully directed attention to activate a specialized brain region has come from Nancy Kanwisher's lab at MIT. She and others had already demonstrated that a specific brain area, located where the temporal and occipital lobes meet, is specialized for processing the appearance of faces. Kanwisher had named this the *fusiform face area*. Does the appearance of a face activate this area automatically, or can you modulate that activity through attention? To find out, Kanwisher's team had eight volunteers view a screen that briefly displayed two faces and two houses simultaneously. Before the images appeared, the researchers told each volunteer to take note of the faces in some trials, or of the houses in others. All four images appeared each time but stayed on the screen for a mere one-fifth of a second. Then the volunteers had to determine whether the cued items (faces or houses) were a matching pair. They were able to do this accurately a little more than three-quarters of the time. The key finding: the brain's specialized face-detecting area was significantly more activated when the subjects were actively looking at faces to see whether they matched than when the faces were only viewed passively because the houses were the cued target. In other words, although both the faces and the houses impinged on the retina and the rest of the visual system (including the fusiform face area), choosing actively to focus attention on the face instantly ramped up activity in the brain's specialized face-recognition area. Its activity, that is, is not strictly automatic, "but depends instead on the allocation of voluntary attention," as the MIT team stated it. Their subsequent work has shown that attention can also ramp up activity in the brain's specialized area for recognizing places, including houses and buildings. And it's not only attention to the outside world that reaches us through our senses that causes such increased activity. Similar activations occur when you conjure up an image in your mind's eye.

Thus the willful act of forming a mental image of a familiar face or place with your eyes closed selectively activates the very same face or place area of the brain that seeing the face or place with your eyes does. "We are not passive recipients but active participants in our own process of perception," Kanwisher summed up.

It is pretty clear, then, that attention can control the brain's sensory processing. But it can do something else, too, something that we only hinted at in our discussion of neuroplasticity. It is a commonplace observation that our perceptions and actions do not take place in a vacuum. Rather, they occur on a stage set that has been concocted from the furniture of our minds. If your mind has been primed with the theory of *pointillism* (the use of tiny dots of primary colors to generate secondary colors), then you will see a Seurat painting in a very different way than if you are ignorant of his technique. Yet the photons of light reflecting off the Seurat and impinging on your retina, there to be conveyed as electrical impulses into your visual cortex, are identical to the photons striking the retina of a less knowledgeable viewer, as well as of one whose mind is distracted. The three viewers "see" very different paintings. Information reaches the brain from the outside world, yes—but in "an ever-changing context of internal representations," as Mike Merzenich put it. Mental states matter. Every stimulus from the world outside impinges on a consciousness that is predisposed to accept it, or to ignore it. We can therefore go further: not only do mental states matter to the physical activity of the brain, but they can contribute to the final perception even more powerfully than the stimulus itself. Neuroscientists are (sometimes reluctantly) admitting mental states into their models for a simple reason: the induction of cortical plasticity discussed in the previous chapters is no more the simple and direct product of particular cortical stimuli than the perception of the Seurat painting is unequivocally determined by the objective pattern of photons emitted from its oil colors: quite the contrary.

In late 1998 I happened on a paper by Mike Merzenich and Rob deCharms that fortified my belief that attention is the mechanism by which the mind effects the expression of volition. The two UCSF scientists noted that when an individual pays attention to some stimulus, the neurons in the cerebral cortex that represent this object show increased activation. But Merzenich and deCharms took this observation further. In addition, they noted, "the pattern of activity of neurons in sensory areas can be altered by patterns of attention, leading to measured shifts in receptive fields or tuning of individual neurons." If individual neurons can be tuned to different stimuli, depending on the mind's attentional state, they concluded, then "entire spatial maps across the cortical surface are systematically distorted by attention . . . [which] implies a rapid remapping of the representational functions of the cortex."

The cortex, that is, is as subject to remapping through attention as it is through the changes in sensory input described in our survey of neuroplasticity. In addition, in all three of the cortical systems where scientists have documented neuroplasticity—the primary auditory cortex, somatosensory cortex, and motor cortex—the variable determining whether or not the brain changes is not the sensory input itself but, crucially, the attentional state of the animal. In 1993 Merzenich showed that passive stimulation alone simply did not cut it. He and his students repeatedly exposed monkeys to specific sound frequencies. When the monkeys were trained to pay attention, the result was the expected tonotopic reorganization of the auditory cortex: the representation of the repeatedly heard frequency expanded. But when the monkeys were distracted by another task, and so were paying little or no attention to the tones piped into their ears, no such tonotopic expansion occurred. Inputs that the monkey does not pay attention to fail to produce long-term cortical changes; closely attended behaviors and inputs do. Let me repeat: when stimuli *identical* to those that induce plastic changes in an attending brain are instead delivered to a nonattending brain, there is no induction of cortical plasticity. Attention, in other words, must be paid.

Since attention is generally considered an internally generated state, it seems that neuroscience has tiptoed up to a conclusion that would be right at home in the canon of some of the Eastern philosophies: introspection, willed attention, subjective state—pick your favorite description of an internal mental state—can redraw the contours of the mind, and in so doing can rewire the circuits of the brain, for it is attention that makes neuroplasticity possible. The role of attention throws into stark relief the power of mind over brain, for it is a mental state (attention) that has the ability to direct neuroplasticity. In so doing, it has the power to alter the very landscape of the brain. "Experience coupled with attention leads to physical changes in the structure and future functioning of the nervous system," Merzenich and deCharms concluded. "This leaves us with a clear physiological fact . . . moment by moment we choose and sculpt how our ever-changing minds will work, we choose who we will be the next moment in a very real sense, and these choices are left embossed in physical form on our material selves."

I had long suspected that attention (especially mindfully directed attention) was the key to the brain changes in OCD patients I was successfully treating with the Four Steps. This was precisely why the Refocusing step was so critical: paying attention to an alternative activity was the means by which the brain changed, quieting activity in the OCD circuit. So it was gratifying to see that Merzenich had collected evidence that focusing attention was the critical action effecting neuroplastic changes in the cortex. And as we saw in Chapter 6, purely mental rehearsal of the kind Alvaro Pascual-Leone and colleagues had volunteers perform with a piano exercise—imagining themselves playing it though not actually doing so—was an early hint of the power of attention. The volunteers may not have been touching the ivories, but their intense concentration on the sequence of notes was enough to increase the representation of those fingers in the motor cortex. They were literally thinking themselves into a new brain.

Similarly, Ed Taub had shown that the more stroke patients

concentrated on their tasks—the more they paid attention—the greater their functional reorganization and recovery. In stroke patients who sustain damage to the prefrontal cortex, and whose attention systems are therefore impaired, recovery is much less likely. Two months after the stroke, a simple measure of attention, such as the patient's ability to count tones presented through headphones, predicts almost uncannily how well the patient will recover motor function. The power of attention, that is, determines whether a stroke patient will remain incapacitated or not. Ian Robertson's research group at Trinity College found much the same thing: "How well people can pay attention just after a right-brain stroke predicts how well they can use their left hands two years later." If the attention circuits in the frontal lobes are damaged by the stroke, the patient recovers less well from injury to other regions of the brain than if the frontal lobes are spared.

The powers of attention being reported by neuroscientists around the world in the late 1990s made me suspect that the process of self-directed brain reorganization I continued to document in my OCD patients might also reflect the workings of attention. In particular, I wondered whether the power of attention to bias brain function might also account for an OCD patient's ability to suppress the neuronal activation caused by obsessive thoughts and strengthen the neuronal activation caused by healthy ones. But even hypothesizing that the specific action an OCD patient chooses to focus attention on (washing hands versus tinkering with the car engine) determines which neuronal representation becomes stronger and which fades away threatens to plunge us down the rabbit hole of Cartesian dualism. In the simplest formulation, do OCD patients—indeed, does any of us?—have a choice about what to pay attention to? Or is attention fully determined by passive brain mechanisms? William James, in the passages I read to Henry Stapp on Christmas Eve, recognized that either was logically possible. If attention is fully determined by a stimulus, then if you knew the precise neuronal wiring and the trillions of synapses in a human brain you could predict precisely what—which stimulus in the environment, or which of the countless thoughts percolating just below the radar of consciousness—a person would pay attention to. The materialist reductionists believe that, under those conditions, we could indeed make such a prediction.

But although we can predict with confidence some of the stimuli that will catch our attention, like the snake that leaps onto the forest path we are hiking or the *boom!* of a building being demolished, we cannot predict others. The *meaning* of experience—how the product of those trillions of synapses will be interpreted by the mind—is inexplicable if you use only materialistic terms. In the case of my OCD patients, whether they attend to the insistent inner voice telling them they left the stove on, or to the voice of mindfulness telling them that message is nothing more (or less) than the manifestation of faulty brain wiring, is *not* predictable. In this case, the ego-dystonic nature of OCD symptoms (the fact that the characteristic intrusive thoughts and urges are experienced as extraneous and alien to the self) enables most patients to distinguish clearly between the competing calls. OCD symptoms can therefore be viewed as painfully amplified versions of the mental events that pass through the mind innumerable times in the course of a day. Most of these mental events are experienced passively, and as outside volitional control; they are often described as "popping into your head." They are thoughts and ideas that may have an identifiable trigger, perhaps a melody that triggers a memory or a sight that prompts a related thought, but feel as if they arise through deterministic mental circuitry over which we have little if any control. They arise unbidden; fleeting, transitory, evanescent, they differ from the thoughts that beset OCD sufferers only in that the latter are much more insistent, discomfiting, and intrusive. OCD thoughts grab the sufferer's attention so insistently that it takes great effort to ignore them. In this way, OCD obsessions illuminate critical differences between mental events that we experience passively and with no apparent effort and those that require significant effort to focus attention on. This aspect of the disease, as I noted earlier, is what attracted me to its study: the hope that such a condition would shed light on the relationship between the mind and the brain and, in particular, on whether mind is causally efficacious in its actions on the brain.

James's dictum "Volitional effort is effort of attention" captures the way OCD patients manage to shift their brain out of pathological thought patterns and into healthy ones. In OCD, two different neural systems compete for attention. One, generated passively and by the pathological brain circuitry underlying the disease, insists you wash your hands again. The other, generated by the active, willful effort characteristic of the Four Steps, beckons to an alternative, healthy behavior, such as gardening. It is the choice of which one to allow into one's attention, which one to hold "steadily before the mind until it *fills* the mind," that shapes subsequent actions. (Even in James's time, OCD was considered a powerful model of when and how something goes wrong with the will. He himself used it as a prime example of a disease of the will.) When my OCD patients succeed in ignoring the siren call of their obsessions, they do so through the power of their attention to hold fast before the mind the image of the healthy alternative to a compulsive act. No one with an ounce of empathy would deny that this requires tremendous effort.

To Henry Stapp, the idea of attention as the motive force behind volition suggested how mind might interact with the quantum brain—how an act of mental effort could focus a stream of consciousness that would otherwise quickly become defocused. Now, for the first time since we began our informal collaboration, Stapp began contemplating a place in his theory for the notion of mental effort. To produce what he would come to call a *quantum theory of consciousness,* he had to reach back through the decades, to his student days at Berkeley in the 1950s. After earning his undergraduate degree in physics from the University of Michigan in 1950, Stapp began work on his Ph.D. thesis at the University of California, Berkeley. His aim was to erect a theoretical framework to analyze the proton-proton scattering experiments being conducted on the cyclotron by Emilio Segrè and Owen Chamberlain (who shared the 1959 Nobel Prize in physics for their discovery of the antiproton). In these experiments, incoming *protons* (the positively charged components of atomic nuclei) caromed off other protons. At first the incoming protons were *polarized* (that is, had their spin vectors aligned) in a certain, known direction. Once they hit the stationary protons they scattered away, with a different polarization. It was logical to expect that this final polarization would have something to do with the initial polarization—or as physicists say, that the polarizations would be correlated. One of Segrè and Chamberlain's bright graduate students, Tom Ypsilantis, happened to be Stapp's roommate. One day, he asked Stapp for help analyzing the scattering result. The work eventually turned into Stapp's thesis and made him familiar with particle correlations. *Correlated particles*—those separated in space or time but sharing a common origin—were soon to trigger a revolution in our understanding of reality.

Through his thesis work, Stapp became one of the first physicists to appreciate what has now become known as Bell's Theorem. John Bell worked at CERN, the sprawling physics lab outside Geneva, Switzerland, designing particle accelerators. He was not paid to do theoretical physics. Yet the soft-spoken, red-bearded Irishman produced what Stapp would, years later, call "the most profound discovery of science." In a 1964 paper, Bell addressed a seeming paradox that had bedeviled physics since 1935. In that year, Albert Einstein and two younger colleagues, Boris Podolsky and Nathan Rosen, had published a paper that had grown out of Einstein's decade-long debate with Niels Bohr about the meaning of the quantum theories that emerged in the 1920s and 1930s. Einstein was convinced that quantum theory was merely a statistical description of a deeper reality that scientists should strive to uncover. He devised countless *thought experiments* (what would happen if . . . ?) to persuade Bohr that quantum theory was inadequate. The paper he wrote with Podolsky and Rosen (the trio became known as EPR) in 1935 proposed one of the most famous thought experiments in modern physics.

"Can Quantum-Mechanical Description of Physical Reality Be Considered Complete?" is about a quality of physical reality called locality. *Locality* means that physical reality in one place cannot be influenced instantaneously by what someone chooses to do at the same time in some faraway place. The core of all classical notions of physical causation, locality holds that all physical effects are caused by local interactions among discrete material particles and their associated fields. Thus if two regions, each bounded in space and time, are separated by a distance so great that even light cannot travel from one to the other, then an action in one region cannot affect anything in the second.

The protagonist of the EPR paper (I am using the simplification offered by the American theoretical physicist David Bohm) is a single quantum particle called a pi meson. It decays into one electron and one positron, which speed off in opposite directions. Quantum mechanics, recall, holds that until an observer observes a property such as the location, momentum, or spin direction of a particle, that property remains undefined. But, as EPR noted, because the positron and electron originated in a single quantum state, their properties remain (according to quantum theory) forever correlated, in a curious and nonclassical state of affairs called *entanglement*. The reality of entanglement has been empirically validated numerous times, but its implications represent one of quantum mechanics' deepest mysteries. Indeed, Schrödinger called entanglement the very essence, "the essential characteristic," of quantum physics. Through entanglement, the spins of two entangled particles, for instance, are not independent. If the spin of the parent particle is, say, 3 up, then the spin of the daughter particles must be something like 1 up and 2 up, or 5 up and 2 down—anything that adds up to the original particle's spin. There is another way of looking at this. If you know the spin of the original particle, and you measure the spin of one of the daughter particles, then you can infer the spin of the other daughter particle. This is the simplest expression of entanglement.

Let's say we make such measurements, proposed EPR. We start with the pi meson's progeny, one electron and one positron. A physicist—we'll call her Alice—measures the spin of the positron after it has flown a great distance. It has flown so far that in the time it takes Alice to measure the positron's spin, not even a signal traveling at the speed of light can reach the electron. The act of measuring the positron should therefore not be able to affect any property of the electron. This is the locality principle at work. But because the positron and electron are correlated, Alice can calculate the spin of the electron, which is based on her measurement of the positron. If her measurement of the positron finds it to have spin up along the x-axis (the horizontal one), then, because the progenitor pi meson has 0 spin, the electron must have spin down along the x-axis. If instead Alice measures the positron's spin along the y-axis (the vertical one) and obtains, say, "left" spin, then she deduces that the electron must have "right" spin along the y-axis.

A problem has arisen. Quantum mechanics insists, as you recall from Chapter 8, that no quantity is a quantity until it is an observed quantity; spin directions do not exist until and unless we measure them. The spin of the electron emerging from the decay of the pi meson is supposed to consist of a superposition of up and down, or right and left. It collapses into a particular spin only if we measure the thing. (We encountered this before, when we noted that the radioactive atom threatening Schrödinger's cat is characterized by a superposition of "decayed" and "intact," collapsing into one or the other only if we peek inside the box.) But Alice, having measured the positron's spin, knows immediately what the electron's spin is. She has therefore defied the rule that quantum properties have no physical reality until they are observed. EPR presented the conflict with quantum theory this way: "If, without in any way disturbing a system, we can predict with certainty the value of a physical quantity, then there exists an element of physical reality corresponding to this physical quantity." Armed with this reasonable criterion for physical reality, EPR asserted that at the instant Alice measures the positron's spin, the electron's spin must have an existence in physical reality even though the electron cannot have been disturbed at that instant by measuring the faraway positron.

EPR were sure they had caught quantum physics in a contradiction. Bohr's Copenhagen Interpretation insisted that properties in the quantum world do not exist until we observe them. Yet in this thought experiment, the spin of the electron is real without our observing it. EPR believed that such properties do have a physical reality independent of our observations of them: location, momentum, and spin direction are elements of reality even if they are not observed. They subscribed to the philosophical position known as *realism,* the belief (antithetical to quantum physics) that quantities exist even without our observing them. To adapt an idea of Einstein's, realism means that the Moon still hangs in the sky whether or not we sneak a peek at it.

EPR had identified an apparent Achilles' heel in quantum theory. Entanglement lets Alice's decision about what to measure *here* instantaneously affect aspects of reality *there*, in violation of locality. Einstein called this long-distance effect "spooky action at a distance," and he thought it absurd—so absurd that it should be enough to sink quantum theory. If the reality of properties of the faraway electron depends on Alice's decision about what measurement to make on the positron, then quantum theory is a philosophical absurdity. Or, as they put it, "The reality of [properties of the second particle] depend upon the process of measurement carried out on the first [particle] which does not disturb the second . . . in any way. No reasonable definition of reality could be expected to permit this."

Physicists inclined to ponder the philosophical implications of quantum theory battled over EPR for almost thirty years before John Bell weighed in with "On the Einstein Podolsky Rosen Paradox." In this paper, he explored whether locality is indeed a property of the physical world. Locality, to repeat, means that physical reality in one place cannot be affected instantaneously by an action in some faraway place. Bell showed that any theory of reality that agrees with the predictions of quantum theory (and, again, all of the predictions of quantum theory are borne out time and again by experiment) must violate locality. Given realism, locality must be discarded as a silly superstition. The universe must be nonlocal. At a deep level, the world is much more closely interconnected than the old physics had let on. The universe must be arranged so that what one freely decides to do *here* must in certain cases influence instantaneously what is true *there*—and *there* is as far away as one would like, from the other side of the laboratory to the other side of the galaxy.

Physicists immediately got busy testing whether the correlations between entangled particles really were in accord with "spooky action at a distance." The experiments typically involve measuring the polarization of pairs of correlated photons, as in the EPR thought experiment. But these experiments were actually carried out, not just cogitated on. Technology had become available for rapidly changing the settings of the measuring device. This development allowed a series of experiments to be performed by Alain Aspect and colleagues at the University of Paris. Published in 1982, they are widely regarded as putting the final nail in the coffin of locality. Aspect's experiments measured correlations between pairs of particles. The particles were separated by so great a distance that no causal connection, no correlation, was possible unless the causation acted faster than the speed of light—instantaneously. And yet Aspect found that the correlations were indeed of the magnitude predicted by quantum mechanics. This seems to show that the physical world is nonlocal: action *here* can instantly affect conditions *there*. (This result disappointed Bell greatly, for, like Einstein, he was uneasy with the weird implications of quantum mechanics and hoped that experiments would shore up realistic, local theories.)

Aspect's conclusions were confirmed in 1997 by Niculus Gisin at the University of Geneva and colleagues. The Swiss team created pairs of entangled photons (quanta of light) and dispatched them through fiber-optic lines to two Swiss villages, Bellevue and Bernex. Aspect had confirmed nonlocality over distances of thirteen meters (the size of his experimental apparatus). Gisin's experiment covered eleven *kilometers*. By the scale of quantum physics, eleven kilometers might as well be 11 billion light-years. Yet Gisin still found that each photon of the pair seemed to know what measurement had been made on its distant partner and to behave accordingly: the photons exhibit the property specified by the measurement made on its partner. Physicists interpret the experiment as indicating that even if tests were conducted across the known universe, they would show that physical reality is nonlocal. Nonlocality appears to be an essential, foundational property of the cosmos. The world, in spite of Einstein's objections, really does seem subject to spooky action at a distance.

At some point in their contemplation of locality, people usually pull up short and ask how being able to predict the polarization of one member of a pair of entangled particles is so weird after all. If you think about it, that prediction seems not much different from a simple, classical case in which you are told that there is a pair of gloves in a drawer. If you blindly grab one and see that it is the left-hand one, you know that the other is the right-hand one. But the gloves differ from the entangled particles in a crucial way. Gloves, being macroscopic and thus describable by classical rather than quantum physics, have an identity independent of our observations. In contrast, as you may remember from Chapter 8, quantum entities have no such identity until we observe them: hence John Archibald Wheeler's conclusion "No phenomenon is a phenomenon until it is an observed phenomenon." According to quantum physics, then, neither the distant quantum particle nor its cousin here in the lab has a spin direction until a measurement fixes that direction. Up until that observation, each particle is a superposition of all possible spin values. If we measure the first particle's spin and find that it is "up," then we have simultaneously determined with equal precision that the spin of the cousin particle that has been flung to the ends of the universe has spin "down." If this property is really brought into existence by the observation that fixes its value, then an observation in one location is directly affecting reality in another, far-off place. Quantum mechanics seems to operate across vast distances instantly in this way. It is nonlocal.

The discovery of nonlocality has shaken our notions of reality and the Cartesian divorce of mind from matter to their very foundations. "Many regard [the discovery of nonlocality] as the most momentous in the history of science," the science historian Robert Nadeau and the physicist Menas Kafatos wrote in their wonderful 1999 book *The Non-Local Universe: The New Physics and Matters of the Mind*. The reason, in large part, is that nonlocality overturns classical ontology. In both classical physics and (as you will recall from Chapter 1) Cartesian dualism, the inner realm of the human mind and the outer realm of the physical world lie on opposite sides of an unbridgeable chasm, leaving mind and physical reality entirely separate and no more capable of meaningful and coherent interactions than different species of salamander on opposite sides of the Grand Canyon. In a nonlocal universe, however, the separation between mind and world meets its ultimate challenge. As Nadeau and Kafatos put it, "The stark division between mind and world sanctioned by classical physics is not in accord with our scientific worldview. When non-locality is factored into our understanding of the relationship between parts and wholes in physics and biology, then mind, or human consciousness, must be viewed as an emergent phenomenon in a seamlessly interconnected whole called the cosmos." An *emergent phenomenon* is one whose characteristics or behaviors cannot be explained in terms of the sum of its parts; if mind is emergent, then it cannot be wholly explained by brain.

Within physics, the implications of nonlocality have generally been downplayed—indeed, have met with almost total silence, like an embarrassing relative at the wedding reception. Why? A good part of the reason is that no practical results seem to arise from the debate about these issues. It might amuse graduate students sitting around after midnight, pondering the meaning of reality and all that, but it didn't provide the basis for the transistor. At bottom, though, the failure to face nonlocality reflects an unease with the implication that the stark divide between mind and world sanctioned by classical physics—in which what is investigated and observed has a reality independent of the mind that observes or investigates—does not accord with what we now know. Almost all scientists, whether trained in the eighteenth century or the twenty-first and whether they articulate it or not, believe that the observer stands apart from the observed, and the act of observation (short of knocking over the apparatus, of course) has no effect on the system being observed. This attitude usually works just fine. But it becomes a problem when the observing system is the same as the system being observed—when, that is, the mind is observing the brain. Nonlocality suggests that nature may not separate ethereal mind from substantive stuff as completely as classical materialist physics assumed. It is here, when the mind contemplates itself and also the brain (as when an OCD patient recognizes compulsions as arising from a brain glitch), that these issues come to a head. In the case of a human being who is observing his own thoughts, the fiction of the dynamic separation of mind and matter needs to be reexamined.

That is what Henry Stapp began to do: explore the physics by which mind can exert a causal influence on brain. To do so, he focused on an odd quantum phenomenon called the Quantum Zeno Effect. Named for the Greek philosopher Zeno of Elea, the Quantum Zeno Effect was introduced to science in 1977 by the physicist George Sudarshan of the University of Texas at Austin and colleagues. If you like nonlocality, you'll love Quantum Zeno, which puts the spookiness of nonlocality to shame: in Quantum Zeno, the questions one puts to nature have the power to influence the dynamic evolution of a system. In particular, repeated and closely spaced observations of a quantum property can freeze that property in place forever, or at least much longer than it would otherwise stay if unwatched.

Consider an atom that has absorbed a photon of energy. That energy has kicked one of the atom's electrons into what's called a higher orbital, kind of like a supermassive asteroid's kicking Mercury into Venus's orbit, and the atom is said to be "excited." But the electron wants to go back where it came from, to its original orbital, as it can do if the atom releases a photon. When the atom does so is one of those chance phenomena, such as when a radioactive atom will decay: the atom has some chance of releasing a photon (and allowing the electron to return home) within a given period. Thus the excited atom exists as a superposition of itself and the unexcited state it will fall into after it has released a photon. Physicists can measure whether the atom is still in its initial state or not. If they carry out such measurements repeatedly and rapidly, they have found, they can keep the atom in its initial state. This is the Quantum Zeno Effect: such a rapid series of observations locks a system into that initial state. The more frequent the observations of a quantum system, the greater the suppression of transitions out of the initial quantum state. Taken to the extreme, observing continuously whether an atom is in a certain quantum state keeps it in that state forever. For this reason, the Quantum Zeno Effect is also known as the watched pot effect. The mere act of rapidly asking questions of a quantum system freezes it in a particular state, preventing it from evolving as it would if we weren't peeking. Simply observing a quantum system suppresses certain of its transitions to other states.

How does it work? Consider this experiment. An ammonia molecule consists of a single atom of nitrogen and three atoms of hydrogen. The arrangement of the four atoms shifts over time because all the atoms are in motion. Let's say that at first the nitrogen atom sits atop the three hydrogens, like an egg nestled on a tripod. (The nitrogen atom has only two options, to be above or below the trio. It cannot be in between.) The wave function that describes the position of the nitrogen is almost entirely concentrated in this configuration: that is, the probability of finding the nitrogen at the apex is nearly 100 percent. Left to its own devices, the wave function would shift as time went by, reflecting the increasing probability that the nitrogen atom would be found below the hydrogens. But before the wave function shifts, we make an observation. The act of observation causes the wave function (which, again, describes the probability of the atom's being in this place or that one) to collapse from several probabilities into a single actuality. This much is standard quantum theory, the well-established collapse of the wave function that follows an observation.

But something interesting has happened. "The wave function has ceased oozing toward the bottom," as Sudarshan and his colleague Tony Rothman explained, "it has been 'reset' to the zero position. And so, by repeated observations at short intervals, . . . one can prevent the nitrogen atom from ever leaving the top position." If you rapidly and repeatedly ask a system, Are you in this state or are you not? and make observations designed to ascertain whether or not the nitrogen atom is where it began, the system will not evolve in the normal way. It will become, in a sense, frozen. As Stapp puts it, "An answer of 'yes' to the posed question [in this case, Is the nitrogen atom on top?] will become fixed and unchanging. The state will be forced to stay longer within the realm that provides a yes answer."

Quantum Zeno has been verified experimentally many times. One of the neatest confirmations came in a 1990 study at the National Institute of Standards and Technology. There, researchers measured the probability that beryllium ions would decay from a high-energy to a low-energy state. As the number of measurements per unit time increased, the probability of that energy transition fell off; the beryllium atoms stayed in their initial, high-energy state because scientists kept asking them, "So, have you decayed yet?" The watched pot never boiled. As Sudarshan and Rothman conclude, "One really can stop an atomic transition by repeatedly looking at it."

The Quantum Zeno Effect "fit beautifully with what Jeff was trying to do," recalls Henry Stapp. It was clear to Stapp, at least in principle, that Quantum Zeno might allow repeated acts of attention—which are, after all, observations by the mind of one strand of thought among the many competing for prominence in the brain—to affect quantum aspects of the brain. "I saw that if the mind puts to nature, in rapid succession, the same repeated question, 'shall I attend to this idea?' then the brain would tend to keep attention focused on that idea," Stapp says. "This is precisely the Quantum Zeno Effect. The mere mental act of rapidly attending would influence the brain's activity in the way Jeff was suggesting." The power of the mind's questioning ("Shall I pay attention to this idea?") to strengthen one idea rather than another so decisively that the privileged idea silences all the others and emerges as the one we focus on—well, this seemed to be an attractive mechanism that would not only account for my results with OCD patients but also fit with everyone's experience that focusing attention helps prevent the mind from wandering. Recall that Mike Merzenich had found that only attended stimuli have the power to alter the cortical map, expanding the region that processes the stimuli an animal focuses on. And recall Alvaro Pascual-Leone's finding that the effort of directed attention alone can produce cortical changes comparable to those generated by physical practice at the piano. It seemed at least possible that it was my OCD patients' efforts at attention, in the step we called Refocusing, that caused the brain changes we detected on PET scans.

In this way, Quantum Zeno could provide a physical basis for the finding that systematic mental Refocusing away from the insistent thoughts of OCD and onto a functional behavior can keep brain activity channeled. Mindfulness and mental effort would then be understood as a way of using attention to control brain state by means of the Quantum Zeno Effect. As Stapp told a 1998 conference in Claremont, California, "The mere choice of which question is asked can influence the behavior of a system. . . . [O]ne's [own] behavior could be influenced in this way by focusing one's attention, if focusing attention corresponds to specifying which question is posed." The Quantum Zeno Effect, he suggested, "could be connected to the psychological experience that intense concentration on an idea tends to hold it in place." Because quantum theory does not specify which question is put to nature or when—the dynamical gap we explored in Chapter 8—there may exist in nature "an effective force associated with mind that is not controlled by the physical aspects of nature," Stapp suggested. "Such a force could control some physical aspect of nature, namely the way a feature of the brain that is directly related to experience deviates, in a way controlled by the observer's focus of attention, from its normal evolution under the influence of physical forces alone."

Stapp began to hammer out the mathematical details by which the Quantum Zeno Effect and nonlocality would allow mental action to be causally efficacious on the brain. He had long recognized, as discussed in Chapter 8, that the Heisenberg choice—What question shall we pose to nature?—provides the basis for a mechanism by which the choice of question determines which of its faces nature deigns to reveal. But the choice of question can be construed as something even more familiar, namely, the choice of what to focus attention on. By the winter of 1999–2000, it was clear to Stapp and me that attention offered an avenue into a scientific understanding of the origin and physics-based mechanism of mental force. It thus offered the hope of understanding how directed mental force acts when OCD patients, by regularly choosing a healthy behavior over a compulsion, alter the gating function of their caudate nucleus in a way that changes the neural circuitry underlying their disease.

What did we know about OCD patients who were following the Four Steps? For one thing, a successful outcome requires that a patient make willful changes in the meaning or value he places on the distressing "error" signals that the brain generates. Only by Relabeling and Revaluing these signals can the patient change the way he processes and responds to them. Once he understands the real nature of these false brain messages, the patient can actively Refocus attention away from the obsessive thoughts. Both the PET scans and the clinical data suggest that the quality of the attentional state—that is, whether it is mindful or unmindful—influences the brain and affects how, and even whether, patients actively process or robotically experience sensory stimuli as well as emotions and thoughts.

A major question now arises. How does the OCD patient focus attention away from the false messages transmitted by the faulty but well-established OCD circuit ("Count the cans in the pantry again!") and toward the barely whispered "true" messages ("No, go feed the roses instead") that are being transmitted by the still-frail circuits that therapy is coaxing into existence? Later on, once the "true" messages have been attended to and acted on for several weeks, they will probably have affected the gating of messages through the caudate and be ever-easier to act on. But early in therapy this process is weak, even nonexistent. It is not at all obvious how a patient heeds the healthy signal, which is just taking shape in his cortex and beginning to forge a new neural pathway through his caudate, and ignores the much more insistent one being generated incessantly by his firmly entrenched and blazingly hyperactive orbital frontal cortex–basal ganglia "error message" circuitry. And once appropriate attention has been paid, how does he activate the motor circuitry that will take him away from the pantry and toward the rose garden? This last is an especially high hurdle, given that movement toward the pantry followed by obsessive counting has been the patient's habitual response to the OCD urge for years. As a result, the maladaptive motor response has its own very well-established brain circuitry in the basal ganglia.

In the buzz of cerebral activity inside the brain, our subjective sense tells us that there arise countless choices, some of them barely breaking through to consciousness. If only for an instant, we hold in our mind a representation of those possible future states—washing our hands or walking into the garden to do battle with the weeds. Those representations have real, physical correlates in different brain states. As researchers such as Stephen Kosslyn of Harvard University have shown, mental imagery activates the same regions of the brain that actual perception does. Thus thinking about washing one's hands, for instance, activates some of the same critical brain structures that actual washing activates, especially at those critical moments when the patient forms the mental image of standing at the sink and washing. "The intended action is represented . . . as a mental image of the intended action, and as a corresponding representation in the brain," says Stapp. In a quantum brain, all the constituents that make up a thought—the diffusion of calcium ions, the propagation of electrons, the release of neurotransmitter—exist as quantum superpositions. Thus the brain itself is characterized by a whole slew of quantum superpositions of possible brain events. The result is a buzzing confusion of alternatives, a more complex version of Schrödinger's alternative (alive or dead) cats. The alternative that persists longer in attention is the one that is caught by a sequence of rapid consents that activates the Quantum Zeno Effect.

This, Henry thought, provided the opening through which attention could give rise to volition. In the brain, the flow of calcium ions within nerve terminals is subject to the Heisenberg Uncertainty Principle. There is a probability associated with whether the calcium ions will trigger the release of neurotransmitter from a terminal vesicle—a probability, that is, and not a certainty. There is, then, also a probability but not a certainty that this neuron will transmit the signal to the next one in the circuit, without which the signal dies without leading to an action. Quantum theory represents these probabilities by means of a superposition of states. Just as an excited atom exists as a superposition of the states "Decay" and "Don't decay," so a synapse exists as a superposition of the states "Release neurotransmitter" and "Don't release neurotransmitter." This superposition corresponds to a superposition of different possible courses of action: if the "Release neurotransmitter" state comes out on top, then neuronal transmission takes place and the thought that this neuron helps generate is born. If the "Don't release neurotransmitter" state wins, then the thought dies before it is even born. By choosing whether and/or how to focus on the various possible states, the mind influences which one of them comes into being.

The more Stapp thought about it, the more he believed that attention picks out one possibility from the cloud of possibilities being thrown up for consideration by the brain. In this case, the choice is which of the superpositions will be the target of our attentional focus. Putting a question to nature, the initial step in collapsing the wave function from a sea of potentialities into one actuality, is then akin to asking, Shall this particular mental event occur? Effortfully attending to one of the possibilities is equivalent to increasing the rate at which these questions to nature are posed. Through the Quantum Zeno Effect, repeatedly and rapidly posing that question affects the behavior of the observed system—namely, the brain. When the mind chooses one of the many possibilities to attend to, it partially freezes into being those patterns of neuronal expression that correspond to the experience of an answer "yes" to the question, Will I do this?

One of the most important, and understandable, quantum processes in the human brain is the migration of calcium ions from the channels through which they enter neuron terminals to the sites where they trigger the release of neurotransmitter from a vesicle. This is a probabilistic process: the ions might or might not trigger that release, with the result that the postsynaptic neuron might or might not fire. Part of this lack of certainty is something even a physicist of the nineteenth century would have understood, it arises from factors like thermal fluctuations and other "noise." But there is, in addition to that lack of certainty, one arising from quantum effects, in particular from the Heisenberg Uncertainty Principle. According to the rules of quantum mechanics, therefore, you get a quantum splitting of the brain into different branches. This occurs in the following manner: since the channel through which the calcium ion must pass to get inside the neuron terminal is extremely narrow (less than one nanometer), it is necessary to apply the Uncertainty Principle. Specifically, since the position of the ion in the channel is extremely restricted, the uncertainty in its velocity must be very large. What this means is that the area in which the ion might land balloons out as it passes from the channel exit to the potential triggering site. Because of this, when the calcium ion gets to the area where it might trigger release of neurotransmitter, it will exist in a superposition of hitting/missing the critical release-inducing site. These quantum effects will generate a superposition of two states: the state in which the neurotransmitter in a vesicle is released, and the state in which the neurotransmitter is not released. Due to this quantum splitting, the brain will tend to contain components that specify alternative possible courses of action. That is, the evolution of the state of the brain in accordance with the Schrödinger equation will normally cause the brain to evolve into a growing ensemble of alternative branches, each representing the neural correlate of some possible conscious experience. Each of these neural correlates has an associated probability of occurring (that is, a probability of being turned from potentiality into actuality by a quantum collapse).

Which possible conscious experience will in fact occur? As noted in Chapter 8, the founders of quantum theory recognized that the mind of the experimenter or observer plays a crucial role in determining which facet of itself nature will reveal. The experimenter plays that role simply by choosing which aspect of nature he wants to probe; which question he wants to ask about the physical world; what he wants to attend to. In this model the brain does practically everything. But mind, by consenting to the rapid re-posing of the question already constructed and briefly presented by brain, can influence brain activity by causing this activity to stay focused on a particular course of action.

Let's take the example of a person suffering from OCD. In this case, one possible brain state corresponds to "Wash your hands again." Another is, "Don't wash—go to the garden." By expending mental effort—or, as I think of it, unleashing mental force—the person can focus attention on this second idea. Doing so, as we saw, brings into play the Quantum Zeno Effect. As a result, the idea—whose physical embodiment is a physical brain state—"Go to the garden" is held in place longer than classical theory predicts. The triumphant idea can then make the body move, and through associated neuroplastic changes, alter the brain's circuitry. This will change the brain in ways that will increase the probability of the "Go to the garden" brain state arising again. (See schematic on pages 362 to 363.)

Mindfulness and its power to help patients effectively Refocus attention seemed to explain how OCD patients manage to choose one thought over a more insistent one. This willful directing of attention can act on the brain to alter its subsequent patterns of activity, for Refocusing on useful behaviors activates the brain circuitry needed to perform them. In this way, brain circuitry is shaped by attentional mechanisms, just as the Quantum Zeno Effect predicts. If it is done regularly, as during the Four Steps, the result is not only a change in behavioral outcome—refusing to accede to the demands of the compulsive urge and instead initiating a healthy behavior—but also a change in the metabolic activity of regions of the brain whose overactivity underlies OCD. Mindfully directed attention is the indispensable factor that brings about the observed brain changes. In consciously rejecting the urge to act on the insistent thoughts of OCD, patients choose alternative and more adaptive behaviors through the willful redirection of attention, with the result that they systematically alter their own brain chemistry, remodeling neuronal circuits in a measurable way. The result is what the late cognitive scientist Francisco Varela recently called "the bootstrapping effect of [mental] action modifying the dynamical landscape" of both consciousness and its neural correlates in the brain.

Once again, William James had sketched the outlines for the emerging theory a century ago. "This coexistence with the triumphant thought of other thoughts . . . would inhibit it but for the effort which makes it prevail," he wrote in *Principles*. "The effort to *attend* is therefore only a part of what the word 'will' covers; it covers also the effort to *consent* to something to which our attention is not quite complete. . . . So that although attention is the first and fundamental thing in volition, *express consent to the reality of what is attended to* is often an additional and quite distinct phenomenon involved." For the stroke victim, the OCD patient, and the depressive, intense effort is required to bring about the requisite Refocusing of attention—a refocusing that will, in turn, sculpt anew the ever-changing brain. The patient generates the mental energy necessary to sustain mindfulness and so activate, strengthen, and stabilize the healthy circuitry through the exertion of willful effort. This effort generates mental force. This force, in its turn, produces plastic and enduring changes in the brain and hence the mind. Intention is made causally efficacious through attention.

Through this mechanism, the mind can enter into the causal

structure of the brain in a way that is not reducible to local mechanical processes—to, that is, electrochemical transmission from one neuron to the next. This power of mind gives our thoughts efficacy, and our volition power. Intention governs attention, and attention exerts real, physical effects on the dynamics of the brain. When I worked out my Four Step treatment of OCD, I had no idea that it would be in line with the quantum mechanical understanding of mind-brain dynamics. I knew that Refocusing and Revaluing make sense psychologically and strongly suspected that these mental/experiential components tap into the power of attention and thus intention to influence brain activity. But which actual physical processes were involved, I had no idea. But now, thanks to Henry Stapp, I do. It is quantum theory that permits willful redirection of attention to have real causal efficacy. In Chapter 1, we explored the conflict between science and moral philosophy posed by Cartesian dualism, and how the two realms Descartes posited—the physical and the mental—have no coherent way to interact. This conflict, devastating in its implications, is now on the verge of being resolved. For quantum theory elegantly explains how our actions are shaped by our will, and our will by our attention, which is not strictly controlled by any known law of nature.

A little humility is in order. Philosophers of an earlier time founded their worldview, materialism, on a set of physical laws associated with Newton and other scientists of the seventeenth century. Those laws turned out to be incomplete and, in their philosophical implications, misleading, especially insofar as they turn the world into a predetermined machine devoid of real moral content. Today, as we derive a scientific worldview from quantum mechanics, we cannot be sure that this theory, too, will not be superseded. For now, however, we are left with the fact that the laws of nature, as Wigner elegantly stated in the epigraph at the beginning of this book, cannot be written without appeal to consciousness. The human mind is at work in the physical universe.

Figure 8

Quantum Effects of Attention

The rules of quantum mechanics allow attention to influence brain function. The release of neurotransmitters requires calcium ions to pass through ion channels in a neuron. Because these channels are extremely narrow, quantum rules and the Uncertainty Principle apply. Since calcium ions trigger vesicles to release neurotransmitters, the release of neurotransmitter is only probabilistic, not certain. In quantum language, the wave function that represents "release neurotransmitter" exists in a superposition with the wave function that represents "don't release neurotransmitter"; each has a probability between 0% and 100% of becoming real. Neurotransmitter release is required to keep a thought going; as a result, whether the "wash hands" or "garden" thought prevails is also a matter of probability. Attention can change the odds on which wave function, and hence which thought, wins:

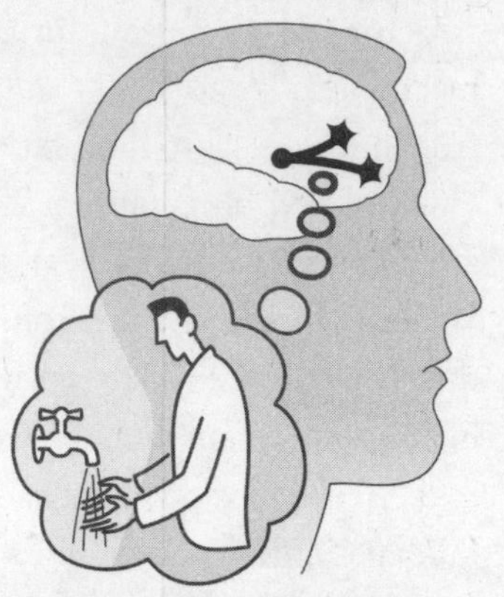

1 In OCD, the brain circuit representing "wash your hands," for instance, fires over and over. This reflects overactivity in the OCD circuit, which includes the orbital frontal cortex, anterior cingulate gyrus, and caudate nucleus.

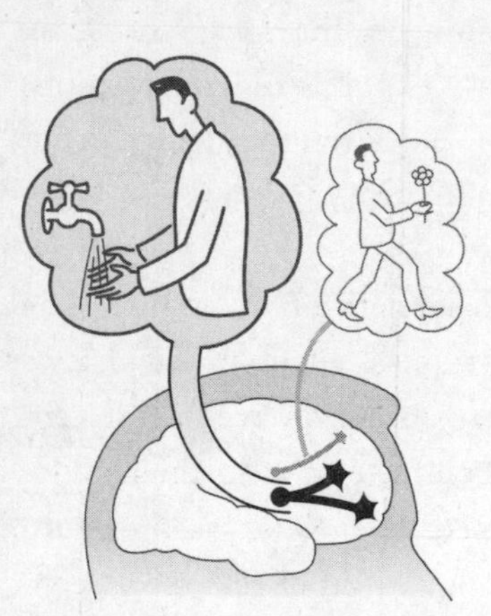

4 The quantum rules allow both states—"release" and "don't release"—to co-exist. Early in therapy, however, the wave representing "release neurotransmitter" in the OCD circuit has a higher probability than the wave representing "release neurotransmitter" in the garden circuit. The patient is much more likely to go to the sink.

5 By expending mental effort and thus unleashing mental force, however, the OCD patient is able, by virtue of the laws of quantum mechanics, to change the odds. Focusing attention on the "garden" thought increases the probability that neurotransmitter will be released in that circuit, not the "wash" circuit.

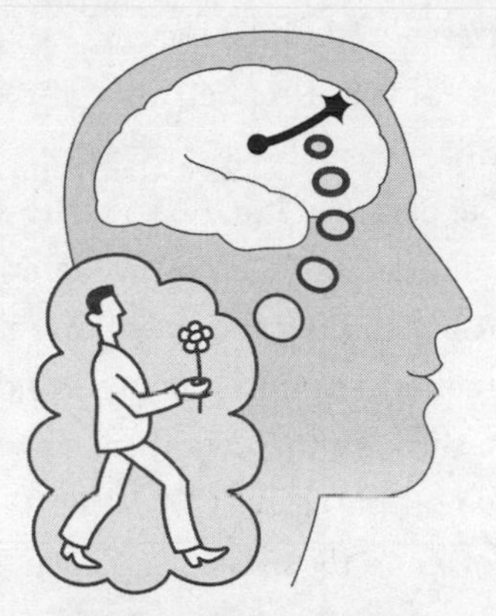

2 Therapy introduces the idea that the OCD patient might go to the garden instead of to the sink. This idea activates planning circuits in the brain's prefrontal cortex. Early in therapy, this circuit is much weaker than the OCD circuit: it has a lower probability of occurring.

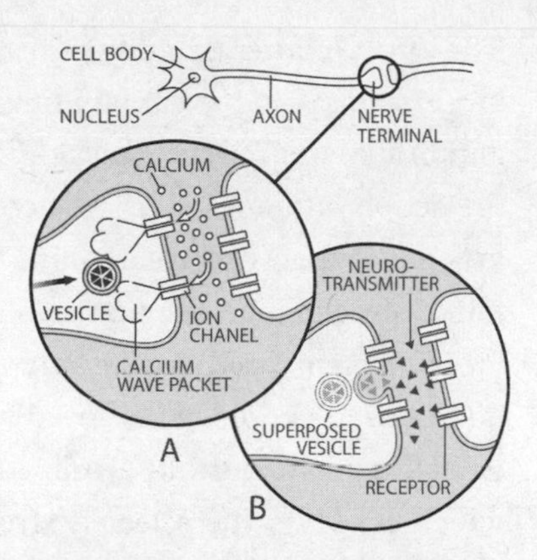

3 The vesicle exists as a superposition of quantum wave functions, one representing "release" and one representing "don't release." This is true in the brain circuit for washing as well as for gardening.

6 The OCD patient can now act on this thought and go to the garden. This increases the chance that, in the future, the "garden" circuit will prevail over the "wash" circuit.

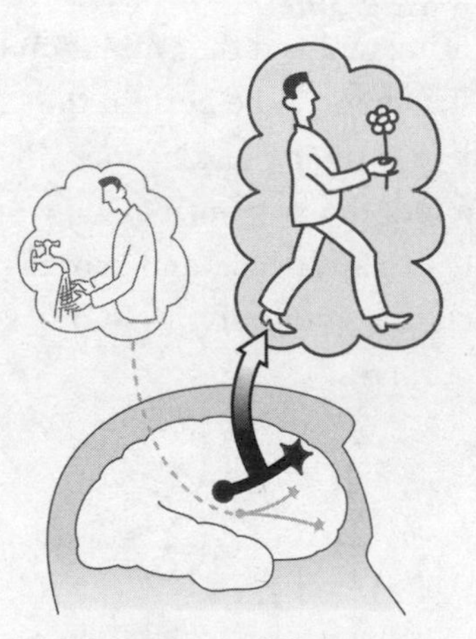

7 If the patient regularly goes to the garden instead of the sink, neuroplasticity kicks in: brain metabolism changes in a way that strengthens the therapeutic circuit. As a result, future OCD urges are easier to overcome.

The effect of attention on the brain offers a rational, coherent, and intuitively satisfying explanation for the interaction between mind and brain, and for the causal efficacy of mental force. It describes the action of the mind as we actually experience it. Consciousness acts on, and acts to create out of an endless universe of predetermined possibilities, the material world—including the brain. Mental effort can speed up the rate at which attention is focused and questions are posed. This speeding up, through the Quantum Zeno Effect, tends to sustain a unified focus on one aspect of reality—which prevents the selected stream of consciousness from losing focus and diffusing. Quantum theory, with the Quantum Zeno Effect, seems to explain how human volition acts in our lives.

For scientifically minded people seeking a rational basis for the belief that truly ethical action is possible, James's epigram—"Volitional effort is effort of attention"—must replace *Cogito ergo sum* as the essential description of the way we experience ourselves and our inner lives. The mind creates the brain. We have the ability to bring will and thus attention to bear on a single nascent possibility struggling to be born in the brain, and thus to turn that possibility into actuality and action. The causal efficacy of attention and will offers the hope of healing the rift, opened by Cartesian dualism, between science and moral philosophy. It is time to explore the closing of this divide, and the meaning that has for the way we live our lives.

EPILOGUE

It is telling that the Decade of the Brain, as (the first) President Bush designated the 1990s, had that name rather than the Decade of the Mind. For it was in the brain rather than the mind that scientists and laypeople alike sought answers, probing the folds and crevasses of our gray matter for the roots of personality and temperament, mental illness and mood, sexual identity and even a predilection for fine food. In my own profession of neuropsychiatry, this attitude is encapsulated in the maxim "For every twisted thought, a twisted molecule." Any mood, preference, or behavior once ascribed to the way we were raised, or even to freely willed volition, instead came to be viewed as the child of our genes and our neurotransmitters, over which we had little, if any, control.

The brain, to be sure, is indeed the physical embodiment of the mind, the organ through which the mind finds expression and through which it acts in the world. Within the brain, ensembles of neurons represent the world beyond, recording both the perceptions of our five senses and the world of mind alone: internally generated imagery produces no less real and measurable a neuronal activation than images of the outside world. But the brain is more than a reflection of our genes. As we saw in Chapter 3, the paltry

35,000 or so genes in the human genome fall woefully short of the task of prescribing the wiring of our 100-trillion-synapse brain. The brain is therefore shaped by and etched with the traces of our experiences—the barrage of sensory stimulation that our peripheral nerves pass along to our brain, the skills we acquire, the knowledge we store, the patterns our thoughts and attention make. All these, and much more, leave their mark.

A mere twenty years ago neuroscientists thought that the brain was structurally immutable by early childhood, and that its functions and abilities were programmed by genes. We now know that that is not so. To the contrary: the brain's ensembles of neurons change over time, forming new connections that become stronger with use, and letting unused synapses weaken until they are able to carry signals no better than a frayed string between two tin cans in the old game of telephone. The neurons that pack our brain at the moment of birth continue to weave themselves into circuits throughout our lives. The real estate that the brain devotes to this activity rather than that one, to this part of the body rather than that one, even to this mental habit rather than that one, is as mutable as a map of congressional districts in the hands of gerrymanderers. The life we lead, in other words, leaves its mark in the form of enduring changes in the complex circuitry of the brain—footprints of the experiences we have had, the actions we have taken. This is neuroplasticity. As Mike Merzenich asserted, the mechanisms of neuroplasticity "account for cortical contributions to our idiosyncratic behavioral abilities and, in extension, for the geniuses, the fools, and the idiot savants among us."

Yet even this perspective assumes a brain more passive than we now understand it to be. It reflects an outdated, classical-physics view of the relationship between mind and matter. For above and beyond the "cortical contributions" to our uniqueness are the choices, decisions, and active will that both propel our actions and, through directed mental force, shape our very brain circuitry.

In the decade since Merzenich's insight, our appreciation of the power of neuroplasticity to reshape the brain has only deepened. We now know that the circuits of our minds change when our fingers fly over the strings of a violin; they change when we suffer an amputation, or a stroke; they change when our ears become tuned to the sounds of our native language and deaf to the phonemes of a foreign one. They change, in short, when the flow of inputs from our senses changes. This much, the Silver Spring monkeys showed us. But the brain—true to its role as the place where Descartes's two realms, the material and the mental, meet and find expression—reflects more than the changing inputs from the body. Neuronal circuits also change when something as gossamer as our thoughts changes, when something as inchoate as mental effort becomes engaged—when, in short, we choose to attend with mindfulness. The power of attention not only allows us to choose what mental direction we will take. It also allows us, by actively focusing attention on one rivulet in the stream of consciousness, to change—in scientifically demonstrable ways—the systematic functioning of our own neural circuitry.

The passive side of mental life, which is generated solely and completely by brain mechanisms, dominates the tone and tenor of our day-to-day, even our second-to-second, experience. During the quotidian business of daily life, the brain does indeed operate very much as a machine does. The brain registers sensory information, processes it, connects it with previously stored sensory experience, and generates an output. It is only a slight exaggeration to say that this much of life is nothing but the brain's going its merry way, running on default awareness. The kind of attention-driven neuroplasticity that Merzenich and his team documented occurs during a mere fraction of our normal experience (more, perhaps, if we are young, and spend many of our waking hours in formal and informal learning); the kind of focused effort that Taub's stroke patients exert is the exception rather than the rule. In general, even the rigorous practice of mindfulness takes up but a few hours in the day of all but the most dedicated practitioners. And even in these cases, when attention is brought to bear, the content of our conscious experience remains largely determined by the inner workings of the brain.

But the content of our character does not, for the amount and quality of attention we focus on this or that aspect of our passive experience as it arises are determined by an active process—one for which brain mechanisms alone provide insufficient explanation. In treating OCD, the capacity to distinguish between passive and active mental processes has turned out to be clinically crucial. When an obsessive thought or compulsive urge enters a patient's mind, the feelings of fear and anxiety it generates are biologically determined. But, as clinical data and PET scans show, patients can willfully change the amount and quality of attention that they focus on those cerebrally generated feelings of anxiety and stress, changing in turn the way the brain works.

The willful focusing of attention is not only a psychological intervention. It is also a biological one. Through changes in the way we focus attention, we have the capacity to make choices about what mental direction we will take; more than that, we also change, in scientifically demonstrable ways, the systematic functioning of neural circuitry. Nowhere is this more clear than among patients with OCD who practice the mindfulness-based Four Step therapy. By Refocusing attention in a mindful fashion, patients change their neurochemistry.

How? By volitional effort, which is effort of attention. Though OCD symptoms may be generated, passively, by the brain, the choice of whether to view those symptoms as "me" or "OCD," whether to become ensnared by them or to focus on a nonpathological behavior, is active. That choice is generated by a patient's mind, and it changes his brain. Mindfulness, as applied in the Four Steps, alters how the connections between the orbital frontal cortex and the caudate nucleus function. The power of attention, and thus the power of mind, reshapes neural circuitry and cortical maps—and does so by means of what I call Directed Mental Force. We now have a scientific basis for asserting that the exercise of the will, the effort of attention, can systematically change the way the brain works. The act of focusing attention has both clinical efficacy (in the treatment of patients besieged by troubling neuropsychiatric symptoms) and biological efficacy (in its power to change the underlying chemistry of the brain). Mind, we now see, has the power to alter biological matter significantly; that three-pound lump of gelatinous ooze within our skulls is truly the mind's brain.

Our will, our volition, our karma, constitutes the essential core of the active part of mental experience. It is the most important, if not the only important, active part of consciousness. We generally think of will as being expressed in the behaviors we exhibit: whether we choose this path or that one, whether we make this decision or that. Even when will is viewed introspectively, we often conceptualize it in terms of an externally pursued goal. But I think the truly important manifestation of will, the one from which our decisions and behaviors flow, is the choice we make about the quality and direction of attentional focus. Mindful or unmindful, wise or unwise—no choice we make is more basic, or important, than this one.

At the end of the nineteenth century, William James recognized that the array of things we can attend to is determined passively by neural conditions—but the amount of attention an aspect of consciousness receives after it has caught our mental eye is determined by active mental processes, by what he called "spiritual force." One's choice of what aspect of experience to focus on is an expression of the active part of mental life. "This strain of attention is the fundamental act of will," James observed in *Psychology: A Briefer Course*. This active component can contribute as much as, and even more than, cerebral conditions in determining where and how attention is directed, and certainly what kind of attention—mindful or unmindful, wise or unwise, diligent or default—is engaged. The feeling that we can make more or less mental effort, as we choose, is not an illusion. Nor is the sense that we have the power to decide, from one moment to another, which aspect of consciousness to attend to. In this critical respect, Jamesian psychology, Buddhist philosophy, and contemporary physics are in total accord. Whereas the contents of consciousness are largely determined by passive processes, the amount and type of attention we pay to those contents are subject to active input via willful mental effort. Cerebral conditions may determine the nature of what's thrown into our minds, but we have the power to choose which aspects of that experience to focus on. The brain may determine the content of our experience, but mind chooses which aspect of that experience receives attention. To repeat: "Volitional effort is effort of attention," James said. And attention—holding before the mind that which, if left to itself, would slip out of consciousness—is the essential achievement of will. This is why effort of attention is, it seems to me, the essential core of any moral act.

What does mind choose to attend to? Buddhist philosophy offers one avenue to understanding this. The traditional practice of Buddhist meditation is based on two broad categories of mental activity: *samatha*, translated as "calmness," "tranquillity," or "quiescence"; and *vipassana*, or "insight." In the beginning stages of training in samatha, attention plays a crucial role by focusing on a single tranquil object, such as the surface of a calm lake or the sensation of breath passing through the nose. The goal is to develop the level of concentration required for attaining a quality of Bare Attention that is steady, powerful, and intense enough to achieve vipassana. Buddhist philosophy teaches that the power of habit can greatly increase the functional effects of the power of karma (which in Buddhist philosophy always means volitional action). Thus the great monk-scholar Ledi Sayadaw (1846–1923) states that "habituating by constant repetition" causes the effects of the subsequent karma to "gain greater and greater proficiency, energy and force—just as one who reads a lesson many times becomes more proficient with each new reading." The will has powers that, at least in the West, have been radically underestimated in an ever more technological and materialist culture. The Law of Karma holds that actions have consequences, and its stress on the vital importance of the state of the will can serve as a counterweight to the materialist bent of Western society, one that has become too credulous about the causal power of material conditions over the human mind. We have been blinded to the power of will to direct attention in ways that can alter the brain. Perhaps, as the discoveries about the power of directed mental effort systematically to alter brain structure and function attract public awareness, we will give greater weight, instead, to the role of volition.

The discovery that the mind can change the brain, momentous as it is both for our image of ourselves and for such practical matters as helping stroke patients, is only the beginning. Finally, after a generation or more in which biological materialism has had neuroscience—indeed, all the life sciences—in a chokehold, we may at last be breaking free. It is said that philosophy is an esoteric, ivory-tower pursuit with no relevance to the world we live in or the way we live. Would that that had been so for the prejudice in favor of biological materialism and its central image, Man the Machine. But biological materialism did, and does, have real-world consequences. We feel its reach every time a pharmaceutical company tells us that, to cure shyness (or "social phobia"), we need only reach for a little pill; every time we fall prey to depression, or anxiety, or inability to sustain attention, and are soothed with the advice that we merely have to get our neurochemicals back into balance to enjoy full mental health. Biological materialism is nothing if not appealing. We need not address the emotional or spiritual causes of our sadness to have the cloud of depression lift; we need not question the way we teach our children before we can rid them of attention deficit disorder. I do not disparage the astounding advances in our understanding of the biochemical and even genetic roots of behavior and illness. Some of those discoveries have been made by my closest friends. But those findings are not the whole story.

Though a substantial majority of the scientists who have done the work leading to those findings agree that there is significantly more to the story than just biology, there has been, up to now, a morbid silence surrounding the moral vacuum created by a worldview dominated by materialist preconceptions. I vividly recall a conversation in which one close and prominent colleague of mine was bemoaning the fact that, according to the dominant materialist view of science, his love for his wife could be explained "solely in terms of the biochemistry of my brain and my viscera." But, because he is a true gentleman who shuns controversy, nothing he does or says in his professional life would give any hint of this demurral. It is my sincere hope that an evolving neurobiology of Directed Mental Force will help rectify this situation.

Human beings are only partially understandable when viewed as the product of material processes. Human beings think, make judgments, and exert effort on the basis of those judgments and in so doing change the material aspects of both their inner and outer worlds in ways that defy the narrow categories of materialist modes of analysis. Understanding our capacity to systematically alter our own neurobiology requires welcoming such concepts as choice and effort into the vocabulary of science. In this new century, questions about the mind-brain interface will become increasingly important as we try to understand how humans function in fields ranging from medicine to economics and political science. Knowing that the mind can, through knowledge and effort, reshape neurobiological processes must powerfully inform that effort.

It is the perspective of what we might call biological humanism, not biological materialism, that fits with the findings of neuroplasticity. It's a mental striving, not a deterministic physical process, that best describes the clinical data on directed neuroplasticity. This may seem to be wishful, even reckless, thinking; after all, to pronounce oneself a skeptic on the subject of biological determinism is to court ridicule, to risk being tarred with the brush of "nonscientific thinking" or even "New Age nonsense." But it seems to me that what we have learned about neuroplasticity and, especially, self-directed neuroplasticity—even this early in our understanding—is that our physical brain alone does not shape our destiny. How can it, when the experiences we undergo, the choices we make, and the acts we undertake inscribe a diary on the living matter of our cortex? The brain continually refines its processing capacities to meet the challenges we present it, increasing the communicative power of neurons and circuits that respond to oft-received inputs or that are tapped for habitual outputs. It is the brain's astonishing power to learn and unlearn, to adapt and change, to carry with it the inscriptions of our experiences, that allows us to throw off the shackles of biological materialism, for it is the life we lead that creates the brain we have. Our new understanding of the power of mind to shape brain can advance not only our knowledge, but also our wisdom. Radical attempts to view the world as a merely material domain, devoid of mind as an active force, neglect the very powers that define humankind. The reality of the mind-shaped brain encourages a cultural climate in which scientific research not only yields advancements in our knowledge, but also adds to our wisdom as an evolving species. By harnessing the power of Directed Mental Force we may yet live up to our taxonomic designation and truly become deserving of the name *Homo sapiens.*

I began, in Chapter 1, with an exploration of the dilemma posed by the notion of a mind's arising from matter, and with Descartes's separation of nature into the material and the mental. Cartesian dualism served science well, at first: by ceding matters of the spirit to men of the cloth, it got the Church off the back of science, which for centuries afterward was perceived as less of a threat to religion's domain than it would otherwise have been (*pace*, Galileo). But Cartesian dualism was a disaster for moral philosophy, setting in motion a process that ultimately reduced human beings to automatons. If all our actions, past and present, can be completely understood as the passive results of machinelike physical mechanisms, without acknowledgment of the existence of consciousness, much less will, then moral responsibility becomes meaningless. If our conscious thoughts matter nothing to the choices we make, and the behavior we engage in, then it is difficult to see how we are any more responsible for our actions than a robot is. That's why the question of whether the mind is capable of real activity (and thus capable of generating a physically effective mental force) is, at its core, an ethical one. "I cannot understand the willingness to act, no matter how we feel, without the belief that acts are really good and bad," James wrote in *The Dilemma of Determinism*. The notion that the mind and the attention it focuses are merely passive effects of material causes, he wrote, "violates my sense of moral reality through and through."

But this conflict between science and moral philosophy vanishes like fog in the light of dawn if, instead of continuing to apply to minds and brains a theory of matter and reality that has been superseded—that is, classical physics—we adopt the most accurate theory of the world advanced so far: quantum theory. In quantum theory, matter and consciousness do not stare at each other across an unbridgeable divide. Rather, they are connected by well-defined and exhaustively tested mathematical rules. "Quantum theory," says Henry Stapp, "rehabilitates the basic premise of moral philosophy. It entails that certain actions that a person can take are influenced by his stream of consciousness, which is not strictly controlled by any known law of nature." A quantum theory of mind, incorporating the discoveries of nonlocality and the Quantum Zeno Effect, offers the hope of mending the breach between science and moral philosophy. It states definitively that real, active, causally efficacious mind operates in the material world.

The shift in understanding inspired by neuroplasticity and the power of mind to shape brain undermines the claim of materialist determinism that humans are essentially nothing more than fleshy computers spitting out the behavioral results of some inescapable neurogenetic program. "The brain is going to do what the brain was always going to do," say the materialists. Both modern physics and contemporary neuroscience reply that they are wrong. The teachings of faith have long railed against the perils of the materialist mind-set. Now neuroscience and physics have joined them at the barricades. The science emerging with the new century tells us that we are not the children of matter alone, nor its slaves. As we establish ourselves in the third millennium, the Law of Karma elaborated so eloquently by Gotama five hundred years before the first millennium still resonates: "All Beings are owners of their Karma. Whatever volitional actions they do, good or evil, of those they shall become the heir."

NOTES

vii *from a psychological-epistemological point of view:* Wigner, E. 1967. *Symmetries and reflections.* Bloomington: Indiana University Press, p. 202.

INTRODUCTION

2 *1913 classic:* Watson, J. B. 1913. Psychology as the behaviorist views it. *Psychological Review, 20,* pp. 158–177.

exposure and response prevention: Baer, L., & Minichello, W. E. 1998. Behavioral treatment for OCD. In: Jenike, M. A., Baer, L., & Minichiello, W. E. (Eds.) 1998. *Obsessive-compulsive disorders: Practical management,* 3rd ed. St. Louis: Mosby.

6 *scientism:* Barzun, J. 2000. *From dawn to decadence: 500 years of Western cultural life.* New York: HarperCollins, p. 218.

10 *Bare Attention:* Nyanaponika Thera. 1973. *The heart of Buddhist meditation.* York Beach, Maine: Samuel Weiser, p. 30.

11 *impartial and well-informed spectator:* Smith, A. 1976. Raphael, D. D., & Macfie, A. L., (Eds.) *The theory of moral sentiments.* New York: Oxford University Press, pp. 112–113.

17 *"the essential achievement of the will":* James, W. 1983. *The principles of psychology.* Cambridge, Mass.: Harvard University Press, p. 1166.

"prolong the stay in consciousness": Ibid., p. 429.

"the utmost a believer in free-will can ever do": Ibid., p. 1177.

18 *"choose and sculpt how our ever-changing minds":* Merzenich, M. M., & deCharms, R. 1996. Neural representations, experience, and change. In: Llinás, R., & Churchland, P. S. (Eds.) *The mind-brain continuum: Sensory processes.* Cambridge, Mass.: MIT Press, pp. 62–81.

19 *"user illusion":* Dennett, D. 1994. In: Guttenplan, S. *A companion to the philosophy of mind.* Oxford, U.K.: Blackwell, pp. 236–243.

CHAPTER ONE

23 *Alcmaeon of Croton:* Burnet, J. 1920. *Early Greek philosophy,* 3rd ed. London: A. & C. Black.

"the brain has the most power for man": Hippocrates, On the sacred disease. Translation from Kirk, G. S., & Raven, J. E. 1963. *The presocratic philosophers: A critical history with a selection of texts.* New York: Cambridge University Press, p. 442.

24 *stimulated tiny spots on the surface:* Penfield, W., & Perot, P. 1963. The brain's record of auditory and visual experience. *Brain, 86,* pp. 595–697.

25 *"The word Mind is obsolete":* Bogen, J. E. 1998. My developing understanding of Roger Wolcott Sperry's philosophy. *Neuropsychologia, 36*(10), pp. 1089–1096.

"understand the brain": Nichols, M.J. & Newsome, W.T. 1999. The neurobiology of cognition. *Nature, 402,* p. C35–38.

27 *"the fundamental features of [the physical] world are as described by physics":* Searle, J. R. 2000. A philosopher unriddles the puzzle of consciousness. *Cerebrum, 2,* pp. 44–54.

27–28 *explanatory gap:* Levine, J. 1983. Materialism and qualia: The explanatory gap. *Pacific Philosophical Quarterly, 6,* pp. 354–361.

Imagine a color-blind neuroscientist: Jackson, J. 1982. Epiphenomenal qualia. *Philosophical Quarterly, 3,* pp. 127–136.

30 *"The problem with materialism":* McGinn, C. 1999. *The mysterious flame: Conscious minds in a material world.* New York: Basic Books, p. 28.

31 *"That one body may act upon another": The Correspondence of Isaac Newton. Volume III, 1688–1694.* Edited by H. W. Turnbull, F.R.S. Cambridge: Published for the Royal Society at the University Press, 1961. Letter 406 Newton to Bentley, 25 February 1692/3.

one version of quantum theory: Von Neumann, J. 1932. *Mathematische Grundlagen der Quanten Mechanik.* English translation from Beyer, R.T. 1953. *Mathematical foundations of quantum mechanics.* Princeton, N.J.: Princeton University Press.

34 *Descartes and La Mettrie:* see the discussion of their work in: Reck, A. J. 1972. *Speculative philosophy: A study of its nature, types, and uses.* Albuquerque: University of New Mexico Press.

36 *As Colin McGinn puts it:* McGinn, 1999, pp. 18–19.

37 *Steven Rose:* Rose, S. 1998. Brains, mind and the world. In: Rose, S. (Ed.) *From brains to consciousness: Essays on the new sciences of the mind.* Princeton, N.J.: Princeton University Press, p. 12.

"That our being should consist of two fundamental elements": Sherrington, C. S. 1947. *The integrative action of the nervous system,* 2d ed. New Haven, Conn.: Yale University Press, p. xxiv.

In 1986 Eccles proposed: Eccles, J. C. 1986. Do mental events cause neural events analogously to the probability fields of quantum mechanics? *Proceedings of the Royal Society of London. Series B: Biological Sciences, 227,* pp. 411–428.

38 *Among the warring theories:* Edelman, G. M., & Tononi, G. A. 2000. *Universe of consciousness: How matter becomes imagination.* New York: Basic Books, p. 6.

"Mentalistic Materialism" as the neurosurgeon Joe Bogen has termed it: Bogen, 1998.

"Mental processes are just brain processes": Flanagan, O. 1992. *Consciousness reconsidered.* Cambridge, Mass.: MIT Press, p. xi.

Churchland and Daniel Dennett: Churchland, P. M., & Churchland, P. S. 1998. *On the contrary: Critical essays, 1987–1997.* Cambridge, Mass.: MIT Press; Dennett, D. C. 1991. *Consciousness explained.* Boston: Little, Brown.

39 *"mind does not move matter":* Herrick, C. J. 1956. *The evolution of human nature.* Austin, Tex.: University of Texas Press, p. 281.

40 *the causal efficacy of mind:* James, W. 1983. The automaton theory. In: *The principles of psychology.* Cambridge Mass: Harvard University Press, Chap. 5.

called such traits spandrels*:* Gould, S. J., & Lewontin, R. C. 1979. The spandrels of San Marco and the Panglossian paradigm: a critique of the adaptationist programme. *Proceedings of the Royal Society of London B, 205,* pp. 581–598.

41 *Emergent materialism:* Sperry, R. W. 1992. Turnabout on consciousness: A mentalist view. *Journal of Mind and Behavior, 13,* pp. 259–280.

42 *As he put it in 1970:* Sperry, R. W. 1970. Perception in the absence of the neocortical commissures. *Research Publications Association for Research in Nervous and Mental Disease, 48,* pp. 123–138.

44 *Agnostic physicalism:* Bogen, 1998.

45 *Process philosophy:* for a useful overview see Reck, A. J. 1972. A useful book on Whitehead's profoundly abstract philosophical system is Sherburn, D. W. (Ed.) 1981. *A key to Whitehead's process and reality.* Chicago: University of Chicago Press.

Dualistic interactionism: Popper, K. R., & Eccles J. C. 1977. *The self and its brain: An argument for interactionism.* New York: Springer International.

46 *the Australian philosopher David Chalmers:* Chalmers, D. J. 1996. *The conscious mind: In search of a fundamental theory.* New York: Oxford University Press.

"started out life as a materialist": Kuhn, R. L. 2000. *Closer to the truth.* New York: McGraw-Hill, p. 21.

47 *"To truly bridge the gap":* Ibid.

48 *"When I first got interested in":* Searle, 2000.

Crick: Crick, F. J. 1994. *The astonishing hypothesis: The scientific search for the soul.* New York: Scribner's.

Edelman: Edelman & Tononi, 2000.

49 *"reductionistic neurobiological explanations":* Singer, W. 1998. Consciousness from a neurobiological perspective. In: Rose (Ed.) 1998, p. 229.

"ambiguous relationship to mind": Rose, 1998.

52 *"mind is but the babbling of a robot":* Doty, R. W. 1998. The five mysteries of the mind, and their consequences. *Neuropsychologia, 36,* pp. 1069–1076.

Chapter Two

54 *obsessive compulsive disorder (OCD):* Two standard general reference books for information on OCD are: Jenike, M. A., Baer, L., & Minichiello, W. E. (Eds.) 1998. *Obsessive-compulsive disorders: Practical management,* 3rd ed. St. Louis: Mosby; Koran, L. M. 1999. *Obsessive-compulsive and related disorders in adults: A comprehensive clinical guide.* New York: Cambridge University Press.

59 *exposure and response prevention:* Foa, E. B., & Wilson, R. 2001. *Stop obsessing! How to overcome your obsessions and compulsions,* rev. ed. New York: Bantam Doubleday Dell; Meyer V., Levy, R., & Schnurer, A. 1974. The behavioral treatment of obsessive-compulsive disorders In: Beech, H. R. (Ed.) *Obsessional states.* London: Methuen, pp. 233–256.

60 *some 25 percent of patients:* Baer, L., & Minichello, W. E. 1998. Behavioral treatment for obsessive-compulsive disorder. In: Jenike, Baer, & Minichiello, 1998, pp. 337–367.

61 *Cognitive therapy—a form of structured introspection—was already*

widely used: Beck, A. T., Rush, A. J., Shaw, B. F., & Emery, G. 1979. *Cognitive therapy of depression.* New York: Guilford Press.

62 *we had studied depression:* Baxter, L. R., Jr., Phelps, M. E., Mazziotta, J. C., Schwartz, J. M., et al. 1985. Cerebral metabolic rates for glucose in mood disorders: Studies with positron emission tomography and fluorodeoxyglucose F 18. *Archives of General Psychiatry, 42,* pp. 441–447.

analysis of the PET scans: Baxter, L. R., Jr., Schwartz, J. M., Mazziotta, J. C., et al. 1988. Cerebral glucose metabolic rates in nondepressed patients with obsessive-compulsive disorder. *American Journal of Psychiatry, 145,* pp. 1560–1563; Baxter, L. R., Jr., Phelps, M. E., Mazziotta, J. C., Guze, B. H., Schwartz, J. M., & Selin, C. E. 1987. Local cerebral glucose metabolic rates in obsessive-compulsive disorder: A comparison with rates in unipolar depression and in normal controls. *Archives of General Psychiatry, 44,* pp. 211–218.

anterior cingulate gyrus: Swedo, S. E., Schapiro, M. B., Grady, C. L., et al. 1989. Cerebral glucose metabolism in childhood-onset obsessive-compulsive disorder. *Archives of General Psychiatry, 46,* pp. 518–523.

63 *elevated metabolism in the orbital frontal cortex:* Rauch, S. L., & Baxter, L. R. 1998. Neuroimaging in obsessive-compulsive disorder and related disorders. In: Jenike, Baer, & Minichiello, 1998, pp. 289–317.

behavioral physiologist E. T. Rolls at Oxford: Thorpe, S. J., Rolls, E. T., & Maddison, S. 1983. The orbitofrontal cortex: Neuronal activity in the behaving monkey. *Experimental Brain Research, 4,* pp. 93–115.

64 *The orbital frontal cortex, it seems, functions as an error detector:* reviews of recent work on this subject are in: Rolls, E. T. 2000. The orbitofrontal cortex and reward. *Cerebral Cortex, 10,* pp. 284–294; O'Doherty, J., Kringelbach, M. L., Rolls, E. T., et al. 2001. Abstract reward and punishment representations in the human orbitofrontal cortex. *Nature Neuroscience, 4,* pp. 95–102; Rogers, R. D., Owen, A. M., Middleton, H. C., et al. 1999. Choosing between small, likely rewards and large, unlikely rewards activates inferior and orbital prefrontal cortex. *Journal of Neuroscience, 15,* pp. 9029–9038.

65 *volunteers play a sort of gambling game:* Bechara, A., Damasio, H., Tranel, D., & Damasio, A. R. 1997. Deciding advantageously before

knowing the advantageous strategy. *Science, 275,* pp. 1293–1295. Bechara, A., Damasio, H., Damasio, A. R., & Lee, G. P. 1999. Different contributions of the human amygdala and ventromedial prefrontal cortex to decision-making. *Journal of Neuroscience, 19,* pp. 5473–5481.

67 *traffic pattern connecting the striatum and the cortex:* For accessible reviews of this see: Schwartz, J. M., 1998. Neuroanatomical aspects of cognitive-behavioral therapy response in obsessive-compulsive disorder: An evolving perspective on brain and behavior. *British Journal of Psychiatry, 173, Supplement 35,* pp. 39–45; Schwartz, J. M. 1997. Obsessive-compulsive disorder. *Science & Medicine, 4(2),* pp. 14–23.

68 *are called* matrisomes: Eblen, F., & Graybiel, A. M. 1995. Highly restricted origin of prefrontal cortical inputs to striosomes in the macaque monkey. *Journal of Neuroscience, 15,* pp. 5999–6013.

neuronal mosaic of reason and passion: Graybiel, A. M., & Rauch, S. L. 2000. Toward a neurobiology of obsessive-compulsive disorder. *Neuron, 28,* pp. 343–347; Graybiel, A. M. & Canales, J. J. 2001. The neurobiology of repetitive behaviors: Clues to the neurobiology of Tourette syndrome. *Advances in Neurology, 85,* pp. 123–131.

tonically active neurons (TANs): Aosaki, T., Kimura, M., & Graybiel, A. M. 1995. Temporal and spatial characteristics of tonically active neurons of the primate's striatum. *Journal of Neurophysiology, 73,* pp. 1234–1252.

70 *serve as a sort of gating mechanism, redirecting information flow:* Graybiel, A. M. 1998. The basal ganglia and chunking of action repertoires. *Neurobiology of Learning and Memory, 70,* pp. 119–136.

role in the development of habits: Jog, M. S., Kubota, Y., Connolly, C. I., Hillegaart, V., & Graybiel, A. M. 1999. Building neural representations of habits. *Science, 26,* pp. 1745–1749.

purposefully alter the response contingencies of their own TANs: Schwartz, J. M. 1999. A role for volition and attention in the generation of new brain circuitry: Toward a neurobiology of mental force. In: Libet, B., Freeman, A., & Sutherland, K. (Eds.) *The volitional brain: Towards a neuroscience of free will.* Thorverton, U.K.: Imprint Academic.

71 *two output pathways: one direct and one indirect:* Baxter, L. R., Jr., Clark, E. C., Iqbal, M., & Ackermann, R. F. 2001. Cortical-subcortical systems in the mediation of obsessive-compulsive disorder: Model-

ing the brain's mediation of a classic "neurosis." In: Lichter, D. G., & Cummings, J. L. (Eds.) *Frontal-subcortical circuits in psychiatric and neurological disorders.* New York: Guilford Press, pp. 207–230.

"worry circuit": Baxter, L. R., Jr., Schwartz, J. M., et al. 1992. Caudate glucose metabolic rate changes with both drug and behavior therapy for obsessive-compulsive disorder. *Archives of General Psychiatry, 49,* pp. 681–689.

72 *"streams of thought and motivation":* Graybiel & Rauch, 2000.

what I came to call Brain Lock: Schwartz, J. M., & Beyette, B. 1997. *Brain lock: Free yourself from obsessive-compulsive behavior.* New York: HarperCollins.

73 *anterior cingulate:* Bush, G., Luu, P., & Posner, M. I. 2000. Cognitive and emotional influences in anterior cingulate cortex. *Trends in Cognitive Sciences, 4,* pp. 215–222.

Researchers at Massachusetts General Hospital: Breiter, H. C., Rauch, S. L., et al. 1996. Functional magnetic resonance imaging of symptom provocation in obsessive-compulsive disorder. *Archives of General Psychiatry, 53,* pp. 595–606; Rauch, S. L., Jenike, M. A., et al. 1994. Regional cerebral blood flow measured during symptom provocation in obsessive-compulsive disorder using oxygen 15–labeled carbon dioxide and positron emission tomography. *Archives of General Psychiatry, 51,* pp. 62–70.

76 *Nyanaponika Thera:* Nyanaponika Thera, 1973.

87 *Ludwig von Mises, who defined* valuing: Von Mises, L. 1962. *The ultimate foundation of economic science: An essay on method.* Kansas City, Kans.: Sheed Andrews & McMeel.

88 *significantly diminished metabolic activity:* Schwartz, J. M., Stoessel, P. W., Baxter, L. R., Jr., et al. 1996. Systematic changes in cerebral glucose metabolic rate after successful behavior modification treatment of obsessive-compulsive disorder. *Archives of General Psychiatry, 53,* pp. 109–113.

92 *Benazon of Wayne State:* Benazon, N. R., Ager, J., & Rosenberg, D. R. 2002. Cognitive behavior therapy in treatment-naïve children and adolescents with obsessive-compulsive disorder: An open trial. *Behavior Research and Therapy, 40,* p. 529–539.

93 *William James posed the question:* Meyers, G. E. (Ed.) 1992. Psychology: Briefer course. In: *William James Writings 1878–1899.* New York: Library of America, p. 417.

Chapter Three

99 *lasting language deficit:* Ratey, J. J. 2000. *A user's guide to the brain.* New York: Pantheon, p. 270.

101 *to process visual information instead:* Eliot, Lise. 1999. *What's going on in there? How the brain and mind develop in the first five years of life.* New York: Bantam, p. 250.

congenitally deaf people: Bavelier, D., & Neville, H. J. 2002. Cross-modal plasticity: where and how? *Nature Reviews Neuroscience,* 3(6), pp. 443–452.

102 *In their breakthrough experiment:* Von Melchner, L., Pallas, S. L., Sur, M. 2000. Visual behaviour mediated by retinal projections directed to the auditory pathway. *Nature, 404,* pp. 871–875.

103 *"the animals 'see'":* Merzenich, M. 2000. Seeing in the sound zone. *Nature, 404,* pp. 820–821.

107 *as a cause of behavioral improvements:* Van Praag, H., Kempermann, G., & Gage, F. H. 2000. Neural consequences of environmental enrichment. *Nature Reviews Neuroscience, 1,* pp. 191–198.

strengthened their synaptic connections: Robertson, I. H., & Murre, J. M. J. 1999. Rehabilitation of brain damage: Brain plasticity and principles of guided recovery. *Psychological Bulletin, 125,* pp. 544–575.

108 *molecular changes:* Kandel, E R. 1998. A new intellectual framework for psychiatry. *American Journal of Psychiatry,* 155(4), pp. 457–469.

111 *an average of 2,500 of these specialized junctions, or synapses:* Gopnik, A., Meltzoff, A. N., & Kuhl, P. K. 1999. *The scientist in the crib: Minds, brains and how children learn.* New York: William Morrow, p. 186.

100 trillion—synapses: Ibid., p. 181.

117 *About half the neurons that form in the fetal brain die before the baby is born:* Ratey, 2000, p. 26.

1.8 million synapses per second: Eliot, 1999, p. 27.

20 billion synapses are pruned every day: Ibid., p. 32.

119 *They literally could not hear any difference:* Gopnik, Meltzoff & Kuhl, 1999, p. 103.

by twelve months they could not: Ibid., p. 107.

rarely learn to speak it like a native: Ibid., p. 192.

forms millions of connections every day: Ibid., p. 1.

120 *in the wilds of New York City:* Ibid., p. 182.

121 *all of the 100 million neurons of the primary visual cortex form:* Eliot, 1999, p. 204.
10 billion per day: Ibid.
visual acuity has improved fivefold: Maurer, D., Lewis, T. L., Brent, H. P., & Levin, A. V. 1999. Rapid improvement in the acuity of infants after visual input. *Science, 286,* pp. 108–109.
sees the world almost as well as a normal adult: Sireteanu, R. 1999. Switching on the infant brain. *Science, 286,* pp. 58–59.

122 *"eliminate addressing errors":* Shatz, C. J. 1992. The developing brain. *Scientific American, 267,* pp. 62–67.

124 *strikes about 1 baby in 10,000:* Sireteanu, 1999, p. 60.
the brain never develops the ability to see normally: Ibid., p. 59.

125 *"visual input was focused on the retina":* Maurer, 1999, p. 108.
as well as normal language development: Sireteanu,1999, p. 61.

127 *led by Elizabeth Sowell:* Sowell, E. R., Thompson, P. M., Holmes, C. J., Jernigan, T. L., Toga, A. W. 1999. In vivo evidence for post-adolescent brain maturation in frontal and striatal regions. *Nature Neuroscience, 2,* pp. 859–861.

128 *increased through age eleven or twelve:* Giedd, J. N., Blumenthal, J., Jeffries N. O., Castellanos, F. X., Liu, H., Zijdenbos, A., Paus, T., Evans, A. C., Rapoport, J. L. 1999. Brain development during childhood and adolescence: a longitudinal MRI study. *Nature Neuroscience, 2,* pp. 861–863.

130 *"In adult centres":* In Lowenstein, D. H. & Parent, J. M. 1999. Brian, heal thyself. *Science, 283,* pp. 1126–1127.
"We are still taught": Ibid. p. 1126.

Chapter Four

133 *As a student at Ohio State:* Guillermo, K. S. 1983. *Monkey business: The disturbing case that launched the animal rights movement.* Washington, D.C.: National Press Books, p. 32.

135 *bought at a toy store:* Guillermo, 1983, p. 25.

136 *The saga of the Silver Spring monkeys:* for an excellent and concise account of the case, see Fraser, Caroline. 1993. The raid at Silver Spring. *The New Yorker, 69,* p. 66.

137 *In 1895 Sherrington:* Reprinted in Denny-Brown, D. (Ed.) 1940. Selected writings of Sir Charles Sherrington. New York: Harper & Bros. pp. 115–119.

138 *Reflecting on the 1895 results:* Sherrington, C.S. 1931. Hughlings Jackson Lecture. *Brain, 54,* pp. 1–28.

139n *By 1947:* Sherington, 1947, pp. xxi–xxiii.

140 *"A negative reinforcer":* Skinner, B.F. 1974. *About Behaviorism.* New York: Alfred A. Knopf, p. 47.

electric shock that lasted up to 3.5 seconds: Taub, E. 1980. Somatosensory deafferentation research with monkeys: Implications for rehabilitation medicine. In Ince, L.P. (Ed.) Behavioral psychology in rehabilitation medicine. Baltimore: Williams & Wilkins, pp. 371–401.

141 *"was to be of long duration, if necessary":* Ibid., p. 374.

142 *the monkey uses it:* Taub, E. 1977. Movement in nonhuman primates deprived of somatosensory feedback. *Exercise and Sports Sciences Reviews, 4,* pp. 335–374.

143 *"except the most precise":* Ibid., p. 368.

144 *"potentially useful":* Ibid., p. 342.

145 *"major difficulties in carrying out deafferentation experiments with monkeys":* Ibid., p. 343.

six out of eleven fetuses died: Ibid., p. 359.

150 *"are not the inevitable consequences of deafferentation":* Guillermo, 1983, p. 133.

152 *"let the monkeys go?":* Kilpatrick, J. 1986. Jailed in Poolesville. *The Washington Post,* May 12, A15.

153 *the fifteen surviving monkeys:* Dajer, T. 1992. Monkeying with the brain. *Discover, 13,* p. 70.

154 *"had been through hell and back":* Ibid.

when they were three or four years old: Pons, T.P., Garraghty, P.E., Ommaya, A.K., Kaas, J.H., Taub, E., & Mishkin, M. 1991. Massive cortical reorganization after sensory deafferentation in adult macaques. *Science, 252,* pp. 1857–1860.

155 *"a couple of millimeters":* Ibid., p. 1857.

156 *Paul stopped eating:* Goldstein, A.A. 1990. Silver Spring monkey undergoes final experiment. *The Washington Post,* January 22, E3.

rejected the advice: Dajer, 1992.

157 *held up experiments on the seven surviving monkeys:* Barnard, N.D. 1990. Animal experimentation: The case of the Silver Spring monkeys. *The Washington Post,* February 25, B3.

158 *"euthanized for humane reasons":* Sullivan, L. W. 1990. Free for all: Morality and the monkeys. *The Washington Post,* March 17, A27.
was denied on April 12, 1991: Okie, S. S., & Jennings, V. 1991. Rescued animals killed: Animal rights group defends euthanasia. *The Washington Post,* April 13, A1.
he never awoke: Ibid.

159 *"advantageous to study the Silver Spring monkeys":* Suplee, C. 1991. Brain's ability to rewire after injury is extensive; "Silver Spring monkeys" used in research. *The Washington Post.* June 28, A3.

Chapter Five

163 *"a distinct and different essence":* Penfield, W. 1975. *The mystery of the mind.* Princeton, N.J.: Princeton University Press, p. 55, 62.
chapter on habit: James, 1983, p. 110.

165 *In 1912 T. Graham Brown and Charles Sherrington:* Graham Brown, T., & Sherrington, C. S. 1912. On the instability of a cortical point. *Proceedings of Royal Science Society of London, 85B,* pp. 250–277.
S. Ivory Franz compared movement maps: Franz, S.I. 1915. Variations in distribution of the motor centers. *Psychological Review, Monograph Supplement 19,* pp. 80–162.
Sherrington himself described "the excitable cortex": Leyton, A. F. S., & Sherrington, C. S. 1917. Observations on the excitable cortex of the chimpanzee, orang-utan and gorilla. *Quarterly Journal of Experimental Physiology, 1,* pp. 135–222.

166 *Karl Lashley, a former colleague of Franz:* Lashley, K. S. 1923. Temporal variation in the function of the gyrus precentralis in primates. *American Journal of Physiology 65,* pp. 585–602.
"plasticity of neural function": Lashley, K. S. 1926. Studies of the cerebral function of learning. *Journal of Comparative Neurology 4,* pp. 1–58.
remodeled continually by experience: Merzenich, M. M., & Jenkins, W. M. 1993. Cortical representations of learned behaviors. In: Andersen, P. et al. (Eds.) *Memory concepts.* New York: Elsevier, pp. 437–454.
Donald Hebb postulated coincident-based synaptic plasticity: Hebb, D. O. 1949. *The organization of behavior: A neuropsychological theory.* New York: John Wiley.

167 *the great Spanish neuroanatomist Ramón y Cajal:* DeFelipe, J., & Jones, E. G. (Eds.) 1988. *Ramón y Cajal Santiago: Cajal on the cerebral cortex: An annotated translation of the complete writings.* New York: Oxford University Press, 1988.
auditory cortex: Disterhoft, J. F., & Stuart, D. K. 1976. Trial sequence of changed unit activity in auditory system of alert rat during conditioned response acquisition and extinction. *Journal of Neurophysiology,* 39(2), pp. 266–281.
"paw cortex": Kalaska, J., & Pomeranz, B. 1979. Chronic paw denervation causes an age-dependent appearance of novel responses from forearm in "paw cortex" of kittens and adult cats. *Journal of Neurophysiology, 42,* pp. 618–633.

168 *amputating a raccoon's fifth digit:* Rasmusson, D. D. 1982. Reorganization of raccoon somatosensory cortex following removal of the fifth digit. *Journal of Comparative Neurology, 10,* pp. 313–326.
somatosensory reorganization in the cortices of raccoons: Kelahan, A. M., & Doetsch, G. S. 1984. Time-dependent changes in the functional organization of somatosensory cerebral cortex following digit amputation in adult raccoons. *Somatosensory Research, 2,* pp. 49–81.
Patrick Wall's prescient suggestion: Wall, P. D. 1977. The presence of ineffective synapses and the circumstances which unmask them. *Philosophical Transactions of the Royal Society of London, Series B: Biological Sciences, 26,* pp. 361–372.

171 *fifteen times as dense as those on, for instance, your shin:* Haseltine, E., 2000. How your brain sees you. *Discover,* September, p. 104.

174 *The poor brain was hoodwinked:* Paul, R. L., Goodman, H., & Merzenich, M. 1972. Alterations in mechanoreceptor input to Brodmann's areas 1 and 3 of the postcentral hand area of *Macaca mulatta* after nerve section and regeneration. *Brain Research, 14,* pp. 1–19.

175 *roughly eight to fourteen square millimeters:* Merzenich, M. M., & Jenkins, W. M. 1993. Reorganization of cortical skin representations of the hand following alterations of skin inputs induced by nerve injury, skin island transfers, and experience. *Journal of Hand Therapy, 6,* pp. 89–104.

176 *neuroscience landmark:* Ibid.

177 *inputs from the radial and ulnar nerves:* Merzenich, M. M., Kaas, J. H., Wall, J. T., et al. 1983. Progression of change following median

nerve section in the cortical representation of the hand in areas 3b and 1 in adult owl and squirrel monkeys. *Neuroscience, 10,* pp. 639–665.

"complete topographic representation": Merzenich & Jenkins, 1993, p. 92.

"completely contrary to a view of sensory systems": Merzenich, Kass, & Wall, 1983, p. 662.

"Hubel and Wiesel's work had shown just the opposite": Hubel, D. H., & Wiesel, T. N. 1970. The period of susceptibility to the physiological effects of unilateral eye closure in kittens. *Journal of Physiology, 206,* pp. 419–436; Hubel, D. H., Wiesel, T. N., & LeVay, S. 1977. Plasticity of ocular dominance columns in monkey striate cortex. *Philosophical Transactions of the Royal Society of London, Series B: Biological Sciences, 197* (26), pp. 377–409.

178 *The representation of the hand:* Merzenich, M. M., Nelson, R. J., Kaas, J. H., et al. 1987. Variability in hand surface representations in areas 3b and 1 in adult owl and squirrel monkeys. *Journal of Comparative Neurology, 258,* pp. 281–296.

"differences in lifelong use of the hands": Merzenich, Nelson, & Kaas, 1987, p. 281.

180 *amputated a single finger in owl monkeys:* Merzenich, M. M., Nelson, R. J., Stryker, M. P., et al. 1984. Somatosensory cortical map changes following digit amputation in adult monkeys. *Journal of Comparative Neurology, 224,* pp. 591–605.

182 *a single, continuous, overlapping representation:* Clark, S. A., Allard, T., Jenkins, W. M., & Merzenich, M. M. 1988. Receptive fields in the body-surface map in adult cortex defined by temporally correlated inputs. *Nature, 332,* pp. 444–445; Allard, T., Clark, S. A., Jenkins, W. M., & Merzenich, M. M. 1991. Reorganization of somatosensory area 3b representations in adult owl monkeys after digital syndactyly. *Journal of Neurophysiology, 66,* pp. 1048–1058.

when the fused digits were separated: Mogilner, A., Grossman, J. A., Ribary, U., et al. 1993. Somatosensory cortical plasticity in adult humans revealed by magnetoencephalography. *Proceedings of the National Academy of Sciences of the United States of America, 90,* pp. 3593–3597.

183 *adult visual cortex seemed just as capable of reorganizing:* Kaas, J. H., Krubitzer, L. A., Chino, Y. M., et al. 1990. Reorganization of

retinotopic cortical maps in adult mammals after lesions of the retina. *Science, 248,* pp. 229–231.

experiment on four of the Silver Spring monkeys: Pons, Garraghty, et al., 1991.

184 *"new direction of research":* Ramachandran, V. S., & Blakeslee, S. 1998. *Phantoms in the brain: Probing the mysteries of the human mind.* New York: William Morrow; Ramachandran, V. S., & Rogers-Ramachandran, D., 2000. Phantom limbs and neural plasticity. *Archives of Neurology, 57,* pp. 317–320.

the term phantom limb: Herman, J. 1998. Phantom limb: From medical knowledge to folk wisdom and back. *Annals of Internal Medicine, 128,* pp. 76–78.

185 *feel the missing appendage:* Ramachandran & Blakeslee, 1998; Ramachandran, V. S., Stewart, M., & Rogers-Ramachandran, D. 1992. Perceptual correlates of massive cortical reorganization. *Neuroreport, 3,* pp. 583–586; Ramachandran, V. S. 1993. Behavioral and magnetoencephalographic correlates of plasticity in the adult human brain. *Proceedings of the National Academy of Sciences of the United States of America, 90,* pp. 10413–10420.

186 *invaded by nerves from the genitals:* Robertson, I. H. 1999. *Mind sculpture: Unlocking your brain's untapped potential.* London: Bantam Press, p. 54. For excellent reviews of the clinical aspects of plasticity, see: Robertson & Murre, 1999, and Robertson, I. H., 1999. Setting goals for cognitive rehabilitation. *Current Opinion and Neurology, 12,* pp. 703–708.

187 *double by 2050:* Taub, E., Uswatte, G., & Pidikiti, R. 1999. Constraint-induced movement therapy: a new family of techniques with broad application to physcial rehabilitation—a clinical review. *Journal of Rehabilitation Research and Development. 36.* pp. 237–251.

188 *speed and strength of movement:* Ibid.

190 *"neurological injury, including stroke":* Taub, E., Miller, N. E., Novack, T. A., et al. 1993. Technique to improve chronic motor deficit after stroke. *Archives of Physical Medicine and Rehabilitation, 74,* pp. 347–354.

In just two weeks: Robertson, I. H., & Murre, J. M. J. 1999. Rehabilitation of brain damage: Brain plasticity and principles of guided recovery. *Psychological Bulletin, 125,* pp. 544–575.

191 *improvement on standard tests of motor ability:* Kunkel, A., Kopp, B., Muller, G., et al. 1999. Constraint-induced movement therapy for motor recovery in chronic stroke patients. *Archives of Physical Medicine and Rehabilitation, 80,* pp. 624–628.

192 *patients who had lost the use of a leg:* Taub, Uswatte, & Pidikiti, 1999.

brain changes in six chronic stroke patients: Liepert, J., Miltner, W. H., Bauder, H., et al. 1998. Motor cortex plasticity during constraint-induced movement therapy in stroke patients. *Neuroscience Letters, 250,* pp. 5–8.

193 *changes in the brain's electrical activity:* Kopp, B., Kunkel, A., Muhlnickel, W., et al. 1999. Plasticity in the motor system related to therapy-induced improvement of movement after stroke. *Neuroreport, 10,* pp. 807–810.

"induced expansion": Taub, Uswatte & Pidikiti, 1999.

196 *left aphasic:* Pulvermuller, F., Neininger, B., et al. 2001. Constraint-induced therapy of chronic aphasia after stroke. *Stroke,* 32(7), pp. 1621–1626.

197 *largely destroyed their Wernicke's area:* Weiller, C., Isensee, C., Rijntjes, M., et al. 1995. Recovery from Wernicke's aphasia: A positron emission tomographic study. *Annals of Neurology, 37,* pp. 723–732.

accompanied by cortical reorganization: Liepert, J., Bauder, H., Wolfgang, H. R., et al. 2000. Treatment-induced cortical reorganization after stroke in humans. *Stroke, 6,* pp. 1210–1216.

reported a similar finding: Buckner, R. L., Corbetta, M., Schatz, J., et al. 1996. Preserved speech abilities and compensation following prefrontal damage. *Proceedings of the National Academy of Sciences of the United States of America, 93,* pp. 1249–1253.

198 *tactile discrimination tasks activate the visual cortex:* Sadato, N., Pascual-Leone, A., Grafman, J., et al. 1996. Activation of the primary visual cortex by Braille reading in blind subjects. *Nature, 380,* pp. 526–528.

superior tactile sense of the congenitally blind: Cohen, L. G., Celnik P., Pascual-Leone, A., et al. 1997. Functional relevance of cross-modal plasticity in blind humans. *Nature, 389,* pp. 180–183; Musso, M., Weiller, C., Kiebel, S., et al. 1999. Training-induced brain plasticity in aphasia. *Brain, 122,* pp. 1781–1790.

199 *for the good of millions of stroke patients:* Taub, E., & Morris, D.M. 2001. Constraint-induced movement therapy to enhance recovery after stroke. *Current Atherosclerosis Reports, 3,* pp. 279–286.

Chapter Six

204 *a route to cortical reorganization that is the polar opposite:* Elbert, T., Candia, V., Altenmuller, E., et al. 1998. Alteration of digital representations in somatosensory cortex in focal hand dystonia. *Neuroreport,* 9, pp. 3571–3575; Byl, N.N., McKenzie, A., & Nagarajan, S.S. 2000. Differences in somatosensory hand organization in a healthy flutist and a flutist with focal hand dystonia: A case report. *Journal of Hand Therapy, 13,* pp. 302–309; Pujol, J., Roset-Llobet, J., Rosines-Cubells, D., et al. 2000. Brain cortical activation during guitar-induced hand dystonia studied by functional MRI. *Neuroimage, 12,* pp. 257–267.

use-dependent cortical reorganization: Jenkins, W.M., Merzenich, M.M., Ochs, M.T., et al. 1990. Functional reorganization of primary somatosensory cortex in adult owl monkeys after behaviorally controlled tactile stimulation. *Journal of Neurophysiology, 63,* pp. 82–104; Nudo, R.J., Jenkins, W.M., Merzenich, M.M., et al. 1992. Neurophysiological correlates of hand preference in primary motor cortex of adult squirrel monkeys. *Journal of Neuroscience, 12,* pp. 2918–2947.

206 *a study of the effects of motor skill learning:* Nudo, R.J., Milliken, G.W., Jenkins, W.M., Merzenich, M.M. 1996. Use-dependent alterations of movement representations in primary motor cortex of adult squirrel monkeys. *Journal of Neuroscience, 16,* pp. 785–807.

208 *"the cortical part of the skill acquisition":* Merzenich, M., Wright, B., Jenkins, W., et al. 1996. Cortical plasticity underlying perceptual, motor, and cognitive skill development: Implications for neurorehabilitation. *Cold Spring Harbor Symposia on Quantitative Biology, 61,* pp. 1–8.

the researchers made small lesions in the monkeys' brains: Xerri, C., Merzenich, M.M., Peterson, B.E., & Jenkins, W. 1998. Plasticity of primary somatosensory cortex paralleling sensorimotor skill recovery from stroke in adult monkeys. *Journal of Neurophysiology, 79,* pp. 2119–2148.

209 *"the reemergence of the representation of functions":* Merzenich, Wright, and Jenkins, 1996, p. 2.

spinning disk experiment: Jenkins, Merzenich, & Ochs, 1990.

210 *flutter-vibration studies:* Recanzone, G. H., Jenkins, W. M., Hradek, G. T., & Merzenich, M. M. 1992. Progressive improvement in discriminative abilities in adult owl monkeys performing a tactile frequency discrimination task. *Journal of Neurophysiology, 67,* pp. 1015–1030.

211 *"cortical representations of the trained hands":* Recanzone, G. H., Merzenich, M. M., Jenkins, W. M., et al. 1992. Topographic reorganization of the hand representation in cortical area 3b of owl monkeys trained in a frequency-discrimination task. *Journal of Neurophysiology, 67,* pp. 1031–1056.

only when the monkeys were attentive: Merzenich, M., Byl, N., Wang, X., & Jenkins, W. 1996. Representational plasticity underlying learning: Contributions to the origins and expressions of neurobehavioral disabilities. In: Ono, T., et al. (Eds.) *Perception, memory, and emotion: Frontiers in neuroscience.* Oxford, U.K., and Tarrytown, N.Y.: Pergamon, pp. 45–61; Merzenich & deCharms, 1996.

212 *discriminate small differences in the frequency of tones:* Recanzone, G. H., Schreiner, C. E., & Merzenich, M. M. 1993. Plasticity in the frequency representation of primary auditory cortex following discrimination training in adult owl monkeys. *Journal of Neuroscience, 13,* pp. 87–103.

"idiosyncratic features of cortical representation": Merzenich, M. M., Recanzone, G. H., Jenkins, W. M., & Grajski, K. A. 1990. Adaptive mechanisms in cortical networks underlying cortical contributions to learning and nondeclarative memory. *Cold Spring Harbor Symposia on Quantitative Biology, 55,* pp. 873–887.

213 *fifteen proficient Braille readers:* Pascual-Leone, A., & Torres, F. 1993. Plasticity of the sensorimotor cortex representation of the reading finger in Braille readers. *Brain, 116, Part 1,* pp. 39–52.

amputation produces extensive reorganization: Flor, H., Elbert, T., Knecht, S., et al. 1995. Phantom-limb pain as a perceptual correlate of cortical reorganization following arm amputation. *Nature, 375,* pp. 482–484.

214 *There was no difference between the string players:* Elbert, T., Pantev, C., Wienbruch, C., Rockstroh, B., & Taub, E. 1995. Increased

cortical representation of the fingers of the left hand in string players. *Science, 270,* pp. 305–307.

215 *tactile stimuli to the fingers changes the maps of the hand:* Wang, X., Merzenich, M. M., Sameshima, K., & Jenkins, W. M. 1995. Remodeling of hand representation in adult cortex determined by timing of tactile stimulation. *Nature, 378,* pp. 71–75.

217 *merely think about practicing it:* Pascual-Leone, A., Dang, N., Cohen, L. G., et al. 1995. Modulation of muscle responses evoked by transcranial magnetic stimulation during the acquisition of new fine motor skills. *Journal of Neurophysiology, 74,* pp. 1037–1045.

218 *Merzenich's group was already suggesting:* Merzenich, Recanzone, & Jenkins, 1990.

219 *simulated writer's cramp in two adult owl monkeys:* Byl, N. N., Merzenich, M. M., & Jenkins, W. M. 1996. A primate genesis model of focal dystonia and repetitive strain injury. I. Learning-induced dedifferentiation of the representation of the hand in the primary somatosensory cortex in adult monkeys. *Neurology, 47,* pp. 508–520.

remap their own somatosensory cortex: Byl, N. N., & McKenzie, A. 2000. Treatment effectiveness for patients with a history of repetitive hand use and focal hand dystonia: A planned, prospective follow-up study. *Journal of Hand Therapy, 13,* pp. 289–301.

220 85 to 98 *percent improvement in fine motor skills:* Byl, N. N., Nagarajan, S., & McKenzie, A. 2000. Effect of sensorimotor training on structure and function in three patients with focal hand dystonia. *Society for Neuroscience Abstracts.*

professional musicians with focal hand dystonia: Candia, V., Elbert, T., Altenmuller, E., et al. 1999. Constraint-induced movement therapy for focal hand dystonia in musicians. *Lancet, 353,* p. 42.

221 *walk in an ever-more constrained way:* Merzenich, Byl, & Wang, 1996, p. 53.

Tinnitus, *or ringing in the ears, is characterized:* Muhlnickel, W., Elbert, T., Taub, E., & Flor, H. 1998. Reorganization of auditory cortex in tinnitus. *Proceedings of the National Academy of Sciences of the United States of America, 95,* pp. 10340–10343.

223 *cortex that used to control the movement of the elbow and shoulder:* Nudo, R. J., Wise, B. M., SiFuentes, F., & Milliken, G. W. 1996. Neural substrates for the effects of rehabilitative training on motor recovery after ischemic infarct. *Science, 272,* pp. 1791–1794.

cortex that used to register when the left arm was touched: Pons, Garraghty, & Ommaya, 1991.

visual cortex that has been reprogrammed to receive and process tactile inputs: Sadato, N., Pascual-Leone, A., Grafman, J., et al. 1998. Neural networks for Braille reading by the blind. *Brain, 121,* pp. 1213–1229.

224 *"most remarkable observations made in recent neuroscience history":* Jones, E. G. 2000. Cortical and subcortical contributions to activity-dependent plasticity in primate somatosensory cortex. *Annual Review of Neuroscience, 23,* pp. 1–37.

"little-attended exercises are of limited value": Merzenich & Jenkins, 1993, p. 102.

Chapter Seven

226 *dyslexia, which affects an estimated 5 to 17 percent:* Temple, E., Poldrack, R. A., Protopapas, A., et al. 2000. Disruption of the neural response to rapid acoustic stimuli in dyslexia: Evidence from functional MRI. *Proceedings of the National Academy of Sciences of the United States of America, 97,* pp. 13907–13912.

processing certain speech sounds—fast ones: Tallal, P. 2000. The science of literacy: From the laboratory to the classroom. *Proceedings of the National Academy of Sciences of the United States of America., 97,* pp. 2402–2404; Poldrack, R. A., Temple, E., Protopapas, A., Nagarajan, S., Tallal, P., Merzenich, M., & Gabrieli, J. D. 2001. Relations between the neural bases of dynamic auditory processing and phonological processing: Evidence from fMRI. *Journal of Cognitive Neuroscience, 13,* pp. 687–697.

227 *a failure to assign neurons to particular phonemes:* Temple, E., Poldrack, R. A., & Salidis, J., et al. 2001. Disrupted neural responses to phonological and orthographic processing in dyslexic children: An fMRI study. *Neuroreport, 12,* pp. 299–307.

228 *"how we might develop a way to train the brain to process sounds correctly":* Tallal, P., Merzenich, M. M., Miller, S., & Jenkins, W. 1998. Language learning impairments: Integrating basic science, technology, and remediation. *Experimental Brain Research, 123,* pp. 210–219.

230 *produce modified speech tapes:* Nagarajan, S. S., Wang, X., Merzenich, M. M., et al. 1998. Speech modification algorithms used

for training language learning–impaired children. *Institute of Electrical and Electronics Engineers Transactions on Rehabilitation Engineering, 6,* pp. 257–268.

233 *the Rutgers and UCSF teams reported their results in the journal* Science: Tallal, P., Miller, S. L., Bedi, G., et al. 1996. Language comprehension in language-learning impaired children improved with acoustically modified speech. *Science, 271,* pp. 81–84; Merzenich, M. M., Jenkins, W. M., Johnston, P., et al. 1996. Temporal processing deficits of language-learning impaired children ameliorated by training. *Science, 271,* pp. 77–81.

235 *Merzenich, Tallal, and colleagues had teamed up with John Gabrieli:* Temple, Poldrack, & Protopapas, et al., 2000.

236 *other scientists:* Beauregard, M., Levesque, J., & Bourgouin, P. 2001 Neural correlates of conscious self-regulation of emotion. *Journal of Neuroscience*, 21(18), p. RC165; Paquette, V., Levesque, J., et al. 2003. "Change the mind and you change the brain": effects of cognitive-behavioral therapy on the neural correlates of spider phobia. *Neuroimage*, 18(2), pp. 401–409; Levesque. J., Eugene, F., et al. 2003. Neural circuitry underlying voluntary suppression of sadness. *Biological Psychiatry*, 53(6), pp. 502–510.

Applied mindfulness could change neuronal circuitry: Schwartz, 1998.

237 *this disease strikes:* Kadesjo, B., & Gillberg, C. 2000. Tourette's disorder: Epidemiology and comorbidity in primary school children. *Journal of the American Academy of Child and Adolescent Psychiatry, 39,* pp. 548–555.

is a biological link between OCD and Tourette's: State, M. W., Pauls, D. L., & Leckman, J. F. 2001. Tourette's syndrome and related disorders. *Child and Adolescent Psychiatric Clinics of North America, 10,* pp. 317–331.

The defining symptoms of Tourette's: For details, see the excellent textbook, Leckman, J., & Cohen, D. J. 1999. *Tourette's syndrome: Tics, obsessions, compulsions: Development, psychopathology and clinical care.* New York: John Wiley & Sons.

The two diseases also seem to share a neural component: Stern, E., Silbersweig, D. A., Chee, K. Y., et al. 2000. A functional neuroanatomy of tics in Tourette syndrome. *Archives of General Psychology, 57,* pp. 741–748.

central role in switching from one behavior to another: Leckman, J. F., & Riddle, M. A. 2000. Tourette's syndrome: When habit-forming systems form habits of their own? *Neuron, 28,* pp. 349–354.

appeared in 1825: Kushner, H. I. 2000. *A cursing brain? The histories of Tourette syndrome.* Cambridge, Mass.: Harvard University Press.

238 *Drugs typically reduce tic symptoms:* Leckman & Cohen, 1999.

239 *That leaves behavioral treatment:* Piacentini, J., & Chang, S. 2001. Behavioral treatments for Tourette syndrome and tic disorders: State of the art. In: Cohen, D. J., Jankovic, J., & Goetz, C. G. (Eds.) *Advances in neurology: Tourette syndrome, 85.* Philadelphia: Lippincott Williams & Wilkins, pp. 319–332.

241 *in a study based on reasoning:* Peterson, B. S., Skudlarski, P., Anderson, A. W., et al. 1998. A functional magnetic resonance imaging study of tic suppression in Tourette syndrome. *Archives of General Psychiatry, 55,* pp. 326–333.

243 *Patients are trained to recognize and label tic urges:* Piacentini & Chang, 2001, p. 328.

244 *"simply as events in the mind":* Teasdale, J. D. 1999. Metacognition, mindfulness and the modification of mood disorders. *Clinical Psychology and Psychotherapy, 6,* pp. 146–155.

in the titles of their research papers: Teasdale, J. D., Segal, Z., & Williams, J. M. G. 1995. How does cognitive therapy prevent depressive relapse, and why should attentional control (mindfulness) training help? *Behavior Research & Therapy, 33,* pp. 25–39.

245 *Teasdale named his approach:* Teasdale, J. D., Segal, Z. V., Williams, J. M. G., et al. 2000. Prevention of relapse/recurrence in major depression by mindfulness-based cognitive therapy. *Journal of Consulting & Clinical Psychology, 68,* pp. 615–623.

cognitive therapy, too, had the power to prevent relapses: Scott, J., Teasdale, J. D., Paykel, E. S., et al. 2000. Effects of cognitive therapy on psychological symptoms and social functioning in residual depression. *British Journal of Psychiatry, 177,* pp. 440–446; Paykel, E. S., et al. 1999. Prevention of relapse in residual depression by cognitive therapy. *Archives of General Psychiatry, 56,* pp. 829–835

Teasdale thought he knew why: Teasdale, J. D. 1999. Emotional processing, three modes of mind and the prevention of relapse in depression. *Behavior Research & Therapy, 37 Supplement 1,* pp. 53–77.

250 *interpersonal therapy:* Brody, A., Saxena, S., Stoessal, P., et al. 2001. Regional brain metabolic changes in patients with major depression treated with either paroxetine or interpersonal therapy. *Archives of General Psychiatry, 58,* pp. 631–640.

how navigation expertise might change the brain: Maguire, E. A., Gadian, D. G., Johnsrude, I. S., Good, C. D., Ashburner, J., Frackowiak, R. S. J., & Frith, C. 2000. Navigation-related structural change in the hippocampi of taxi drivers. *Proceedings of the National Academy of Sciences of the United States of America, 97,* pp. 4398–4403.

252 *exploring the cellular and molecular mechanisms:* Van Praag, Kempermann, & Gage, 2000.

the actual creation of new neurons: Kempermann, G., & Gage, F. H. 1999. New nerve cells for the adult brain. *Scientific American, 280,* pp. 48–53.

the formation and survival of new neurons: Kempermann, G., Kuhn, H. G., & Gage, F. H. 1997. More hippocampal neurons in adult mice living in an enriched environment. *Nature, 386,* pp. 493–495.

directly related to learning tasks: Gould, E., Beylin, A., Tanapat, P., Reeves, A., & Shors, T. J. 1999. Learning enhances adult neurogenesis in the hippocampal formation. *Nature Neuroscience, 2,* pp. 260–265.

253 *exercising on a wheel:* van Praag, H., Christie, B. R., Sejnowski, T. J., & Gage, F. H. 1999. Running enhances neurogenesis, learning, and long-term potentiation in mice. *Proceedings of the National Academy of Sciences of the United States of America, 96,* pp. 13427–13431; van Praag, H., Kempermann, G., & Gage, F. H. 1999. Running increases cell proliferation and neurogenesis in the adult mouse dentate gyrus. *Nature Neuroscience 2,* pp. 266–270.

newly generated neurons: Shors, T. J., Miesegaes, G., et al. 2001. Neurogenesis in the adult is involved in the formation of trace memories. *Nature*, 410, pp. 372–376.

neurogenisis occurs in the adult human hippocampus: Eriksson, P. S., Perfilieva, E., Bjork-Ericksson, T., et al. 1998. Neurogenisis in the adult human hippocampus. *Nature Medicine, 4,* pp. 1313–1317.

254 *"it is for the science of the future":* Cajal, S. R., & Mays R. T. 1959. Degeneration and regeneration of the nervous system. New York: Hafner, p. 750.

Chapter Eight

255 *"actual philosophy":* Born, M. 1968. *My life and my views.* New York: Scribner.

"from the object": Heisenberg, W. 1958. The representation of nature in contemporary physics. *Daedalus, 87,* pp. 95–108.

my second book: Schwartz, J. M., Gottlieb, A., & Buckley, P. 1998. *A return to innocence.* New York: Regan Books/HarperCollins.

256 *a book published the year before:* Chalmers, D. 1996. *The conscious mind: In search of a fundamental theory.* New York: Oxford University Press.

258 *a justice on the Australian Supreme Court:* Hodgson, D. 1991. *The mind matters: Consciousness and choice in a quantum world.* New York: Oxford University Press.

259 *purchased Stapp's 1993 book:* Stapp H. 1993. *Mind, matter and quantum mechanics.* New York: Springer-Verlag.

"not coercive": James, 1983, p. 1177.

"the feeling of effort": Ibid., p. 1142.

"active element": Ibid., p. 428.

260 *"the brain is an instrument of possibilities":* Ibid., p. 144.

261 *"the ordinary laws of physics and chemistry":* Wigner, E. 1969. Are we machines? *Proceedings of the American Philosophical Society, 113,* pp. 95–101.

"as important as that of Newton": Motz, L., & Weaver, J. H. 1989. *The story of physics.* New York Perseus, pp. 194–195.

262 *Planck's talk:* Zeilinger, A. 2000. The quantum centennial. *Nature, 408,* pp. 639–641.

"a deep understanding of chemistry": Greenberger, D. M. 1986. Preface. In: Greenberger, D. M. (Ed.) New techniques and ideas in quantum measurement theory. *Annals of the New York Academy of Sciences, 480,* pp. xiii–xiv.

"gift from the gods": Ibid., p. xiii.

263 *"the silliest is quantum theory":* Kaku, M. 1995. *Hyperspace. A scientific odyssey through parallel universes, time warps and the tenth dimension.* New York: Anchor.

264 *"Any other situation in quantum mechanics":* Feynman, R. 1965. *The character of physical law.* Cambridge, Mass.: MIT Press.

266 *physicists in Paris:* Gribbin, J. 1995. *Schrödinger's kittens and the search for reality: solving the quantum msyteries.* New York: Little, Brown.

267 *Hitachi research labs:* Ibid., p. 7.

270 *"God does not play dice":* Einstein's exact words, in a letter to Cornel Lanczos on March 21, 1942, were "It is hard to sneak a look at God's card's. But that he would choose to play dice with the world . . . is something I cannot believe for a single moment."

physicist John Bell showed: Bell, J. 1987. *Speakable and unspeakable in quantum mechanics.* New York: Cambridge University Press. For a lucid discussion of this challenging subject, see: Stapp, H. 2001. Quantum theory and the role of mind in nature. *Foundations of Physics, 31,* pp. 1465–1499. Available online at: http://www-physics.lbl.gov/~stapp/vnr.txt.

271 many-worlds view: Everett, H., III. 1957. Relative state formulation of quantum mechanics. *Review of Modern Physics, 29,* p. 454–462.

272 *Copenhagen Interpretation:* Stapp, H. 1972. The Copenhagen interpretation. *American Journal of Physics, 40,* pp. 1098–1116.

273 *"objective existence of an electron":* Pagels, H. 1982. *The cosmic code: Quantum physics as the language of nature.* New York: Simon & Schuster.

274 *the mind of an observer:* Wigner, 1967 (see especially chapters 12 and 13).

"One aim of the physical sciences": Bronowski, J. 1973. *The ascent of man.* Boston: Little, Brown.

"has thus evaporated": Heisenberg, 1958.

"It is wrong": In: Nadeau, R., & Kafatos, M. 1999. *The non-local universe: The new physics and matters of the mind.* New York: Oxford University Press, p. 96.

275 *Schrödinger's cat:* Schrödinger's original 1935 description of the cat is translated in: Jauch, J. M. 1977. *Foundations of quantum mechanics.* Reading, Mass.: Addison-Wesley, p. 185.

276 *"shrouded in mystery":* Wigner, 1967.

280 *his book on the foundations of quantum theory:* von Neumann, J. 1955. *Mathematical foundations of quantum mechanics.* Princeton,

N.J.: Princeton University Press. Translation from the 1932 German original.

283 *"the content of consciousness":* Wigner, 1969.

286 *"biologists are more prone":* Wigner, E. P. 1964. Two kinds of reality. *The Monist, 48,* pp. 248–264.

287 *"A brain was always going to do":* Dennett, 1994.

Chapter Nine

290 *a talk on how my OCD work:* Schwartz, J. M. 2000. First steps toward a theory of mental force: PET imaging of systematic cerebral changes after psychological treatment of obsessive-compulsive disorder. In: Hameroff, S. R., Kaszniak, A. W., & Chalmers, D. J. (Eds.) *Toward a science of consciousness III: The third Tucson discussions and debates.* Cambridge, Mass.: MIT Press, pp. 111–122.

294 *final version of my "Volitional Brain" paper:* Schwartz, J. M. 1999. A role for volition and attention in the generation of new brain circuitry: Toward a neurobiology of mental force. In Libet, Freeman, & Sutherland, 1999, pp. 115–142.

295 *"mind as a force field":* Lindahl, B. I. B., & Århem, P. 1994. Mind as a force field: Comments on a new interactionist hypothesis. *Journal of Theoretical Biology, 171,* pp. 111–122.

"conscious mental field": Libet, B. 1996. Conscious mind as a field. *Journal of Theoretical Biology, 178,* pp. 223–224.

296 *In his own* JCS *paper, Stapp argued:* Stapp, H. P. 1999. Attention, intention, and will in quantum physics. In: Libet, Freeman, & Sutherland, 1999, pp. 143–164.

297 *our strongest argument yet:* Ibid., pp. 140–142.

298 *Kant, in fact, succumbed to the same temptation:* Ibid., p. ix.

299 *In 1931, Einstein had declared:* Ibid., p. xii.

Carl Rogers wrote: Rogers, C. R. 1964. Freedom and commitment. *The Humanist 29,* pp. 37–40.

conditioned responses to stimuli: Skinner, B. F. 1971. *Beyond freedom and dignity.* New York: Alfred A Knopf.

risk taking: Benjamin, J., Li, L., Patterson, C., et al. 1996. Population and familial association between the D4 dopamine receptor gene and measures of novelty seeking. *Nature Genetics, 12,* pp. 81–84.

and hence obesity: Barinaga, M. 1995. "Obese" protein slims mice. *Science, 269,* pp. 475–476.

dopamine imbalances with addiction: Koob, G. F., & Bloom, F. E. 1988. Cellular and molecular mechanisms of drug-dependence. *Science, 242,* pp. 715–723.

300 *eternity is impossible:* James, William. 1992. The dilemma of determinism. In: *William James Writings 1878–1899,* p. 570.

"is not imagined to be ultimately responsible for itself": Libet, Freeman, & Sutherland, 1999, pp. ix–xxiii.

"needed for a particular movement": Doty, R. W. 1998. The five mysteries of the mind, and their consequences. *Neuropsychologia,* 36, pp. 1069–1076.

301 *"effortless volitions":* James, William. 1992. Psychology: Briefer course. In: *William James Writings 1878–1899,* p. 423.

"Actualities": James, Ibid, p. 570.

302 *as the theorist Thomas Clark puts it:* Clark, T. W. 1999. Fear of mechanism. In Libet, Freeman, & Sutherland, 1999, p. 277.

"a benign user illusion": Dennett, D.C. 1991. *Consciousness explained.* Boston: Little, Brown.

303 *"the owner" of the state of your will:* Anguttara Nikåya V, 57. Translated in: Nyanaponika Thera & Bhikkhu Bodhi. 1999. *Numerical discourses of the Buddha.* Walnut Creek, Calif.: AltaMira Press, p. 135.

work reported in 1964: Kornhuber, H. H., & Deecke, L. 1964. Brain potential changes in man preceding and following voluntary movement, displayed with magnetic tape storage and time-reversed analysis. *Pflugers Archiv für die gesamte Physiologie des Menschen und der Tiere, 281,* p .52.

304 *he reported in 1982 and 1985:* Libet, B., Wright, E. W., Jr., & Gleason, C. A. 1982. Readiness-potentials preceding unrestricted "spontaneous" vs. pre-planned voluntary acts. *Electroencephalography and Clinical Neurophysiology, 54,* pp. 322–335; Libet, B. 1985. Unconscious cerebral initiative and the role of conscious will in voluntary action. *Behavioral & Brain Sciences, 8,* pp. 529–566.

"produced the movement": Libet, B. 1999. Do we have free will? In: Libet, Freeman, & Sutherland, 1999, pp. 47–57.

305 *"how we could view free will":* Libet in ibid., p. 49.

306 *"allow[s] enough time":* Libet in ibid., p. 51.

307 *as Libet wrote in 1998:* Libet, B. 1998. Do the models offer testable proposals of brain functions for conscious experience? In: Jasper,

H. H., Descarries, L., Castellucci, V. F., & Rossignol, S. (Eds.) *Advances in neurology, 77: Consciousness: At the frontiers of neuroscience.* Philadelphia: Lippincott-Raven, p. 215.

"free won't": Claxton, G. 1999. Who dunnit? Unpicking the "seems" of free will. In: Libet, Freeman, & Sutherland, 1999, pp. 99–113.

Experiments published in 1983: Libet, B., Wright, E. W., Jr., & Gleason, C. A. 1983. Preparation- or intention-to-act, in relation to pre-event potentials recorded at the vertex. *Electroencephalography and Clinical Neurophysiology 56,* pp. 367–372; Libet, B., Gleason, C. A., Wright, E. W., & Pearl, D. K. 1983. Time of conscious intention to act in relation to onset of cerebral activity (readiness-potential): The unconscious initiation of a freely voluntary act. *Brain, 106, Part 3,* pp. 623–642.

almost two full seconds: Deecke, L., & Lang, W. 1996. Generation of movement-related potentials and fields in the supplementary sensorimotor area and the primary motor area. *Advances in Neurology 7,* pp. 127–146.

"veto the process": Libet in Libet, Freeman, & Sutherland, 1999, pp. 51–52.

308 *Ten Commandments:* Libet in ibid., p. 54.

all five of the basic moral precepts: Saddhatissa, H. 1987. *Buddhist ethics.* London: Wisdom Publications.

"Restraint everywhere": Dhammapada, Verse 361.

"I've always been able to avoid that question": Horgan, John. 1999. *The undiscovered mind: How the human brain defies replication, medication and explanation.* New York: Free Press, p. 234.

309 *"volition is nothing but attention":* James, 1983, p. 424.

310 *"our conscious veto may not require":* Libet in Libet, Freeman, & Sutherland, 1999, p. 53.

"attention is the fundamental act of will": James, 1983, p. 1168.

311 *led by the Swedish physiologist David Ingvar:* Ingvar, D. H., & Philipson, L. 1977. Distribution of cerebral blood-flow in dominant hemisphere during motor ideation and motor-performance. *Annals of Neurology, 2,* pp. 230–237.

activated during the willful mental activity: Frith, C. D., Friston, K., Liddle, P. F., & Frackowiak, R. S. J. 1991. Willed action and the prefrontal cortex in man: A study with PET. *Proceedings of the Royal Society of London. Series B: Biological Sciences, 244,* pp. 241–246; Passingham, R. 1993. *The frontal lobes and voluntary action.* Oxford: Oxford University Press; Miller, E.K. 2000. The prefrontal cortex and cognitive control. *Nature Reviews Neuroscience, 1,* pp. 59–65.

In schizophrenics: Spence, S.A., Brooks, D.J., Hirsch, S.R., et al. 1997. A PET study of voluntary movement in schizophrenic patients experiencing passivity phenomena (delusions of alien control). *Brain, 120,* pp. 1997–2011; Frackowiak, R.S.J., Friston, K.J., Frith, C., & Dolan, R. 1997. Human brain function. San Diego: Academic Press; Spence, S.A., Hirsch, S.R., Brooks, D.J., & Grasby, P.M. 1998. Prefrontal cortex activity in people with schizophrenia and control subjects: Evidence from positron emission tomography for remission of "hypofrontality" with recovery from acute schizophrenia. *British Journal of Psychiatry, 17,* pp. 316–323.

In depression: Drevets, W.C. 1998. Functional neuroimaging studies of depression: The anatomy of melancholia. *Annual Review of Medicine, 49,* pp. 341–361; Mayberg, H.S., Liotti, M., Brannan, S.K., et al. 1999. Reciprocal limbic-cortical function and negative mood: Converging PET findings in depression and normal sadness. *American Journal of Psychology, 156,* pp. 675–682.

what Ingvar calls "action programs for willed acts": Ingvar, D.H. 1999. On volition: A neurophysiologically oriented essay. In Libet, Freeman, & Sutherland, 1999, pp. 1–10.

primary role for the prefrontal cortex: Seitz, R.J., Stephan, K.M., & Binkofski, F. 2000. Control of action as mediated by the human frontal lobe. *Experimental Brain Research, 133,* pp. 71–80.

and associated brain regions: Libet, Freeman, Sutherland, 1999, p. 16.

312 *accompanied by activity in the dorsolateral prefrontal cortex:* Jahanshahi, M., Jenkins, I.H., Brown, R.G., et al. 1995. Self-initiated versus externally triggered movements. I. An investigation using measurement of regional cerebral blood flow with PET and movement-related potentials in normal and Parkinson's disease subjects. *Brain, 118,* pp. 913–933; Jenkins, I.H., Jahanshahi, M., Jueptner, M., et al. 2000. Self-initiated versus externally triggered movements. II. The effect of movement predictability on regional cerebral blood flow. *Brain, 123,* pp. 1216–1228.

unable to stifle inappropriate responses: Spence, S.S., & Frith, C. 1999. Towards a functional anatomy of volition. In: Libet, Freeman, & Sutherland, 1999, 11–29.

313 *testing how volition affects conscious perception:* Silbersweig, D. A., & Stern, E. 1998. Towards a functional neuroanatomy of conscious perception and its modulation by volition: Implications of human auditory neuroimaging studies. *Philosophical Transactions of the Royal Society of London, Series B: Biological Sciences, 353,* pp. 1883–1888.

fortune sent Stern and Silbersweig a young man known as S.B.: Engelien, A., Huber, W., Silbersweig, D., et al. 2000. The neural correlates of "deaf-hearing" in man: Conscious sensory awareness enabled by attentional modulation. *Brain, 123,* pp. 532–545.

315 *experiments in the late 1990s:* Birbaumer, N., Ghanayim, N., Hinterberger, T., et al. 1999. A spelling device for the paralysed. *Nature, 398,* pp. 297–298; Kübler, A., Kotchoubey, B., Hinterberger, T, et al. 1999. The thought translation device: A neurophysiological approach to communication in total motor paralysis. *Experimental Brain Research, 124,* pp. 223–232.

317 *"Volitional effort":* James, 1992, pp. 417–418.

"master of course of thought": Majjhima Nikåya, Sutta 20. Translated in: Nyanaponika Thera & Bhikku Bodhi. 1995. *The Middle Length Discourses of the Buddha.* Somerville, MA: Wisdom Publications, p. 213.

319 *"strange arrogance":* James, 1983, pp. 429–430.

321 *"It is volition, monks, that I declare to be Karma (Action)":* Anguttara Nikåya VI, 63. *Numerical Discourses,* p. 173.

"Volition becomes the chief": Ledi Sayadaw. 1999. *The Manuals of Dhamma.* Maharastra, India: Vipassana Research Institute, p. 95.

322 *"[One] branch of these bifurcations":* James, 1992, p. 593.

Chapter Ten

323 *copy of James's:* Page references for William James are to the following editions: James, William. 1992. Psychology: Briefer course, In: *William James Writings 1878–1899.* New York: Library of America, p. 272,278. James, William. 1983. *The principles of psychology.* Cambridge, Mass.: Harvard University Press, p. 429.

326 *"pivotal question":* James, 1890/1983, p. 424.

327 *"'footlights'":* James, 1890/1983, p. 426.

"limited processing resources": Kastner, S., & Ungerleider, L. G. 2000. Mechanisms of visual attention in the human cortex. *Annual Review of Neuroscience, 23,* pp. 315–341.

328 *Selectively focusing attention on target images:* Kastner, S. & Ungerleider, L. G. 2001. The neural basis of biased competition in human visual cortex. *Neuropsychologia, 39,* pp. 1263–1276.

"biasing the brain circuit for the important stimuli": Desimone, R. 1998. Visual attention mediated by biased competition in extrastriate visual cortex. *Philosophical Transactions of the Royal Society of London, Series B: Biological Sciences, 353,* pp. 1245–1255.

329 *fascinating series of experiments:* see the papers referenced in the two preceding notes.

331 *activity spikes in human brains:* Kastner, S., Pinsk, M. A., De Weerd, P., Desimone, R., & Ungerleider, L. G. 1999. Increased activity in human visual cortex during directed attention in the absence of visual stimulation. *Neuron, 22,* pp. 751–761.

332 *In 1990, researchers:* Corbetta, M., Miezin, F. M., Dobmeyer, S., et al. 1990. Attentional modulation of neural processing of shape, color, and velocity in humans. *Science, 248,* pp. 1556–1559; Corbetta, M., Miezin, F. M., Dobmeyer, S., et al. 1991. Selective and divided attention during visual discriminations of shape, color, and speed: Functional anatomy by positron emission tomography. *Journal of Neuroscience, 11,* pp. 2383–2402.

333 *during the directing of such selective attention:* Rees, G. & Lavie, N. 2001. What can functional imaging reveal about the role of attention in visual awareness? *Neuropsychologia, 39,* pp. 1343–1353; de Fockert, J. W., Rees, G., Frith, C. D., & Lavie, N. 2001. The role of working memory in visual selective attention. *Science, 291,* pp. 1803–1806; Vandenberghe, R., Duncan, J., Arnell, K. M., et al. 2000. Maintaining and shifting attention within left or right hemifield. *Cerebral Cortex, 10,* pp. 706–713.

paying attention to the vibrations: Meyer, E., Ferguson, S. S., Zatorre, R. J., et al. 1991. Attention modulates somatosensory cerebral blood flow response to vibrotactile stimulation as measured by positron emission tomography. *Annals of Neurology, 29,* pp. 440–443.

334 *"can sculpt brain activity":* Robertson, 1999, p. 43.

fascinating experiment, Dick Passingham: Jueptner, M., Stephan, K. M., Frith, C. D., et al. 1997. Anatomy of motor learning. I. Frontal cortex and attention to action. *Journal of Neurophysiology, 77,* pp. 1313–1324; Toni, I., et al. 1998. The time course of changes during motor sequence learning: A whole-brain fMRI study. *NeuroImage, 8,* p. 50.

336 *willful selection of self-initiated responses:* Jenkins, I. H., Jahanshahi, M., Jueptner, M., et al. 2000. Self-initiated versus externally triggered movements. II. The effect of movement predictability on regional cerebral blood flow. *Brain, 123,* pp. 1216–1228.

fusiform face area: Kanwisher, N., McDermott, J., & Chun, M. M. 1997. The fusiform face area: A module in human extrastriate cortex specialized for face perception. *Journal of Neuroscience, 17,* pp. 4302–4311; Kanwisher, N., & Wojciulik, E. 2000. Visual attention: Insights from brain imaging. *Nature Reviews Neuroscience, 1,* pp. 91–100.

as the MIT team stated it: Wojciulik, E., Kanwisher, N., & Driver, J. 1998. Covert visual attention modulates face-specific activity in the human fusiform gyrus: fMRI study. *Journal of Neurophysiology, 79,* pp. 1574–1578.

an image in your mind's eye: O'Craven, K. M. & Kanwisher, N. 2000. Mental imagery of faces and places activates corresponding stimulus-specific brain regions. *Journal of Cognitive Neuroscience, 12,* pp. 1013–1023.

337 *"active participants in our own process of perception":* Kanwisher, N., & Downing, P. 1998. Separating the wheat from the chaff. *Science, 282,* pp. 57–58.

338 *"altered by patterns of attention":* Merzenich & deCharms, 1996, p. 62.

tonotopic reorganization of the auditory cortex: Recanzone, G. H., Schreiner, C. E., & Merzenich, M. M. 1993. Plasticity in the frequency representation of primary auditory cortex following discrimination training in adult owl monkeys. *Journal of Neuroscience, 13,* pp. 87–103.

339 *"Experience coupled with attention":* Merzenich & deCharms, 1996, p. 77.

339–340 *the more stroke patients concentrated on their tasks:* Taub, E., & Morris, D. M. 2001. Constraint-induced movement therapy to enhance recovery after stroke. *Current Atherosclerosis Reports, 3,* pp. 279–286; Taub, E., Uswatte, G., & Pidikiti, R. 1999. Constraint-induced movement therapy: A new family of techniques with broad application to physical rehabilitation; a clinical review. *Journal of Rehabilitation Research and Development, 36,* pp. 237–251.

"just after a right-brain stroke": Robertson, 1999, p. 108.

342 *"steadily before the mind":* James, 1983, p. 1169.

He himself used it: Ibid., p. 1152.

343 *Bell's Theorem:* Bell, 1987; Stapp, 2001, pp. 1475–1479.
Albert Einstein and two younger colleagues: Einstein, A., Podolsky, B., & Rosen, N. 1935. Can quantum-mechanical description of physical reality be considered complete? *Physical Review, 47,* pp. 777–780.

344 *David Bohm:* Gribbin, John. 1995. *Schrödinger's kittens and the search for reality: Solving the quantum mysteries.* New York: Little, Brown.
Schrödinger called entanglement: Zeilinger, A. 2000. The quantum centennial. *Nature, 408,* pp. 639–641.

347 *experiments by Alain Aspect:* Aspect, A., Dailbard, J., & Roger, G. 1982. Experimental test of Bell inequalities using time-varying analyzers. *Physical Review Letters, 49 (25),* pp. 1804–1807.

348 *Aspect's conclusions were confirmed:* Tittle, W., Brendel, J., Zbinden, H., & Gisin, N. 1998. Violation of Bell inequalities by photons more than 10 km apart. *Physical Review Letters, 81 (17),* pp. 3563–3566. See also Stapp, H. A Bell-type theorem without hidden variables. *American Journal of Physics,* in press. Appears at www-physics.lbl.gov/~stapp/stappfiles.html, where Stapp shows that nonlocality holds within an orthodox quantum perspective.

349 *"most momentous in the history of science":* Nadeau, R., & Kafatos, M. 1999. *The non-local universe: The new physics and matters of the mind.* New York: Oxford University Press.

351 *called the Quantum Zeno Effect:* Misra, B., & Sudarshan, E. C. G. 1977. Zeno's paradox in quantum-theory. *Journal of Mathematical Physics, 18 (4),* pp. 756–763. An approachable description of Quantum Zeno is in: Milonni, P. W. 2000. *Nature, 405,* p. 526.

352 *"The wave function has ceased oozing":* Rothman, T., & Sudarshan, G. 1998. *Doubt and certainty.* Reading, Mass.: Perseus Books, p. 290.

353 *the probability that beryllium ions would decay:* Casti, J. L. 2000. *Paradigms regained.* New York: William Morrow, p. 233.

356 *activates the same regions of the brain:* Kosslyn S. M., Ganis, G., & Thompson, W. N. 2001. Neural foundations of imagery. *Nature Reviews Neuroscience, 2* (9), pp. 635–642.
the migration of calcium ions: Stapp, 2001, p. 1485.

360 *"bootstrapping effect":* Varela, F. 1999. Present-time consciousness. *Journal of Consciousness Studies, 6* (2–3), pp. 111–140.
"coexistence with the triumphant thought of other thoughts": James, 1983, p. 1172.

INDEX

Page numbers in *italics* refer to illustrations.